Sustainability, Energy and Architecture

Case Studies in Realizing Green Buildings

Sustainability, Energy and Architecture

Case Studies in Realizing Green Buildings

Ali Sayigh

ELSEVIER

AMSTERDAM • BOSTON • HEIDELBERG • LONDON
NEW YORK • OXFORD • PARIS • SAN DIEGO
SAN FRANCISCO • SINGAPORE • SYDNEY • TOKYO

Academic Press is an imprint of Elsevier

Academic Press is an imprint of Elsevier
The Boulevard, Langford Lane, Kidlington, Oxford OX5 1GB, UK
225 Wyman Street, Waltham, MA 02451, USA

First edition 2014

British Library Cataloguing in Publication Data
A catalogue record for this book is available from the British Library

Library of Congress Cataloging-in-Publication Data
A catalog record for this book is availabe from the Library of Congress

ISBN: 978-0-12-397269-9

For information on all Academic Press publications visit our web site at books.elsevier.com

Printed and bound in United States of America

14 15 16 17 10 9 8 7 6 5 4 3 2 1

Contents

Preface

Sustainability, Energy and Architecture has been produced to serve architects, students and practitioners alike; a reference book for all those interested in buildings and their relationship to energy and the environment. Each of the 17 chapters has been written by a leading authority in their field. They highlight the practise of building design throughout the world, followed by critical analysis.

It is hoped that architects and builders will use and benefit from this worldwide experience and expertise. Both vernacular and modern architecture are addressed, and the text sets out the principles for architects to follow when designing buildings, giving full consideration to energy; whether embedded in the building materials or required to make the buildings comfortable for work or residence.

Special attention has been given to the GCC and the Middle East, where recent building development has, by and large, been erected for no rhyme or reason to solve the accommodation requirements of a large immigrant population of skilled and non-skilled workers, without thought for esthetic or planning constraints. Thankfully there are signs that this attitude is changing for the better.

The book also looks at the speed at which some regions of the world have changed from traditional buildings to modern skyscrapers and their neglect of the principles of sustainability.

The book includes more than 50 design case studies from the UK, The Netherlands, Germany, Italy, Sweden, Romania, New Zealand, Canada, the GCC, Iraq, Iran, Cuba and few selected countries from central and southern Africa.

As editor-in-chief, I am proud and honored to have worked with these eminent architects and building experts in putting forward this work of insight into building design, operation and experience.

Ali Sayigh
BSc, DIC, PhD, FIE, FIEE, CEng
Chairman of World Renewable Energy Congress
Director General of WREN, Chairman of IEI
WREN, P O Box 362, Brighton BN2 1YH, UK
T: 44- (0) 1273 625 643, F: 44- (0) 1273 625 768
Email: asayigh@wrnuk.co.uk http://wrenuk.co.uk

Prof Mohsen Aboulnaga

He is an expert on Sustainable Development, Sustainability, Built Environment and Green Building. Dr Aboulnaga is a Professor of Sustainable Built Environment. His areas of competency including: strategy planning, policy development and tools, strategic environmental assessment, low-carbon society and scenarios, eco-friendly cities, green building policies and guidelines, renewable energy and climate change. In 2009, he was appointed Strategy & Policy Advisor – Environment & Infrastructure at The Prime Minister's Office of the UAE (February 2009 – March 2010). Prior to that, he held a position in the capacity of Strategy & Policy Advisor at The Executive Council, Government of Dubai from May 2007 to 2009. He was involved in developing Dubai Government Strategic Green Building Policy and Dubai Heat Island and Orthophoto Flyover Project in collaboration with Lawrence Berkeley National Laboratory. Dr Aboulnaga is author and co-author of more than 60 published int'l refereed journal and conference papers and holds a Ph.D. in sustainable building Environments from The University of Leeds, U.K.

Professor Khalid A Al-Sallal

Dr Al-Sallal received a Master of Environmental Planning from Arizona State University in 1988 and a PhD from Texas A&M University at College Station in 1995. He has been teaching architectural engineering at the UAE University since 1996 and currently holds a professor rank. His area of expertise is architectural design with emphasis on building energy. His research has focused on building performance and simulation, carbon-neutral design and zero energy buildings. He is member of several organizations and societies and editor of few journals in the area of architecture and environment and produced more than 40 publications in international refereed journals and specialized conferences. He designed several residential and public buildings including the Yemeni Ministry of Education in Sana'a and won architectural competition prizes for the design of the Yemen Kuwait Bank and the design of a commercial residential complex for the Yemeni Government in Sana'a. He received several awards for his work and design.

Arch Rahman Azari

He is a PhD candidate in Built Environment at the University of Washington. With a background in architecture, he has researched on various aspects of green buildings including green project delivery, energy-efficiency in buildings, environmental life-cycle assessments of buildings, etc. The outcomes of his research have been published as research papers in proceedings of many international conferences.

Professor George Baird

Dr George Baird is a Professor of Building Science at the School of Architecture, Victoria University of Wellington, New Zealand where he specialises in building environmental science and engineering services, building performance generally, and the energy efficient design and operation of buildings. At the Victoria University of Wellington School of Architecture he has been variously Director of Energy Research Group, Dean of Faculty of Architecture, Director of the Centre for Building Performance Research, and Associate Dean Research. He is currently a Fellow of the Chartered Institution of Building Services Engineers (UK), the Institution of Professional Engineers (NZ) and of the Institute of Refrigerating Heating and Air Conditioning Engineers (NZ), and a Foundation Member of the Energy Management Association (NZ). He was recipient of the 1999 NZ Science and Technology Bronze Medal "For singular contribution to energy efficiency of New Zealand buildings and to building performance research....”; ‘Pioneers of WREN Chairman’s Award’ of the World Renewable Energy Network for ‘contribution to the world of renewable energy through publications, teaching and promotion of renewable energy’. Author and co-authors of innumerable technical papers and case studies.

Professor Dania González Couret

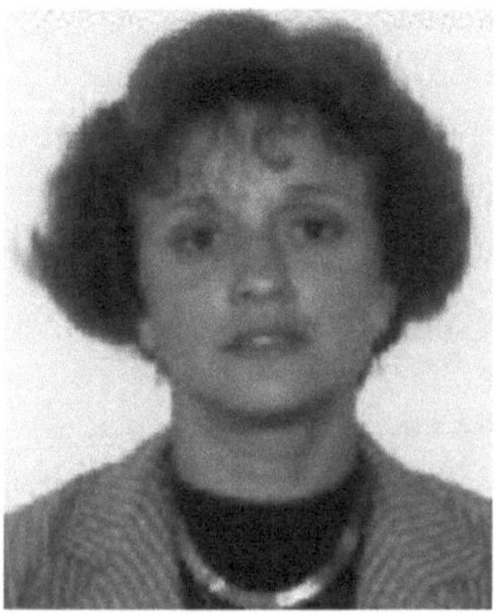

She is the Dean of Research, Faculty of Architecture, Havana, Cuba. Professor Couret she obtained her first degree in Architecture (ISPJAE, 1979); PhD (ISPJAE, 1994); Post-Doctoral Studies in Lund University (Sweden, 1997 and 1999); Doctor in Science (ISPJAE, 2007). Now, she is titular professor, and Vice Dean for Research and Postgraduate Education in the Faculty of Architecture in Havana. Prof Couret is President of the Academic Committee for PhD Program in Architecture and the Master Course in Social Housing. Vice President of the National Tribunal for PhD in Architecture and Urbanism, reviewer in Renewable Energy Journal and INVI (Housing Institute, University of Chile), and Member of the Jury in several international design competitions.

Dr Ruxandra Crutescu

Born in Bucharest, Romania, Ruxandra Crutescu is a University Lecturer at the Faculty of Architecture, “Spiru Haret” University, Bucharest, and is the Head of the Research-Development-Innovation Department at Passivehaus Institut, Bragadiru, whose founder she is. Her contribution to the scientific activity consists of approx. 80 articles published in various scientific magazines or on the occasion of national and international scientific conferences and a number of books having as main subject the ecological architecture, the durable development in architectural matters, the use of renewable energies in buildings and architecture, in order to reduce the greenhouse gas emissions and protect nature. As member of Architects’ Order of Romania. Active participant to the life of the Romanian and the international scientific community, she is Reviewer and Associate Editor for different Romanian and international publications.

Dr Nada El-Zein

Nada is a naturalized US citizen who was born and raised in Beirut, Lebanon. She received her BS, MS, and PhD from the University of Illinois in Champaign-Urbana, IL and received the highest honors possible. While at U of I she was among a very select group of PhD candidates to study under the direct mentorship of Dr Nick Holonyak, the inventor of the LED. Nada subsequently worked at Motorola for 7 years where she was the R&D Manager for nanotechnologies. After Motorola she joined EpiWorks where she led the overall R&D effort for this epitaxial wafer manufacturer which produces HBTs/FETs (compound semi-conductor transistors for wireless applications), and high-power lasers and detectors for telecom and military applications. After EpiWorks she left the R&D world and joined the international company AkzoNobel, In 2007 Nada left AkzoNobel to start LED Light Energy LLC, in the USA and is now heavily involved in the design, specification and installation of LED fixtures in various projects.

Prof Dr Manuel Correia Guedes

Dept of Civil Engineering & Architecture Instituto Superior Tecnico, Lisbon, Portugal. Professor, Director of the Architectural Research Centre (ICIST-N8), Portugal. He is Director of the Architectural Research Centre of the Instituto Superior Tecnico (ICIST-Group 8). He is Responsible for several Disciplines of the courses of Architecture, Civil Engineering and Territorial Engineering. He is supervising several PhD and MSc students.

Participation in various research projects: Chief Coordinator of a COOPENER E.U. project (SUREAFRICA), National Coordinator of an ASIA-LINK project, Dr Guedes, participated in various international and national conferences, seminars and workshops. Published many papers and books. Since 1985 he participated in various projects, namely the Portuguese Pavilion in Seville's EXPO 92, two residential buildings in Vila Real, the competition for the National Assembly building, and the building of the Agronomy Faculty (UTL). He worked as an architect in several Portuguese architectural companies.

He published many articles and papers in the fields of bioclimatic architecture and the built environment.

Dr Neveen Hamza

Dr Hamza first degree was in Architecture and a PhD in Building Science. She has been involved in teaching and researching building environmental design since 1997. Her research transcends the fictitious barriers between building performance as a science, and its socio-cultural values which should ideally fuse to express its architecture. She published several papers in conferences and journals on environmental simulation of buildings including the thermal performance double skin facades in hot arid areas, as well as researching pragmatic issues of how building regulations affect the design team and setting policy for energy conservation in the mature built environments in the UK. She is a scientific reviewer in scientific journals including Energy, and Renewable Energy and also contributes on a number of scientific conference committees.

Ms Shawna Henderson

Shawna Henderson, CEO of Bfreehomes, has been working in the field of energy-efficiency and housing since 1992. Her experience with the R-2000 and EnerGuide for Houses (ecoENERGY) programs, coupled with research carried out for Canada Mortgage and Housing Corporation (CMHC) and Natural Resources Canada, provides the backbone of Bfreehomes consulting services. Shawna has worked successfully with eight to fifteen home design clients each year since 1992, on such widely varied projects as load-bearing straw bale homes, double-wall new construction, standard stick framing and gut rehabs of older houses. In addition, she worked using renewable energy in buildings. In 2010, Shawna participated on the Information Subcommittee of the Ener Guide for Houses Rating Service Upgrade Process. In 2007, she participated on the Selection Committee for CMHC's EQuilibrium House Initiative. Shawna has published books and papers in about building technology.

Professor Andrew Miller

Andrew Miller is Professor of Building Sustainability and Head of the Centre for Sustainability of the Built Environment (CSBE) in the School of Environment and Technology at the University of Brighton in the UK. He is also adjunct professor in the School of the Built Environment at Curtin University of Technology in Perth Australia. He is a chartered Building Services Engineer with over 30 years of experience in research related to the environmental performance of buildings. His research includes the embodied energy of materials, thermal performance of buildings in operation as well as recycling of materials at the end of the useful life of the building. He is particularly interested in passive building design and in low carbon refurbishment of existing buildings.

Prof Bahram Moshfegh

He studied at the Linköping University, Sweden, where he received a PhD degree in Energy Systems in 1992. He was appointed Professor of energy systems there in 2000. Professor Moshfegh is the Chairman of Division of Energy Systems at the Linköping University since 2000. He has been involved as expert for Swedish parliament and funding research council both nationally and internationally, member of the scientific committee for the Swedish Research School Energy Systems, member of the scientific committee and organization committee as well as invited speaker for many *International Conferences.* Referee for many international journals and international conferences in the field of energy and building. Professor Moshfegh was the Chairman for the *World Renewable Energy Congress (WREC-2011) on May 8–13, 2011* in Linköping, Sweden. Professor Moshfegh has authored or co-authored more than 150 papers and research reports mainly presented in international journals or at international conferences with referee.

Professor Dr.-Ing. Helmut F.O. Müller, Architect

1966–1971, Study of Architecture at University of Hanover and Stuttgart (Diploma), 1971–1972 DAAD scholarship at the London University College (School of Environmental Studies/Bartlett School) 1972–1982 Professional activity in design and research 1979-Doctor's Degree at the University of Stuttgart (Dr.-Ing.) 1982–1993 Professor at the Polytec of Cologne, Department of Architecture 1991–1997-Foundation and director of the Institute of Light and BuildingTechnology at the Polytec of Cologne (ILB), 1993–2009 Univ. Professor, Chair of Environmental Architecture, Department of Building, University of Dortmund 1997–2005 General Manager, GLB, Gesellschaft für Licht und Bautechnik mbH, Dortmund since 2006, he was deputy chairman of FiTLicht e.V. Development Association Innovative Daylight Utilization. Since 2008, a director of Green Building R&D, Düsseldorf, Partner of office consortium, 4greenarchitecture, Düsseldorf. Member of board of directors, Schuermann Spannel AG, Bochum.

Professor Marco Sala

Marco Sala, graduated from University of Florence in Architecture in 1972 is registered architect since 1973, Chairman of Marco Sala Associates, Architecture and Energy Consultants, he specialises in passive solar systems in architecture, environmental design and energy conscious building design. Full Professor in Architectural Technology at the University of Florence. Founder and Director of ABITA Interuniversity Research Centre based in Florence University, Director of European Master ABITA in Sustainable Architecture, has extensive experience in European research projects, as group leader and coordinator (JOULE, ALTENER, SAVE, VII FW, EIE) Fonder member of TIA (Teaching in Architecture Energy and Environment World Network). Published many papers and books in Architecture Practice.

Professor Ali Sayigh, BSc, DIC, AWP, PhD, F Inst E, F IET, CEng

Chairman and founder of World Renewable Energy Congress and Council which been held in 23 different countries up to now, Director General of World Renewable Energy Network (WREN) since 1990, Chairman, Founder of the Arab Solar Energy Society, Past Chairman of the UK Solar Energy Society, Director of Solar Seminars at ICTP-Trieste, Italy between 1977–1995, Professor and head of Solar Energy in Saudi Arabia, Kuwait, and Reading Universities from 1969 to 1994. Presently he is professor at the University of Hertfordshire. Founding member of the Arab Science and Technology Foundation (ASTF), Shaijah, UAE, Fellow of the Institute of Energy; Fellow of the Institution of Electrical Engineers; and Chartered Engineer. He published more than 400 papers and contributed and edited more than 30 books. He is editor and editor in chief of several international journals including the one which he founded in 1983 Solar and Wind Technology. This journal in 1990 is called The International Journal of Renewable Energy which is published by Elsevier Science Ltd. He is Editor-in-chief of Comprehensive Renewable Energy, will be published in March 2012, by Elsevier Company. Prof Sayigh established with Sovereign Publications the annual Renewable Energy Magazine since 2000, Co Editor-in-chief of ICPSR Journal "ISESCO Vision of Science and Technology". Editor of International Journal of Environmental Sciences and Technology. He was Professor at King Saud University, Reading University and University of Hertfordshire (1978–2004). He graduated 32 PhD Students in various Engineering, Building Technology and renewable Energy. Prof Sayigh is chairman of Iraq Energy Institute. Editor-in-chief of Comprehensive Renewable Energy with 154 contributors.

Arch Nazar Sayigh

He is Director of Architecture, Glas Architects. Nazar is a fully qualified architect and chartered member of the RIBA with over 14 years of professional experience. Prior to setting up Glas, he worked for a number of award-winning design firms including Buschow Henley Architects and Mary Thum Associates. More recently his firm has been responsible for some of South London's most distinctive and successful mixed use housing developments, including 1 Druid Street and 134–144 Southwark Bridge Road. Glas currently undertake a wide range of design led projects from small commercial fit out schemes to large scale master planning exercises. Nazar is a member of Southwark Council's highly acclaimed Design Review Panel, a steering committee member of The World Renewable Energy Congress, an RIBA competition assessor and has been an invited guest critic at the Bartlett School of Architecture and East London University school of Architecture.

Dr Maryam Singery

Maryam Singery is an assistant professor in Urban Studies at Islamic Azad University, Iran. She was born on February 02, 1976 the city of Tabriz, Iran. Presently Ph D, Urbanism (Candidate) at the Faculty of Art and Architecture, Science and Research Campus, Islamic Azad University, Tehran, Iran. She obtained her M.A in Architecture during (1996–2003) on Islamic Azad University, Tabriz Branch in Iran. Full time Staff Member of Architecture and art Department at Islamic Azad University, Tabriz Branch, central Instructor from 2002 up to this date. She participated in two International Congresses: the World Renewable Energy Congress IX, Aug 2006, Florence, Italy and the 2nd PALENC Conference and 28th AIVC Conference, September 2007, Greece

Professor Wim Zeiler

Wim Zeiler studied Mechanical Engineering (Design and construction; Methodical Design) at the Technische Universiteit Twente. In 1983 he started working in Kropman Building Services contracting, in which he currently holds the position of specialist building services technology. In 2001 he became full professor Building Services at the faculty of Architecture, Building and Planning at the Eindhoven University of Technology. His current research is on Integral Design, Agent technology, Climatic Design and Renewable Energy in the Built Environment. He is member of the board of advice of ISSO (Dutch society for stimulation of research on building services in the Netherlands) and participated in many different committees and boards within the TVVL (the Dutch Society for Building Services in the Built Environment).

Chapter 1

Dutch Efforts Towards a Sustainable Built Environment

Wim Zeiler
TU Eindhoven, Faculty of the Built Environment, Eindhoven, The Netherlands

Chapter Outline

1.1 INTRODUCTION

Until the nineteen sixties in The Netherlands, almost all houses used a coal stove that only heated one room, in which everyone gathered in winter. The growing need for a healthier indoor climate and the discovery of enormous amounts of natural gas in The Netherlands led to the widespread installation of gas-fired central heating systems in Dutch homes [1]. While this initially led to a big increase in energy use, step by step measures were introduced to reduce the energy demand. Sustainable building has a respectable history in The Netherlands especially since the early nineteen seventies. The report 'The limits to growth' produced by the Club of Rome in 1972, and the 1973 energy crises shortly afterwards, showed the downside of the long period of growing prosperity experienced since the early nineteen fifties. In October 1973, Middle Eastern OPEC nations stopped exporting oil to the US and other Western nations. The embargo forced Western governments to consider alternative sources and strategies for energy, such as their cost and supply, which up to 1973 no one had

Sustainability, Energy and Architecture. ISBN: 9780123972699 worried about. Governmental regulations before this period were concerned primarily with the indoor environment and health, but after the energy crises, the focus shifted to saving energy. In new government energy policies introduced between 1973 and 1990, energy efficiency and diversification became the key subjects. The first solar energy projects were completed in Boxmeer and Oss (1975) and Zoetermeer (1977) [2].

One of the oldest Dutch Photo Voltaic (PV) hybrid systems and the first 'large scale' PV system in Europe was installed in 1983 on the island of Terschelling [3]. At the 'Willem Barentsz' Higher Maritime School, a 43 kWp PV system was coupled to a 40 kW wind turbine and a large battery bank. Later a new 75 kW wind turbine was installed, and in 1995 a diesel engine was added [4]. In 1988 the first off-grid solar house was opened in Castricum, having a 2,5 KWp PV array with a 10 kWh storage capacity. It was an autonomous system, dimensioned for winter demand and functioned very well and reliably until 2012. These projects were the first large scale applications of PN in the Netherlands and were folloed by many ohers:

- 1991: 10 houses with a grid-connected PV-system were built in Heerhugowaard.
- 1994: 66 houses were built in the district Nieuw Sloten in Amsterdam (250 kWp) using PV.
- 1996–2000: project Nieuwland: 1,3 MWp in Amersfoort, 500 houses equipped with PV roofs were constructed.

The publication in 1987 of the Brundtland committee's report 'Our Common Future' made policy makers realize that sustainable development required as high a priority as a healthy indoor environment and energy savings. This led to the RIVM (National Institute of Public Health and the Environment) report Zorgen voor Morgen (Looking after Tomorrow), and in 1989 the government responded with the National Environmental Policy Plan (Nationaal Milieubeleidsplan – NMP). The 1995 'Sustainable Building Plan: Investing in the Future' marked the beginning of a programmatic approach by central government and the energy performance coefficient (EPC) was incorporated in the building code. The government published the 2000–2004 Sustainable Building Policy Program in order to embed sustainable building in policy and practice. The Dutch pavilion at the Hanover World Expo 2000, (see Figure 1.1) designed by the Dutch architectural firm MVDRV, demonstrated trends in sustainable building on land use by multi-level function, integration of renewable energy (5th floor), preserving greenery (3rd floor) and reducing environmental impact, within a natural setting [1].

An interesting project from this period is the ecological building project EVA Lanxmeer, Culemborg, completed in 2004. The project, consisting of around 200 houses, is an example of integrated sustainable building and innovative urban landscape design features [1].

Energy performance certification for homes was introduced in The Netherlands in January 2008, one year before the introduction date prescribed by the

FIGURE 1.1 **Dutch paviloen World Expo Hannover.**

European Union. The Netherlands was an early adopter of the EPC initiative, and all transactions in the Dutch housing market now need to be accompanied by an energy performance certificate. The energy performance grading ranges from 'A++' for exceptionally energy-efficient dwellings, to 'G', for highly inefficient buildings.

The rationale of energy performance certification schemes of the EU energy label is that it will cause buyers to favor houses which have higher levels of energy efficiency. However, in the absence of objective data that measures to what extent such schemes actually do have an impact on homebuyer behavior, it is difficult to determine whether they are a meaningful intervention in seeking to address climate change. Agentschap NL, an agency of the Dutch Ministry of Economic Affairs, administers the quality control and registration of the certificates. As of September 2009, more than 100,000 residential homes (rental and owner-occupied) had been certified. The report by Dirk Brounen of Erasmus University, Rotterdam, and Nils Kok of Maastricht University, based on data from The Netherlands, provides some of the first evidence regarding the market adoption and financial impact of energy performance certificates.

Energy performance certification is not fully mandatory in The Netherlands: homebuyers are allowed to sign a waiver that removes the seller's obligation to provide a certificate for the dwelling. This semi-mandatory choice for energy performance certification creates a natural experiment to study the adoption and market effects of the energy label in the residential housing market. During the first three months of 2008, more than 25% of all housing transactions had an energy label. However, soon after this the adoption rate for Dutch energy labels started to decline, eventually reaching a rate of less than 7% of the 150,000 homes on the market in September 2009. The empirical results show that the choice of certification is driven partially by the quality of a dwelling, with duplex homes constructed during the 1970s and 1980s and located in high-density, low-income areas being significantly more likely to obtain an energy performance certificate. There is a premium associated with properties that demonstrate high levels of energy efficiency, with a 2.8% higher sale price for properties with an A, B or C certificate. The price increment achieved varies with the grade of the certificate, and mostly reflects the financial benefits obtained from lower energy costs in more energy-efficient buildings.

Zero or nearly zero impact buildings are designed and built in every country. In this context, zero energy means that the building does not use fossil fuels, but only renewable energy. Energy 0 projects were already in development in the late nineteen thirties – e.g. the 1939 MIT Solar House 1 [5], in The Netherlands. An overview and evaluation of the early Dutch projects are given by Gilijamse [6] and Hoiting et al. [7]. In the last Now zero or nearly zero impact buildings are designed and built in The Netherlands as a matter of course. The energy consultant Kroon, built a zero energy house in Woudbrugge in 1993 (see Figure 1.2).

However at that time the necessary technology had to be developed. The house was the first net zero energy building that had a fully integrated PV implementation on the roof combined with thermal solar collectors. The roof of the house carries 3.4 kWp in photovoltaic cells, connected to the public grid, and a 12 m^2 active (thermal) solar collector [8,9]. In 1995, a measurement program showed that the energy consumption of 1,070 m^3 natural gas equivalent was more than balanced by the energy production of 1,146 m^3 natural gas equivalent and proved that the house met the designer's 'zero-energy' target. Figure 1.2 shows the energy concept of the house: more information can be found in references [8,10].

After a few experiments with zero-energy houses in 2000, the first Net Zero Energy Building (NZEB) School in The Netherlands was completed [11,12] (see Figure 1.3). The school's electricity requirement of around 14.650 kWh is supplied by 145 m^2 PV-panels on the roof, while its heating needs of around 4000m^3 gas (~ 16.000 kWh) are supplied by participation in a wind turbine park. To investigate the results of this ZEB school approach, thermal comfort and indoor air quality were measured.

Over a week, different measurements in schools were made, and questionnaires issued to determine the indoor air quality and thermal comfort, and it was found that the ZEB school does not perform better on all aspects [13].

Although the projects proved that is possible to build a zero-energy house and zero-energy school, the investment proved too costly for large scale application. For example the photovoltaic system of the NZEB house in Woudbrugge

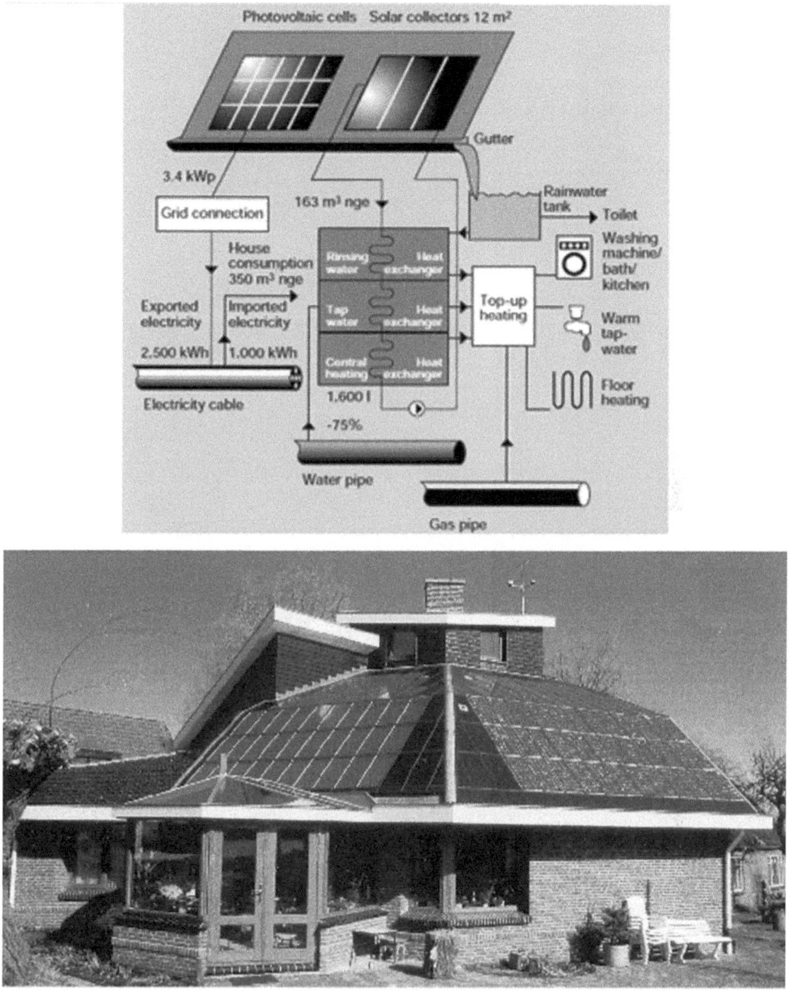

FIGURE 1.2 Schematic of the photovoltaic and active (thermal) solar heating systems and zero-energy house.

amounted to € 52,000, whereas the revenues from the annual yield of 2,905 kWh at a rate of € 0.113/kWh, giving an income of € 329 [8]. This would not produce a realistic pay-back period. However, the purpose of the project was to demonstrate the technical feasibility of the zero-energy concept. The project concluded that the required equipment will become cheaper and energy prices will rise to the point where the zero-energy concept would become economically feasible.

A new direction in the development of zero energy buildings was also followed in The Netherlands, due to developments in other countries: Passive Houses.

FIGURE 1.3 **First Dutch NZEB School.**

1.2 PASSIVE HOUSES

An important first step to reach NZEB is the application of the Passive House strategy. This leads to energy savings in heating of 80%, compared to conventional standards for new buildings. The general definition of a Passive House is that its energy consumption is limited to a maximum of around 15 kWh/m^2 for space heating, and a maximum total of around 120 kWh/m^2 for heating, domestic hot water and electricity for electrical equipment and lighting. To meet these criteria, the Passive House concept focuses first and foremost on reducing the energy demand of the building. Passive House solutions give high priority to the performance of the thermal envelope: this means high grade insulation for walls, roofs, floors and windows/doors, thermal bridge-free construction and air tightness [14]. For well-insulated, low energy houses the heating energy needed for the ventilation is around 50 to 65% of the total heat demand. This is the reason that the ventilation is often significantly reduced; some mention values as low as a ventilation rate of 0.4. There is a competition between energy saving on the one hand and good indoor air quality on the other.

Designing and building Passive Houses in a country is not a straightforward matter, as can be seen from the experience of building such houses in Germany and Austria. Each country has its own building tradition, architecture, building technologies, climate and culture. Architects and builders are familiar with local construction materials and solutions which have been developed to meet the specific building codes and standards of that country [15]. In The Netherlands, a number of houses have been built according to Passivhause Projectierungs Paket (PhPP), but the actual performance of these houses has not yet been thoroughly investigated. A study at the Technische Universiteit Eindhoven investigated whether the ventilation levels achieved by the installed mechanical ventilation systems in these Dutch Passive Houses were sufficient [16]. For the study, three Passive Houses in The Netherlands were chosen. Case 1 was of one of the 12 houses that were built against the inner slope of a river dike. Cases 2 and 3 were villas in which the technology used was also based on Passive House technology. Case 4 was a three story house that was investigated by DHV [17]. This house has special, large night ventilation facilities. The results of all the measurements are given in Table 1.1. The average airflow found in Cases 1 to 3 was lower than recommended by the Dutch Building Code, while the capacity installed is sufficient. Another remarkable fact is that the energy use is 13% higher than calculated, presumably due to the higher average temperature recorded in the living room compared to the program of requirements (22 °C instead of 20 °C). This 'rebound' effect is presumably due to people preferring a higher comfort level if this is possible: something to be careful about.

1.3 TYPES OF CASE STUDIES

Sustainable building continues to increase in importance, not only for new buildings but also in renovation. Over the past decades, several concepts have been developed to reduce the environmental impact of dwellings [18]. A fairly new concept in this area is the greenhouse residence concept (Dutch: kaswoning): which is a combination of a normal building and a greenhouse [18]. *A greenhouse residence (GHR) is a dwelling covered by a transparent building envelope, the size of which is the same order of magnitude of the dwelling, and which is not used for commercial horticultural activities [19].* This approach tries to optimize the use of energy from the sun during the cold season. Different versions of greenhouse residences have been built in The Netherlands (Culemborg and Almere), Germany, Sweden, France and Japan. Three series of greenhouse residences have been built in Culemborg (The Netherlands) [20–22]. They were designed by Arjan Karssenberg and Peter Wienberg of the architectural firm 'KWSA' [20]. In May 2002 the first series (Figure 1.4), consisting of one row of six greenhouse dwellings, was completed. This was followed in 2006 by the second series, a row of five dwellings and two workspaces [21], and by the third series in 2009. The residences of the first series are almost completely covered by the greenhouse and those of the second series have an adjacent greenhouse [20]. Preheated air from the greenhouse is used to ventilate the residence [20]. The residences are also very well insulated. Photovoltaic–panels are assembled on the greenhouse and solar collectors are applied for low temperature heating. Single glazing is used for the greenhouse [22]. This causes the temperature inside the greenhouse to rise by approximately 3–5°C over the outside tempera-

TABLE 1.1 Information and Measurement Results for Dutch Passive Houses [8]

	Case 1	Case 2	Case 3	Case 4

Town	Sliedrecht	Dalem	Duiven	Roosendaal
Type of house	Terraced house	Villa	Villa	Terraced house
Year of completion	2004	2000	2004	2008
Bedroom 1 [2 persons]*	800	750	950	600
Bedroom 2*	800	700	1650	600

**Average night time level CO_2 [ppm]*

FIGURE 1.4 **Greenhouse residences Culemborg 1st series (2002), 2nd series (2006) and 3rd series (2009) [24].**

ture [23]. In order to keep temperatures inside the dwelling acceptable during the summer season, the doors and windows of the dwelling and the windows of the greenhouse are opened during the night. Sun screens in the greenhouse are not necessary, according to the architects [21]. Nevertheless, at least two of the six greenhouses are provided with sun screens in the form of cloths or a parasol [22]. An automated system is installed to control the opening of the greenhouse windows. This can also be done manually, however when it rains or when high wind speeds occur the windows are closed automatically. The architects of the greenhouse residences in Culemborg claim that the greenhouse functions as a buffer between the indoor and outdoor environment, and therefore a greenhouse residence has a very low energy demand and a pleasant indoor climate. Extra space for living due to the indoor terrace, light and openness, lack of weathering of the materials of the shell of the residence, low maintenance and cheap construction are mentioned as other advantages of the greenhouse residence [20].

At the moment the technology exists [25], but unfortunately very few designers have the necessary skills or experience needed to design zero impact buildings. This task requires a profound knowledge of energy and technology. An integrated design approach is needed, not only by the architect but also by the structural consultant, building services consultant and building physics consultant. A method needs to be developed to allow other, largely engineering, building disciplines to be integrated into the design process of NZEB from the outset in a meaningful way. This is not something done by merely putting the different designers together around a table, but needs greater support. This leads to the necessity of an integral design approach not only by the architect but from the structural consultant, building services consultant and building physics consultant. In this context, the traditional approaches to organizing and planning these complex processes are no longer sufficient [26]. First and foremost, a method needs to be developed to allow other, largely engineering, building disciplines to become integrated into the design process of NZEB from the outset in a meaningful way. In the last few years, some design tools have been tested in different design workshops for NZEB design, an exercise in which more than 100 experienced professionals, including architects and consulting engineers, participated. One of the participating architects, Carl-Peter Goossen, became so enthusiastic about the design approach that he made it the leading principle in the design management of the projects of his company, BouwQuest [27]. A good example of his work is the Veldhuizerschool in Ede. It is the first new Passive House primary school in The Netherlands.

1.4 THE VELDHUIZERSCHOOL EDE

The construction of the school mentioned above was completed in the middle of 2011. The building is extremely airtight and highly insulated (Rc = 10 m^2K/W). It also has triple glazing. The air conditioning is performed by individual room ventilation systems with heat exchangers. The air distribution is done by textile ducts. Energy is only needed at the start of lessons; during the rest of the time the students generate enough heat to warm the room. The school has a green sedum roof and is a pilot project for an integral Building Information Model (BIM) approach [28] (see Figure 1.5).

1.5 CHRISTIAAN HUYGENS COLLEGE: AN ENERGY PLUS SCHOOL

Christiaan Huygens College is the first CO_2-neutral and energy plus school in The Netherlands. The new building of the Christiaan Huygens College has a compact construction in order to limit the surface area of the façade, thus reducing excess heat absorbed in summer and heat loss in winter by as much as possible. At the same time, well-insulated windows allow natural daylight to flood the building without causing overheating. Thanks in part to the Energy Roof (see Figure 1.6) used in its construction, the school building designed by architect Thomas Rau (RAU) generates more energy than the school needs for its own use.

The Energy Roof was developed by the Schiebroek roofing company in cooperation with Volantis and Technische Universiteit Eindhoven. It has a thermal system based on solar power, which uses an evacuated tube collector with a heat exchanger which is invisibly incorporated into the insulating layer of the roof construction. The tube collector is covered in synthetic roofing material with an integrated layer of photovoltaic cells that generate electricity (see Figure 1.7).

During peak hours, the system produces more energy than the school, the sport hall and the adjoining homes need. That surplus is stored in an underground aquifer storage system. This surplus energy will be used in the adjoining sport hall and nearby apartments by the Trudo housing corporation. The energy saved by all three buildings together amounts to 130,000 euros a year. In the

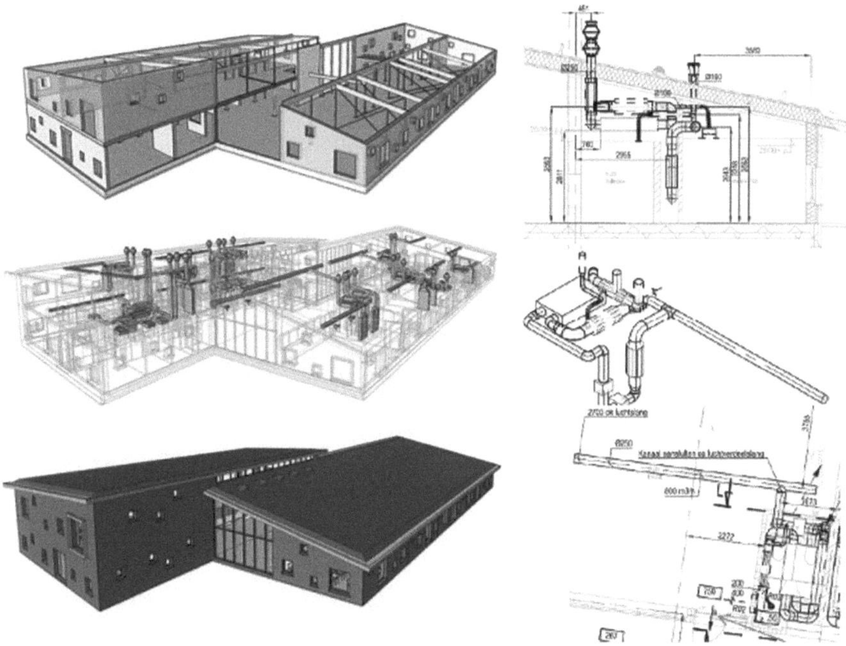

FIGURE 1.5 The Veldhuizerschool in Ede [Goosen 2010b].

FIGURE 1.6 Christian Huygens College Eindhoven with energy roof.

winter, this energy can be brought back up to heat the buildings – see Figure 1.8 for the total energy concept of the building. The building was completed in November 2010.

1.6 CONVENTIONAL DUTCH BUILDING DESIGN

Dutch households are responsible for 19% of the total primary energy use of The Netherlands. The Dutch housing stock has a poor energy performance, with over 46% of the total having an energy label of E or worse on a scale that reaches from A++ to G (energy neutral to very energy inefficient). Therefore, substantial energy savings can be achieved in the Dutch housing sector.

Post-war (1946–1966) multilevel residences represent an important group in Dutch housing stock, comprising about 11% of the total, or around 740,000 households. These homes traditionally have a poor energy performance and belong to the F and G energy efficiency categories unless substantial renovation has taken place. In addition, several other important

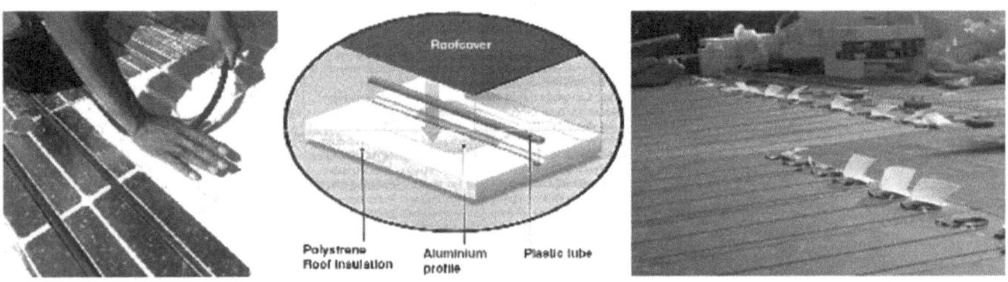

FIGURE 1.7 Energy roof construction.

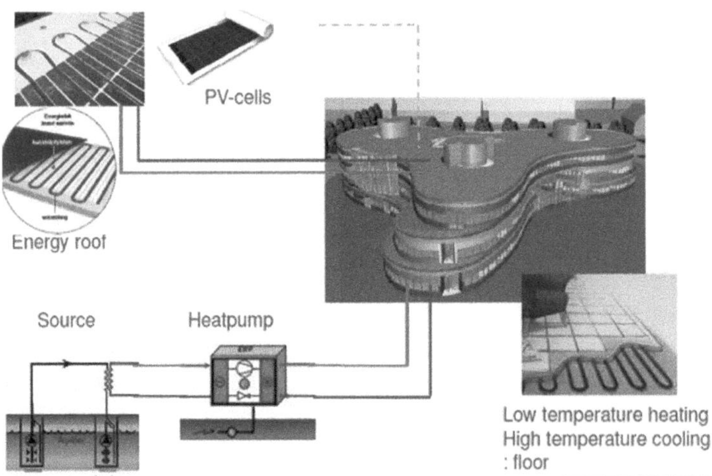

FIGURE 1.8 Total energy concept of Christiaan Huygens college Eindhoven.

problems occur with this specific type of housing: the flats are characterized by a monotonous quality in their architecture and functionality. Furthermore, they are usually poorly maintained, and degradation is visible in many cases. As a consequence, The Netherlands now has to deal with a lack of suitable housing that is still affordable for those with low incomes. Therefore a general consensus exists on the need for a (physical) transformation of these flats.

1.7 ENERGY SAVING TECHNIQUES

In The Netherlands in the early nineteen nineties, a three-step schema was developed which ranked sustainable measures for the building industry in order of sustainability preference. The first three-waymodel was introduced by Lysen [29], under the name 'Trias Energica', after Charles de Montesquieu's 'Trias Politica' from 1752. The following steps for reaching a sustainable energy supply were suggested by Lysen [30]:

1. Continuing improvement of energy efficiency,
2. Greater use of sustainable energy sources,
3. Cleaner use of the remaining fossil fuels.

Duijvestein [31], introduced a more structured approach, aimed at the building industry, which placed the three factors in order of their impact on sustainability. The most favorable measure was put at the top and the least favorable became the last step. This aligning process has led to the 'Trias Energetica' now commonly used in The Netherlands, which has three steps [32]:

1. Prevent the use of energy by reconsidering the energy use (prevention);
2. Use sustainable energy sources as widely as possible (renewable);
3. When an energy demand still remains, then use fossil fuels as efficiently as possible (efficiency).

Energy saving policy in The Netherlands has traditionally concentrated on the heating of houses and offices [30]. However, it is of course necessary to minimize energy use in both heating and cooling, and to use an energy source that is as sustainable as possible. The 'Trias Energetica' refers to three categories of measures which can achieve a sustainable solution for the energy demand of houses and offices:

1. Take measures which decrease the building's energy use, such as insulation and efficient ventilators;
2. Use as sustainable a source for the energy as is possible. The use of solar power in an active or passive way, and the use of wind energy, are examples in this category;

3. If there aren't enough sustainable sources, then it is necessary to use fossil fuels. In this case use the supplies as effectively as possible. The use of highly efficient boilers is a way to turn gas into heat.

Based on Duijvestein [31] and Lysen [29] the Trias Energetica refers to three categories of measures, which can achieve a sustainable solution for energy use in houses and offices (see Figure 1.9):

1. Take energy saving measures
2. Use sustainable energy sources
3. Use fossil fuels as efficiently as possible

1.8 NOVEL DESIGN AND EXAMPLES

Strongly fluctuating and rising energy prices, the depletion of fossil fuels and growing awareness of global warming have led to actions intended to reduce carbon dioxide emissions. As the built environment is responsible for nearly 40% of CO_2 emissions, new approaches are necessary. In July 2009 the G8, the eight most important industrial countries, agreed on an 80% reduction in CO_2 emissions by 2050. In The Netherlands, stakeholders agreed with the government to reduce step by step the emissions of new houses with a first reduction of CO_2 with 25% in 2012 and a 50% reduction of CO_2 in 2016.

In 2009, the Dutch government started the so-called UKP-NESK program to stimulate innovation for energy neutral buildings. UKP means 'unique chances projects', and NESK means 'towards energy neutral schools and offices' (Naar Energieneutrale Scholen en Kantoren). In 2010, this program funded projects which showed exceptional innovation in the area of energy conservation, sustainability or organization within the building industry (see Table 1.2).

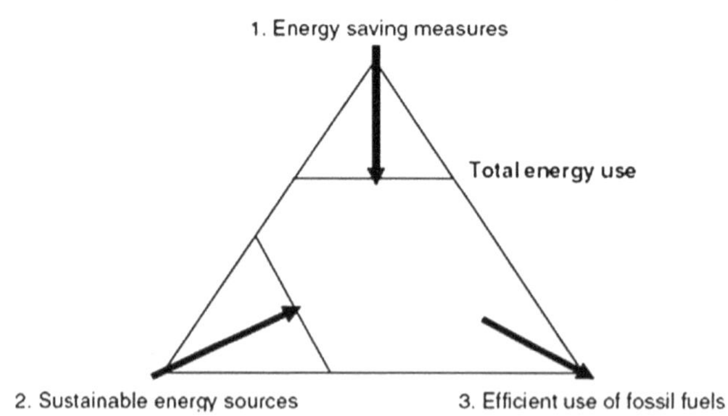

FIGURE 1.9 Visual model towards a sustainable implementation of the power consumption according to the Trias. [32]

TABLE 1.2A

	Project	Type	Location	Year	Special features of project
	TNT Green Office	New office	Hoofddorp	2010	Cooperation between principle and project developer, bio heat power combination, heat pump, aquifer
	Villa Flora	New office	Venlo	2011	Technology from green houses applied
	Zeswegen	New office	Heerlen	2012	Heatpump uses mijnwater as heat and cold source.

TABLE 1.2B Overview of Dutch UKP NESK Projects Offices and Schools

	Project	Location	Year	Special features of project	PV m^2
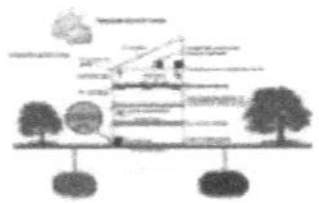	Baken Poort	Almere, 6800 m^2	2013	Energy neutral high level insulation, low temperature heating, HR ventilation, HP, aquifer	3000
	Hart van Oijen	Lith, 2447 m^2	2012	Energy neutral by applying a biogas-CHP, low temperature heating, HP, aquifer	150
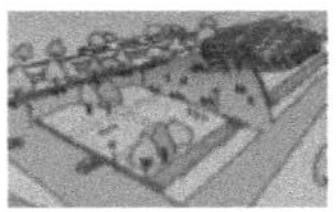	VMBO Huygens College & De Polsstok	Heerhugowaard, 4257 m^2	2012	Energy neutral based on applying Passive House-concept, R=10 insulation, HR ventilation, HP, aquifer, solar boiler	2000
	SO/VSO OdyZee	Goes, 2458 m^2	2011	Energy neutral based on applying Passive House-concept, R=10 insulation, HP, solar boiler, low temperature heating	499
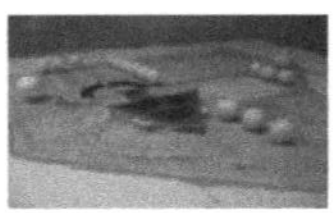	MFC Brede School	Kollumerland, 1787 m^2	2012	Energy neutral by applying a combination of sustainable energy technologies, R=6 insulation, HP, floor heating/cooling	1190
	Klimaatneutraal DSK-II	Haarlem, 2735 m^2	2012	Energy neutral, uses the heat of a computer server room for heating and hot water, HP, aquifer	1000
	Het Klaverblad	Amsterdam, 3177 m^2	2012	Energy neutral by applying PV-panels which are installed by the local energy distribution company without extra costs for the school, connected to district heating	1000

These are the most interesting projects in the field of NZEB in The Netherlands. Unfortunately none of the projects are finished, and therefore only a list of the projects is presented. In future our intention is to investigate the design process of the projects as well as the actual performance outcome to comfort, health and energy.

1.9 THE TNT GREEN OFFICE

The TNT Group, a global express delivery service headquartered in The Netherlands, recently announced its intention to move its operations to a green office building that would be newly developed. TNT Green Office project shows how important it is that a client has a strong, clear vision about the sustainability goals they want to achieve in their buildings. Despite current trends in housing, sustainable office development is still far from being mainstream. One of the causes is the circle of blame: a vicious circle in which the stakeholders blame each other for not initiating either demand or supply of sustainable buildings. One of the intentions of TNT was to break this circle of blame. The office building of TNT in Hoofddorp (NL) is their first Green Office. Green Offices is part of the ambitious Planet-Me program, through which TNT wants to achieve its ambition to become the first emission-free mail and express delivery company in the world. The most important key success factors for sustainable office buildings are: 'commitment to sustainability from the involved persons', 'willingness of the end-user to invest in sustainability', 'focus on long term value creation', 'early involvement of all stakeholders in the project' and 'clear definition of sustainability goals'. All these requirements were met in the new Green Office in Hoofddorp, and it should meet the highest standards regarding sustainability: it is CO_2 emission free, the design should also achieve more than 1,000 points under the Dutch green building certification GreenCalc+ and it is LEED (Leadership in Energy & Environmental Design) Platinum certified.

The development of GreenCalc started in 1997. The GreenCalc+ assessment method is a questionnaire which allows you to estimate how much land it takes to run and maintain your office. The results can be then used to calculate what developers call the environmental index of a building. This is done by calculating the environmental impact of the building by a Life Cycle Analysis (LCA). The GreenCalc+ software consists of four modules, each representing a different aspect of the building's characteristics; mobility, materials, water and energy. The input values for this program are divided into four groups; materials, energy, water and travel to and from work.

LEED is an American methodology for assessing the sustainability of a building. This was developed by the US Green Building Council (USGBC) for the US Department of Energy. The pilot version (LEED 1.0) for new construction was first launched at the USGBC Membership Summit in August 1998. The current LEED Reference Guide presents detailed information on how to achieve the scheme's credits, which are divided into the following six groups; sustainable site, water efficiency, energy and atmosphere, materials and resources, indoor environmental quality and innovation and design process.

To guarantee that their goal for a highly sustainable building would be achieved, TNT entered into cooperation with an experienced consortium in the field of sustainability: Triodos Bank and OVG Development. The sustainability ambition is an essential part of the contract between TNT and the consortium which had not only to develop the building but also had to own the office for 10 years. TNT Real Estate has entered into an innovative form of contract that is similar to a DBFMO (Design-Build-Finance-Maintain-Operate) agreement with the consortium [34]. TNT Real Estate has a contract for 10 years, under which it will pay a fixed price for water, electricity, heating and cooling. This gives TNT, as the consortium's principal, hard guarantees in the areas of energy and other sustainability measures. The consortium and the end-user (TNT) are both encouraged to reduce energy consumption throughout the lease period of the building. The consortium acts as an Energy Service Company (Esco) for the tenant and all investments in additional sustainability measures are taken within a ten-year payback period [34].

Architect Paul de Ruiter is responsible for the design of the Green Office. He has a clear vision about sustainability and architecture [33].

'After the release of Al Gore's 'An Inconvenient Truth', Dutch politicians are prepared to put the environment on the political agenda. However, the Dutch government primarily views sustainable construction in terms of energy-efficient construction, and to this end has decided to implement a more stringent Energy Performance Norm, without giving much thought to the architectural quality, let alone the quality of the building's interior. The results of this approach are smaller windows, more isolation and balanced ventilation – resulting in, for example, airtight homes and schools that are actually very unhealthy for their users. Not enough oxygen; excessive CO_2 levels; insufficient daylight and restricted views; buildings that are too ugly – this can hardly be what was intended. The quality norms relating to comfortable homes, a healthy life and a properly functioning professional and residential environment are far more important than mere energy efficiency. It makes far more sense to base the theme of sustainability on the viewpoint of the end user. Architecture needs to develop a radical service attitude. In other words, it should not only satisfy the demands of the client, but actually exceed them, delivering added value as a result. We can only achieve real improvements by radically addressing the actual needs of the end users, clients and their environment. Only this way, we can create an obligation to really get to grips with innovative solutions that work and as a result have a tangible impact on the world of tomorrow.' [33]

The design of the TNT Green Office is characterized by sustainability, transparency and connectivity. First a volume study was done to test different volumes regarding criteria such as compactness, flexibility, daylight factor, view, building costs and the highest LEED results. The design consists of two rectangular parallel volumes, each six stories high. On the Westside (the Geniedijk) the lower three stories of these two volumes are connected by terrace-like volumes and the upper three stories by connective bridges [34]. These connections also offer great meeting places for the employees. On the Eastside both volumes are connected by a third 'floating' volume. For the TNT headquarters, 16-meter long concrete floor slabs were used, which were made out of recycled rubble and granulate. Due to the long span, fewer supports were needed, which saves material and generates spaces that can be divided up freely [34]. See Figure 1.10.

1.10 SUSTAINABILITY

The atrium has been designed in such a way that as much daylight as possible can enter and it offers the employees a beautiful view. The atrium and the entrance are clearly connected, and the terrace-like volumes encourage employees to take the stairs instead of the elevator, thereby serving both a health and social purpose [33,34] (see Figure 1.11).

The presence of daylight in living and working areas is of crucial importance to making the working and living environment feel good, and also for the health of the occupants. Daylight gives energy, generates happiness and stimulates productivity [33]. Daylight was the leitmotif in the design of the

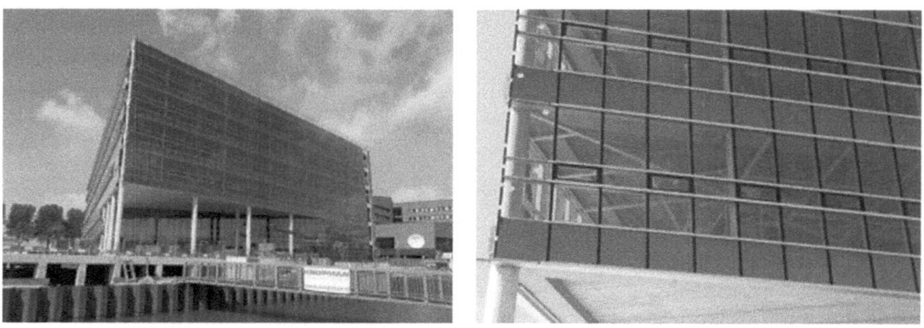

FIGURE 1.10 Facade of the TNT Green Office building Hoofddorp.

FIGURE 1.11 Roof and area of the atrium.

building, which has a completely glazed north façade. In addition to an optimum interior climate, daylight incidence was at the center of Paul de Ruiter's sustainability philosophy, which goes beyond purely technical aspects:

'It has to be about the people who work in the building. You can erect a building that saves energy, yet is still bad for employees. That's not sustainable.'

1.11 DIVERSE SUSTAINABILITY MEASURES

Several sustainability techniques have been applied in the construction of the Green Office: high levels of insulation, intelligent awning, hybrid ventilation (natural if possible, mechanical if necessary), heat recovery from the ventilated air, energy-saving equipment and lighting, long term cold/heat storage, generation of electricity through the use of bio-CHP (Green Machine) and an advanced building management system.

The surplus of heat in the summer and the surplus of cold in the winter are stored below ground level through cold/heat storage. The stored heat is used to warm the building in the winter, and the stored cold to cool it down in the summer. The electricity for the two heat pumps of 332 kW is delivered by the bio-CHP.

All electricity for the Green Office (on a yearly basis) is generated on location in a sustainable way using a bio-CHP. For the remainder of the peak demand, green energy is purchased. This way, the TNT Green Office operates completely CO_2 emission free. The produced heat is supplied to nearby (yet to be realized) office buildings. The surplus of heat in the summer and the surplus of cold in the winter are stored below ground level through cold/heat storage. The stored heat is used to warm the building in the winter, and the stored cold to cool it down in the summer. A solar hot water heating system has been added to the building. The system includes two solar collectors with an aperture size of 2.4 m^2 each. The system contributes heat to the Domestic Hot Water (DHW) system.

1.12 RESULTS OF GREENCALC+ AND LEED ASSESSMENT

Based on the design and the actual realization the Milieu Index Gebouw has been calculated for the TNT Green office. The building will have an Environmental Index of at least 1,000 points, calculated by GreenCalc methodology. This is more than 1.5 times better than the highest Building Environmental Index for any currently existing building. This index is determined from the materials and the quantities used in construction. During the design and preparation for the construction of the building, continuous attention is devoted to assessing whether the choice of materials is the most environmentally friendly, and whether the quantities used remain within the initial estimates, in order to guarantee that the building achieves an ultimate index of at least 1,000. It is not just materials that count in this measure, however, but water and energy treatment also contribute to a higher index. The TNT Green Office has a bio-CHP (heat-power coupling to biofuel) for the purpose of generating power. The high score for energy is due to the compensation effect of applying the bio-CHP, without that the score for energy would have been around 220. This would lead to a GreenCalc+ score of 481, still among the best existing buildings – for example, one of the other most environmentally friendly office buildings in The Netherlands, the 2004 Rijkswaterstaat building in Terneuzen, has a score of 323.

The LEED assessment considers the areas of design, implementation, ultimate use and management. This is tracked in five categories: materials, energy consumption, efficient water use, interior environment and the environment. By way of example: the location is important, therefore including the proximity of the building to the public transportation network. The methodology even extends to the need for documenting the specific properties of the paints used. These may not contain more harmful substances than prescribed by LEED. In addition, a prescribed minimum quantity of recycled materials must be used and a large percentage of the materials used must be 'regional'. The highest certificate that can be issued under this methodology is the LEED Platinum Certificate and this will indeed be the certificate awarded to the TNT Green Office building.

Kropman has in cooperation with Octalix and Nijeboer-Hage won a grant in the tender UKP NESK for a demonstration project of building an energy-0 office building for CBW-MITEX. A Building Energy Management System in combination with wireless technology is one of the key technological elements of the innovative project. Octalix® is an integral Energy and Comfort Management System (control switch is shown below) for commercial buildings: new buildings and existing buildings. This system is directed at energy saving by making use of a monitoring system to monitor energy use at a detailed level, i.e. per room, and in this way gaining insight into waste.

The most important distinguishing factor of Octalix in regard to a Building Management System (BMS) solution is that Octalix® works on the principle of integral operation on demand and that it is also possible to monitor all energy flows at a detailed level for each room. Octalix® incorporates advanced sensors that monitor the indoor climate in a room and supply the underlying installations with information. This will, for instance, prevent the air-conditioning in the whole building from running in summer, but will instead only activate it in those rooms in which there are actually people. During a meeting, ventilation will be increased in response to CO_2 measurement, whereas the energy consumption in empty rooms will be limited to a minimum. This is operating (on demand in optima format), which optimally spreads the available capacity of the climate control installations across the premises and thus keeps energy consumption as low as possible.

The CBW Mitex head office has a gross floor area of 2900 m^2 and was completed in 2011. It was designed to be CO_2 neutral and net zero energy (see Figure 1.12).

The fully triple-glazed façades are important in reducing energy consumption. Heat pumps in combination with Long Term Energy Storage (LTES) aquifers, an underground heating and cooling storage facility, is combined with concrete core activation. A special feature of the building's management is an integral

FIGURE 1.12 View of the main entrance and the back of the building [Rau].

system of sensors produced by Octalix, which ensures that the building is only ventilated, warmed, cooled and lit in those parts of the building where people are actually present. This system helps to save up to 40% of the building's energy needs. The system also responds to the number of people present in the building, thus improving efficiency even further (see Figure 1.13).

Kropman and Octalix want to develop their approach further and present it to a broad audience. The expected use of instrumentation is based on a concept for the use of different wireless sensors and infra-red cameras, and is shown in Figure 1.14. Such a grid of low cost sensors would make it possible to control and manage the comfort of an individual and minimize the energy necessary for this. Figure 1.14 shows how the actual experimental setting of the Smart Energy Building grid could look, based on a recent design by Kropman.

This chapter presents a few of the current Net Zero Emiision Buildings (NZEB) planned and already built in 2010. It shows that the step can be taken from Passive House or NZEB to Plus Energy buildings.

FIGURE 1.13 Interior with view of the courtyard, foyer and reception desk [Rau].

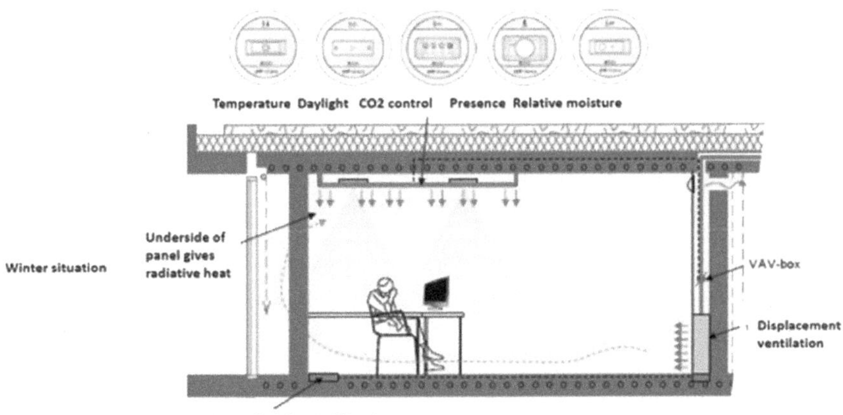

FIGURE 1.14 The ZEB Smart grid concept: individual-workspace-room level.

REFERENCES

[1] R. Rovers, Sustainable Housing Projects, Implementing a Conceptual Approach, Techne Press, Amsterdam, 2008.
[2] D. Turrent, A review of solar houses in Europe, Energy and Buildings 3 (3) (1981) 229–245.
[3] E.H. Lysen, The International Energy Agency PVPS Program – PV hybrids and the future, PV Hybrid Power Systems Conference Aix-en-Provence, France, 2000.
[4] E. Lysen, Fifty years of solar PV in the Netherlands, Dutch Solar Cell R&D Seminar, Utrecht, 2006, http://www.energieplatform.nl/fileadmin/node/documents/Zon-PV/lysen_50yearssolarpvnetherlands.pdf (accessed 15 december 2012).
[5] P. Hernandez, P. Kenny, From net energy to zero energy buildings: Defining life cycle zero energy buildings (LC-ZEB0), Energy and Buildings 22 (6) (June 2010) 815–821.
[6] W. Gilijamse, Energie-neutrale woningen: mogelijkheden in Nedrland, Proceedings Duurzame Energie: Met het oog op morgen, Nederlandse Duurzame Energie conferentie, 25–26 November 1999, Noordwijkerhout.
[7] H. Hoiting, F. Boer, G. Donze, L. Verhoef, M. Dingenouts, Evaluatie energienulwoningen, W/E-rapportnummer 5789, Gouda, August 2001.
[8] CADDET, 1996, A Zero-energy House in the Netherlands, IEA-OECD Technical Brochure No.27, 1996.
[9] T.J. Schoen, Building-integrated PV instalaltions in the Netherlands: Examples and operational experiences, Solar Energy 70 (6) (2001) 467–477.
[10] A. Kroon, J.H. Boumans, Met de zon naar realisatie nul-energiewoning Woubrugge, 2008, http://www.energienulhuis-kroon.nl/p3inhoud.htm (accessed 15 december 2012).
[11] J.C. Weenen, W.J.M. van, Dettmers, M.S.J. Overtoom, H.G.J.M. Poldermans, Development of the First Sustainable Elementary School of The Netherlands, 2000, Proceedings International Conference Sustainable Building 2000, 22–25 October, Maastricht, the Netherlands, pp. 651–653.
[12] Wenen H. van, Passive Solar Schools International Expertise in Support of the First Sustainable Elementary School of The Netherlands International Workshop, 3–5 March 2000, Castricum, The Netherlands, 2000, Organized within the framework of the National Demonstration Project of the Dutch Governmental Building Agency (RGD): 'Sustainable Decision-Making'.
[13] W. Zeiler, G. Boxem, Sustainable schools: better than traditional schools? Proceeding ROOMVENT, Buxan, Korea, 2009.
[14] I. Strom, L. Joosten, C. Boonstra, Passive House Solutions, final version Working paper 1.2 PEP Promotion of European Passive Houses, 2006, rapport nr. DHV_WP1.2, 23-05-2005, May 2006.
[15] H.F. Kaan, B.J. Boer, de, Passive Houses: Achievable concepts for low CO2 housing, Proceedings ISES Conference 2005, 2006, Orlando, USA, September 2005, ECN-RX–;06–019.
[16] J.R. Balvers, G. Boxem, M.H. Wit, de, Indoor air quality in low-energy houses in the Netherlands, Does mechanical ventilation provide a healthy indoor environment? Indoor Air, Copenhagen, 2008.
[17] R.M. Cloquet, L.A.H. Joosten, Wonen in een passiefhuis, voor het eerst binnenmilieu in passiefhuizen gementen, Bouwfysica 2, (2010).
[18] S.J. v Velzen, Greenhouse residence – feasibility of greenhouse residence concept for renovation, MSc thesis, Technische Universiteit Eindhoven, 2010, http://www.bwk.tue.nl/bps/hensen/research.html (accessed 15 december 2012).
[19] L. F. v Ree, Building performance evaluation of newly built greenhouse residences in terms of heating energy and thermal comfort, MSc thesis, Technische Universiteit Eindhoven, 2010, http://www.bwk.tue.nl/bps/hensen/research.html (accessed 15 december 2012).

[20] J. Bakkers, Wonen in een kas, energiebewust, Bouwkavels.nl 1 (2008) 7–11.
[21] SenterNovem, Praktijkvoorbeelden energiezuinige woningbouw, Juni 2007.
[22] I.R. Vollebregt, De serre heruitgevonden, 2003.
[23] Vries, G.d. and P. Teeuw, 13 in een dozijn. Voorbeelden van duurzame gebouwen in Nederland. 2007, Boxtel: Aeneas, uitgeverij van vakinformatie.
[24] KWSA. Zeven woningen/ateliers (nieuwe kaswoningen) eerste fase, te Eva Lanxmeer, Culemborg, 2003–2006. 2003 [cited April 15th, 2011]; Available from: http://www.kwsa.nl/nieuwekaswoning.html (accessed 15 december 2012).
[25] Charron R., 2005, A review of low and net-zero solar home initiatives, CanmetENERGY, CETC number: 2005-111/2005/07-12.
[26] Aken J.E. van, Valid knowledge for professional design of large and complex design processes, Design Studies 26 (4) (2005) 379–404.
[27] C.P. Goossen, a, Morfologisch ontwerpen, 2010, http://www.bouwquest.nl/ontwerpmanagement/morfologisch-ontwerpen (accessed 15 december 2012).
[28] C.P. Goossen, b, Veldhuizerschool goes Passief!, 2010, http://www.bouwquest.nl/architectuur/veldhuizerschool.html (accessed 15 december 2012).
[29] E. Lysen, The Trias Energica: Solar Energy Strategies for Developing Countries, 1996.
[30] H.J.H. Brouwers, A.G. Entrop, New triplet visions on sustainable building, The 2005 World Sustainable Building Conference, Tokyo, 27–29 September 2005 (SB05Tokyo).
[31] C.A.J. Duijvestein, Ecologisch bouwen (In Dutch), 1997.
[32] B. Entrop, J. Brouwers, A. Van der Veen, Model voor duurzaam ruimtegebruik gemeenten: Journal of Building Appraisal 5 (4): 293-310.
[33] P. de. Ruiter, Philosophy, 2010, http://www.paulderuiter.nl/bureau/item/Philosophy.html (accessed 15 december 2012)
[34] OVG, 2010, TNT Centre, http://www.ovg.nl/index.php?pageID=3411&projectID=376993 (accessed 15 december 2012)..

Chapter 2

Low Energy Approaches to Design-Led Schemes – Five Case Studies

Nazar Sayigh
Glas Architects, London, UK

Chapter Outline

Sustainability, Energy and Architecture. ISBN: 9780123972699

2.1 INTRODUCTION

Glas place low energy and passive/sustainable design at the heart of all their schemes. However, we are ever mindful of the constraints and limitations placed on such design strategies when operating within a restricted economic environment, where affordability, practicality and value for money are key considerations for building procurers.

The following case studies illustrate our approach to this important element of building design, the strategies employed and techniques proposed in order to deliver low energy solutions while meeting the needs of our clients. Each scheme has been chosen to illustrate a particular approach, and while most of the schemes embody several aspects of sustainable design, it is the key feature of each scheme that marks them out. To be of additional use we felt it was worth grouping the schemes broadly into large, medium and small. Case studies 1 and 2 cover the largest of our projects, while 4 and 5 illustrate two of our smallest. I hope they also illustrate that such elements need not be an afterthought, or bolt-on addendum to a building, but rather can be seen as intrinsic to the design solution.

2.2 CASE STUDIES 1 AND 2 – OVERVIEW

The first case studies cover two large mixed use projects located in central South East London. They were chosen because, far from employing a host of renewable add on technologies, they made the best possible use of site and context to deliver improved energy use and higher levels of user comfort and amenity. The key to this success lay primarily in careful site analysis, efficient space planning. [including the use of standardized dwelling layouts] and ultimately with the clever use of orientation.

2.2.1 33–134 Webber Street – Case Study 1

Glas were invited to propose a master plan for this large triangular plot, situated just north of Elephant and Castle, South East London. Planning guidance required a mixture of uses to replace the existing industrial warehousing, but with an emphasis on the provision of both affordable and private housing.

Given the size of the site and the likely density of development, initial thoughts turned to the possibility of including renewable energy options, from extensive photovoltaic (PV) and wind to the provision of a site-wide combined heat and power supply via a centralized Combined Heat and Power (CHP) or similar boiler.

However all these options were discounted by our client as the site was likely to be split up into various tenures with a range of landlords and end users. Attention turned to simple, passive methods of reducing the energy use for the proposed buildings as well as the possibility of improving the environment for the end users.

Note: Above numbers refer to photos on following page.

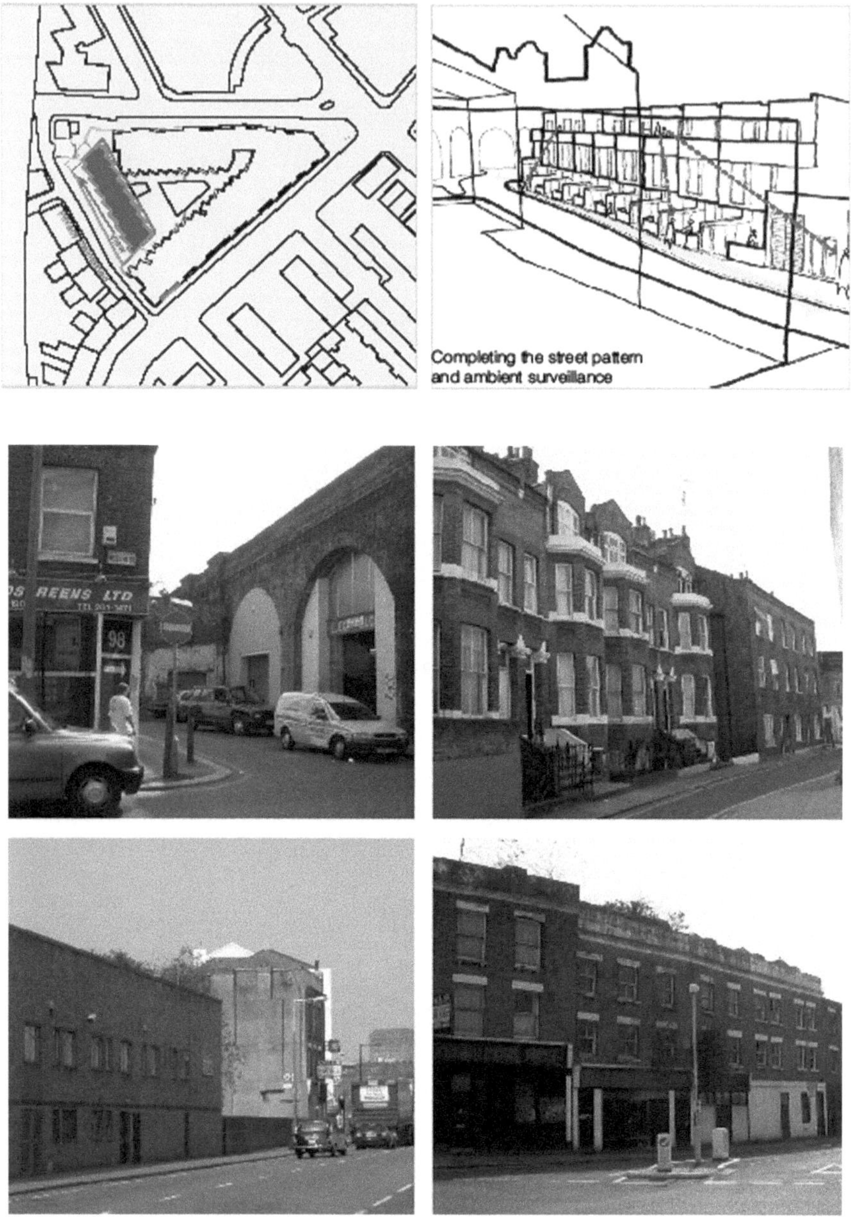
Completing the street pattern
and ambient surveillance
REENS LTD
98

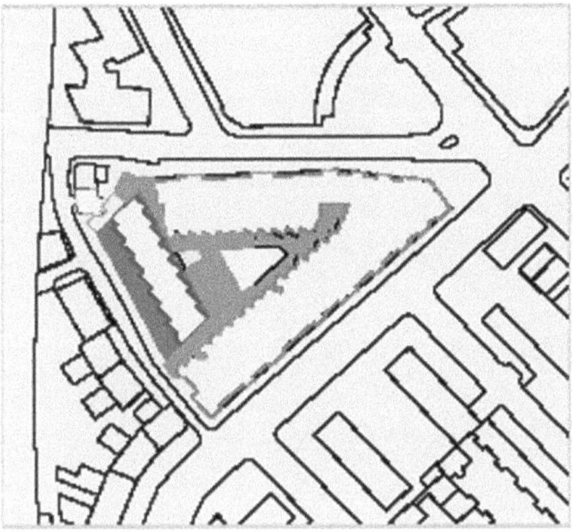

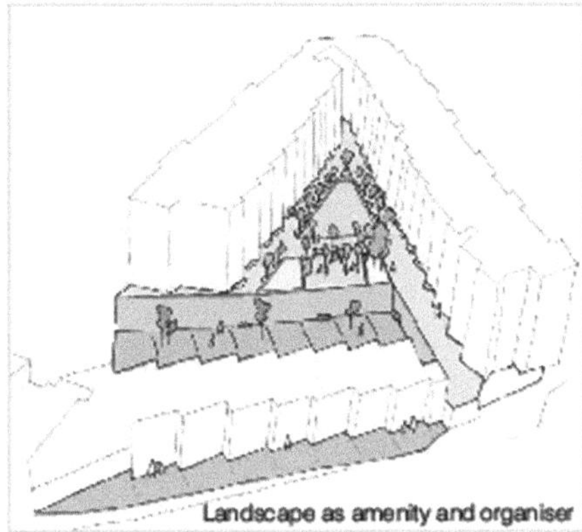

Once the overall target density for the site had been agreed with the planning authority, basic massing and envelope studies indicated that the scheme was likely to deliver upwards of 150 units, with the possibility of extensive ground floor commercial space.

Planning restrictions imposed a maximum height of seven stories, with a requirement for the blocks to be stepped. Initial studies explored several options for the siting, height, massing and form of the blocks; from a centrally located 'cluster' of small towers, to a large wide rectangular 'slab' running the entire length of the site.

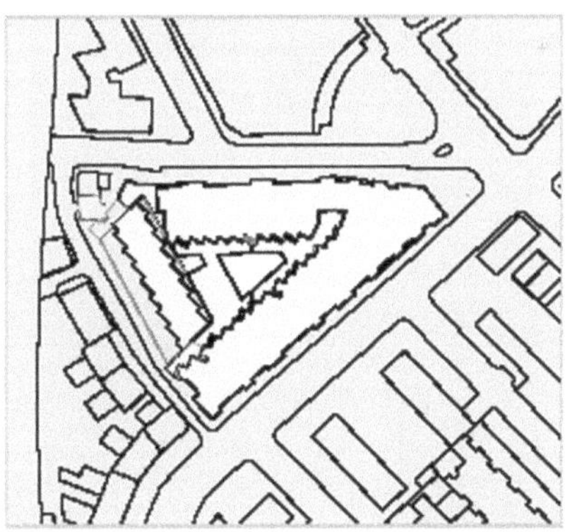

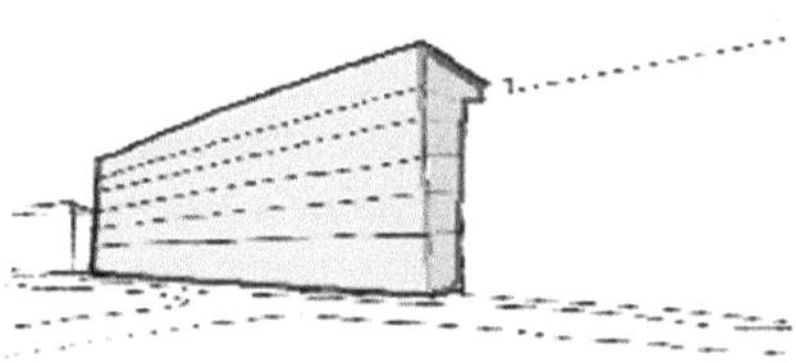

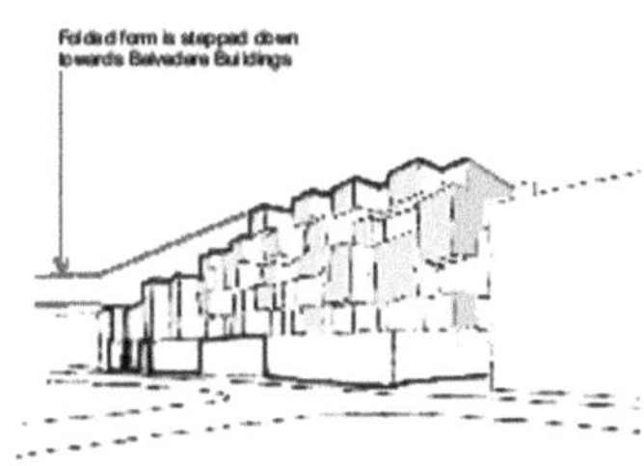

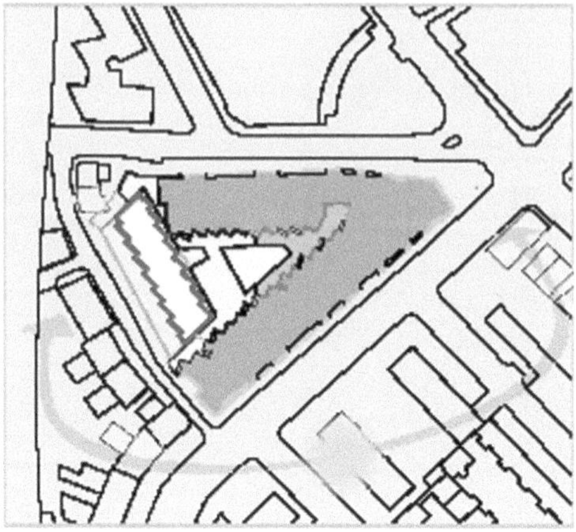

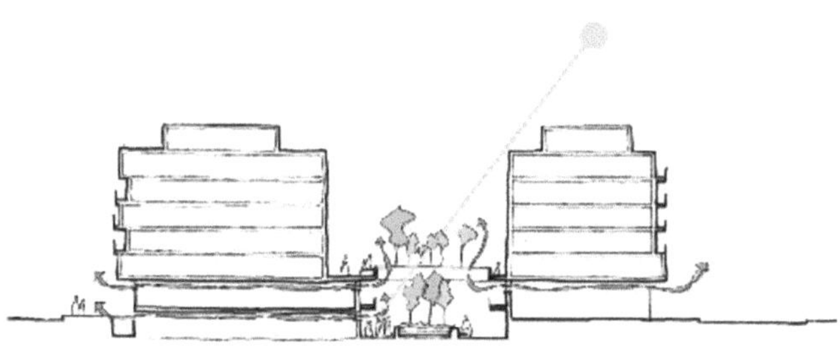

Sun path analysis revealed the south eastern and south western sides of the site benefited from good levels of natural light throughout the day, with the south western side in particular benefiting from excellent levels of daylight. This data was utilized to generate a form that would maximize access to daylight, while allowing the site to be efficiently planned.

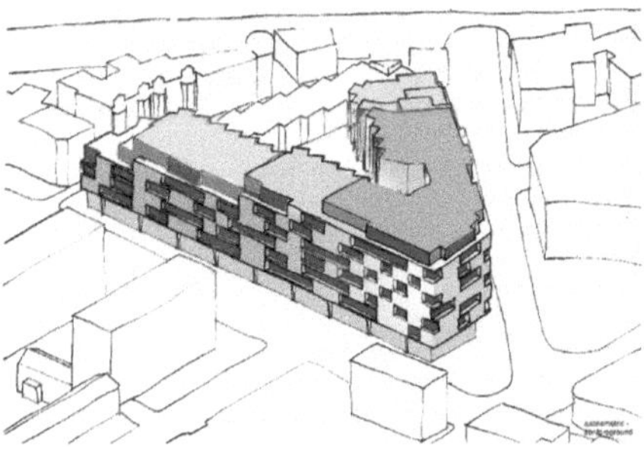

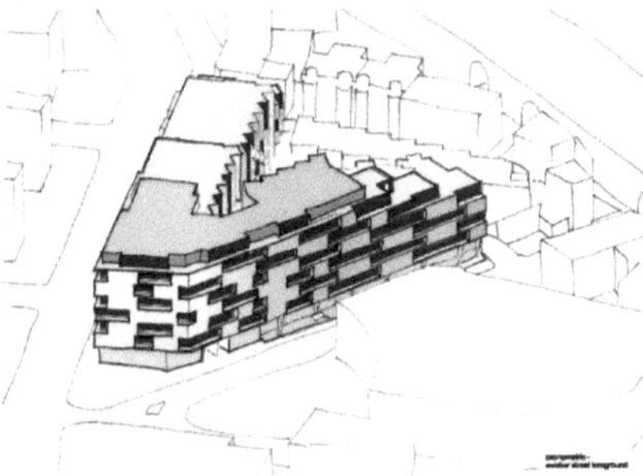

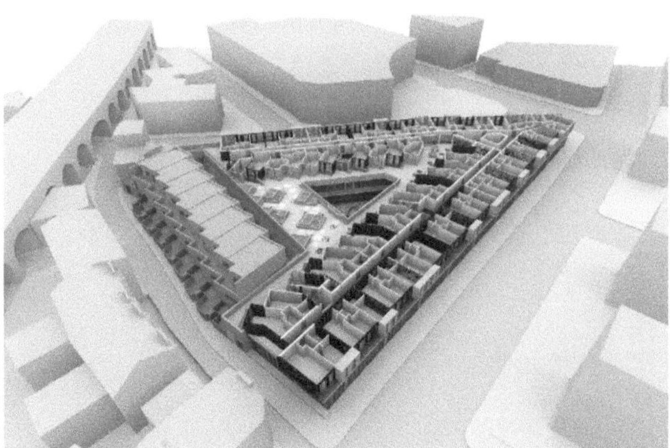

A 'ribbon' block was proposed to run round three of the sides, leaving the south west side open. The north east corner of the site, lying directly opposite a busy junction, was the most physically constrained. It was decided to place the highest element of the ribbon block here in order to address this prominent junction. The scheme then stepped down along the south eastern and north western sides to the open end opposite Belvedere Terrace.

On-site car parking for the residential block necessitated a basement level above which commercial floor space was required. Provision for these provided the final element to our scheme, an expansive 'deck' off which the ribbon block could rise. A central courtyard delivering valuable private amenity space was now possible.

Detailed planning of the ribbon block provided an opportunity to further enhance the residential unit's amenity, by way of orientation. All individual units were to be given their own balconies, and were required to be dual aspect. An angled chamfered plan form was adopted for the units, allowing for the courtyard inward facing units to be orientated towards the open south western end; thus maximizing the levels of natural daylight. In addition privacy was maintained between the dwellings. The staggered plan form of the units also allowed for some cross ventilation.

A staggered approach was also adopted for the final element of the scheme along Belvedere Terrace.

Seven new town houses were provided that referenced an existing Victorian terrace opposite. By utilizing a similar orientation to the ribbon block behind, these houses could benefit from improved daylight, cross ventilation and heightened privacy.

2.2.2 Stead Street Development, Southwark, London – Case Study 2

This development was part of an invited competition to provide early decanting sites for the residents of housing estates at Elephant and Castle; the lessons learnt from delivering our Webber Street scheme were applied here, from the study and use of orientation to developing a more functional and sustainable façade treatment. A large former car park was given as the site, with the brief requiring a mixed use scheme to provide affordable and private housing, as well as community facilities. Initially site-wide renewable energy options were investigated, however the client's intention was for the scheme to be plugged into a new district-wide combined heat/power supply located off site.

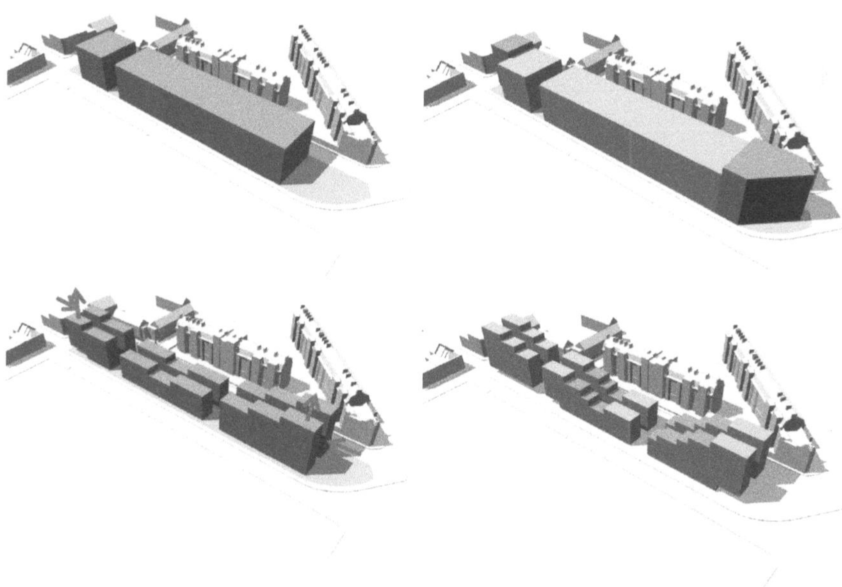

Our key focus became the provision of low tech, passive measures that would be integral to the scheme design and enhance both the energy performance and comfort/amenity of the dwellings.

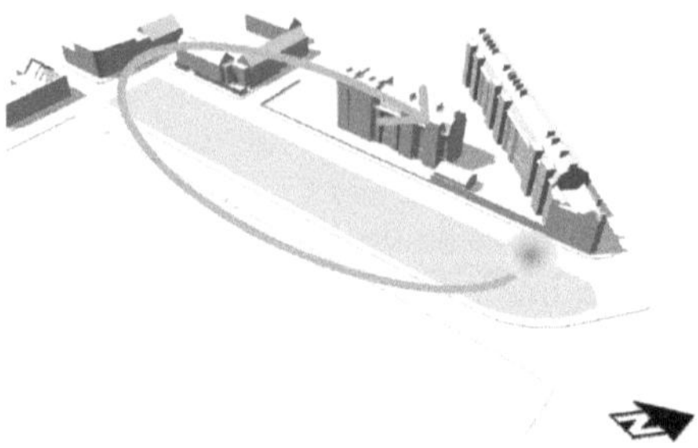

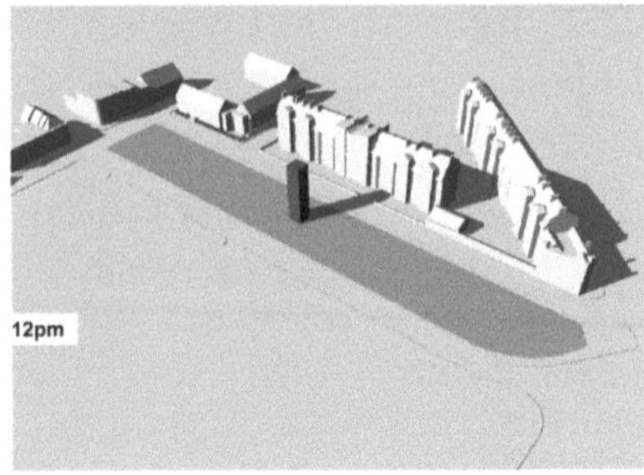

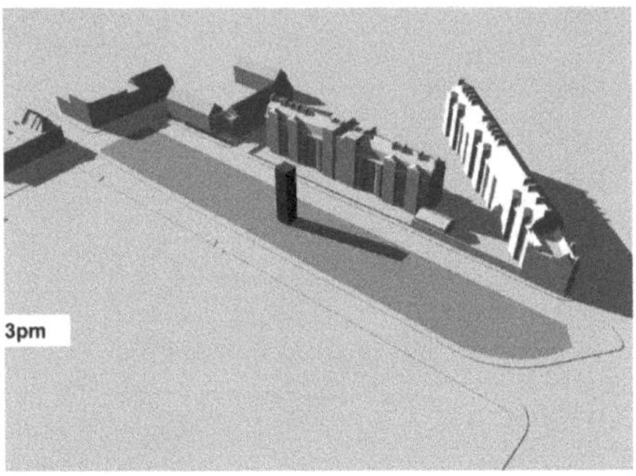

As with Webber Street, initial site investigation included aspect, orientation and sun path analysis; the results of which were used to guide early block massing studies. Height restrictions imposed by the planning department allowed for the tallest blocks to be 6–7 stories and the presence of a large Victorian terrace block adjacent to the site provided further height restriction due to potential loss of light.

Our solution proposed a series of stepped blocks, ranging from two taller ‘bookend’ sections to a lower central area adjacent to the Victorian terrace. A range of standardized unit types was devised to best meet the needs of future occupiers, optimizing access to fresh air, sunlight/daylight and amenity space. The units were based on a regular simple plan form module, enabling the procurement of a range of standardized components, from window units to entire kitchen suites.

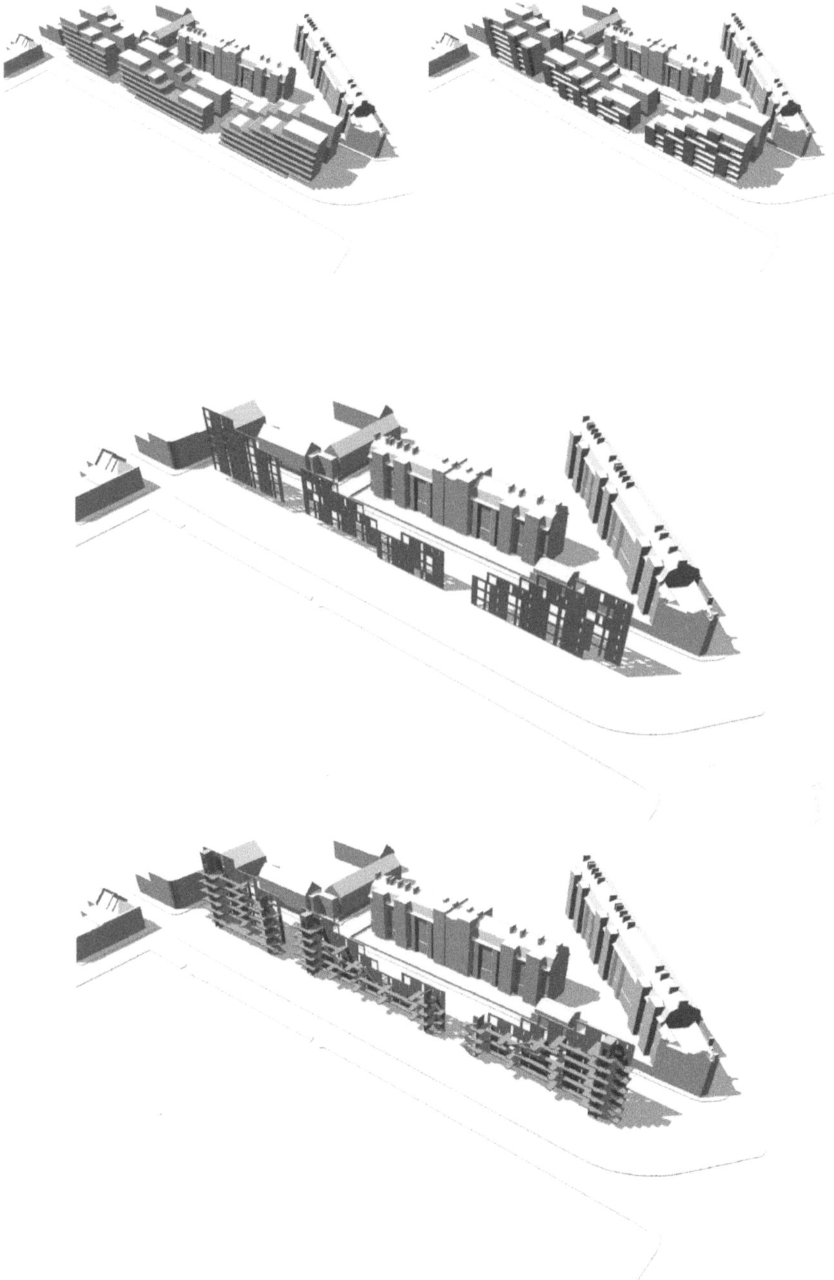

Further refinement of the blocks included the distribution of the unit types throughout the scheme to meet the client's brief. Key requirements for each unit included adequate external amenity space, dual aspect, high thermal comfort levels and good ventilation.

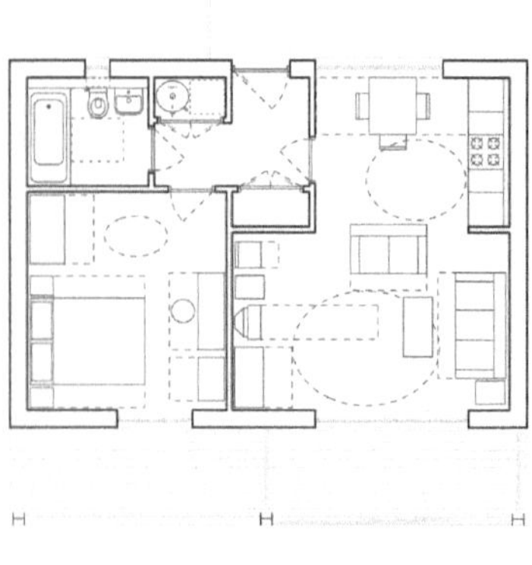

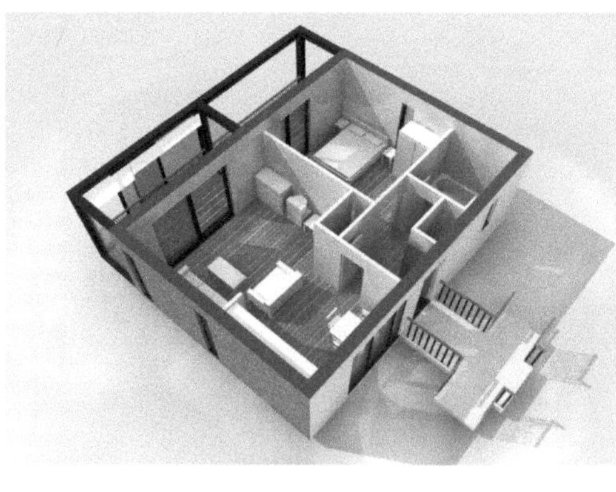

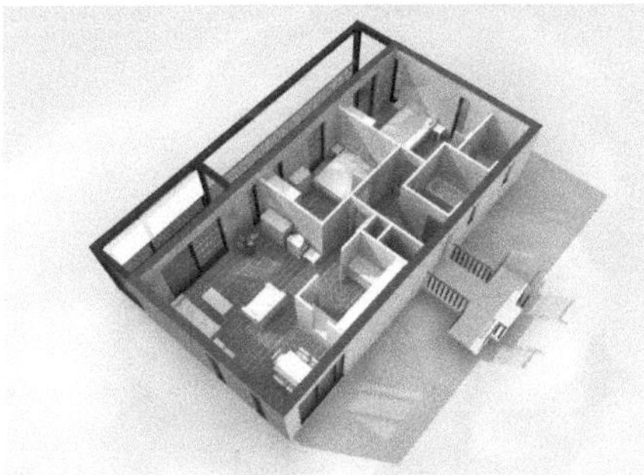

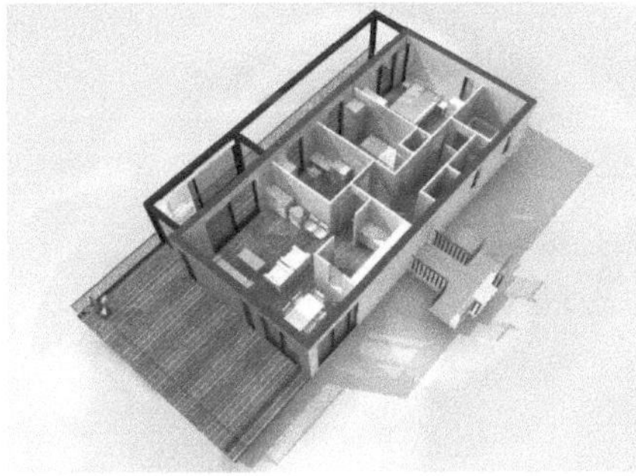

To meet these needs a central 'foyer' was devised, providing secure, covered access to the units as well as secondary amenity space. Passive cross ventilation via the central space would also be a realistic option to achieve, and so a system allowing fresh external supply air to be supplied to the central foyer was incorporated. By providing roof level, openable vents, a natural stack effect could be encouraged, helping to draw in fresh external air to the foyer.

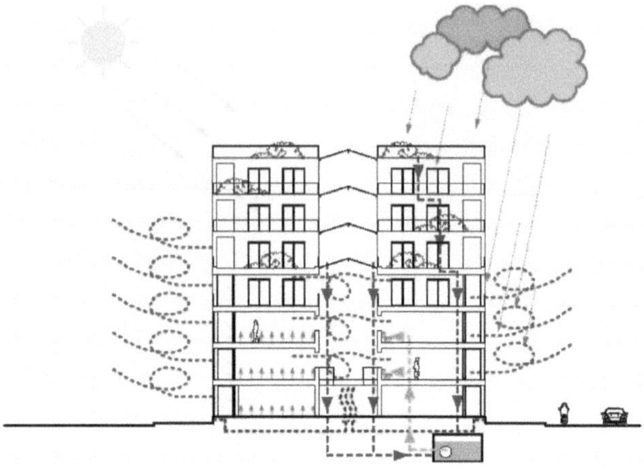

An external lattice-work of terraces was to provide the required private amenity space. This presented an opportunity to further develop an idea considered for Webber Street; the winter garden enclosure. Our scheme at Webber Street included in-set private terraces to the main elevation's adjacent living rooms. In the case of Stead Street, semi-enclosed, glazed areas could be provided opposite each unit's living room doors, allowing for a flexible semi-enclosed outdoor space.

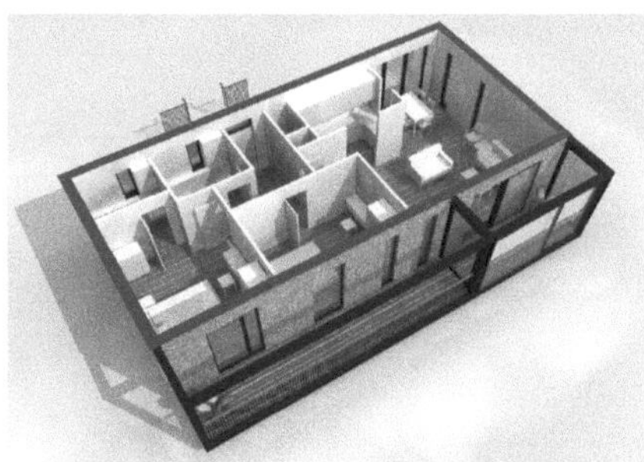

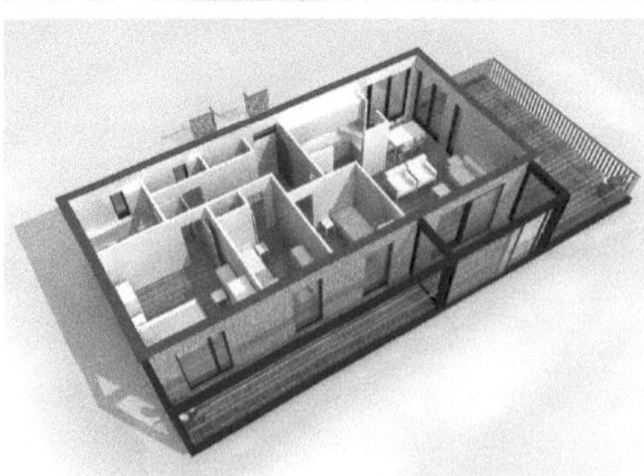

Temperature differentials between these glazed semi-outdoor areas and the internal living rooms could thus provide a good source of heating via passive solar gain and cross ventilation.

A final element to the scheme was the provision of rooftop private terraces for the larger family units, enabling much needed urban vegetation to be introduced at these high levels.

2.3 CASE STUDY 3 – OVERVIEW

This third case study explores our most sustainable scheme to date; won after a limited invited competition, the new, multi-purpose hall at Tower House School was required to deliver three distinct functions under the same roof – assembly/dining/performance – while also combining a music school, large flexible stage and a catering kitchen for the preparation of school meals.

The triangular plan, with three distinct wings surrounding a large, roofed hall space incorporated a unique ground source, passive ventilation system that utilized a network of underground large diameter concrete supply pipes.

In addition, high levels of insulation, natural daylight and low energy lighting ensured that the building's energy use remained far lower than that of comparable conventional building types. Materials were also carefully selected for their excellent life-cycle credentials, recyclability and robustness/fitness for purpose.

Despite the scheme employing a host of eco-friendly features, it was chosen for this chapter for the unique way in which it provided a low cost, maintenance free, passive air ventilation system to a building that would otherwise have made do with a standard mechanical alternative.

2.3.1 Multi-Purpose Hall, Tower House Scholl, Sheen, Richmond, London – Case Study 3

An invited competition brief called for a small multi-purpose hall, on a tight triangular plot at the far corner of a constrained playground, shoehorned into the grounds of a suburban former Victorian detached house.

The school governors outlined two key criteria for the winning commission: firstly that the scheme be as 'green' as possible; secondly that it achieve the brief within the maximum budget of £500K.

It became clear from the outset that to provide the accommodation the school desired – a new music school, a dedicated stage/performance space, an assembly and dining hall complete with a kitchen facility; and all 'under one roof' – almost the entire plot would need to be utilized.

Our solution proposed a triangular plan. This offered the best compromise between the various functions, and suited the site's restricted shape – allowing us the space to keep the structure below two stories in height; itself a key constraint, as the site was bounded on all sides by the gardens of three separate dwellings.

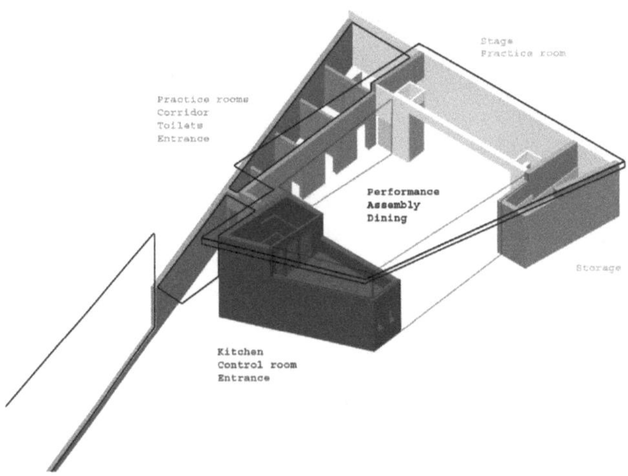

Clients frequently have pre-conceived ideas of what a 'green' building means: that the building uses no energy; that it requires zero cooling/heating, that it is made from entirely recyclable materials all obtained from clean, ethical non-polluting sources; and even that it looks 'eco'.

As a project progresses however, external factors act to alter, shift and undermine the original aspirations. Cost is almost always one of these.

To deliver a genuinely 'green' scheme and to avoid the cost trap, we decided to focus on one aspect of the building's design, the ventilation. It was essential that such an approach should be 'hard wired' into the building rather than be tacked on as an extra.

Given the site's orientation and the possibility of a large roof area, PV was considered, but it was the focus on providing a sustainable, low energy approach to the ventilation that ultimately kept our solution simple, cost effective, elegant and deliverable.

The key space within the project was the multi-purpose hall able to accommodate 100 pupils for morning assembly, fully seated lunchtime dining, and evening performances alongside visiting parents and guests.

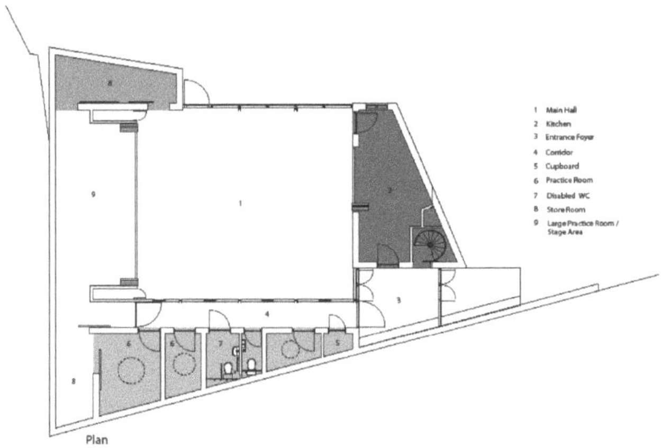

The need to change use throughout the day meant light control was important, so a system of retractable full height curtains was proposed which could be easily deployed to provide enclosure, sound attenuation and blackout. Allowing for these curtains however presented challenges with regards to the ventilation supply and the cooling/heating of the hall, especially with the varying temperature demands placed on the space by the multiple uses.

The hall assumed a central position within the plan, leaving three zones for the remaining functions.

A long, narrow 'wing' to the south of the hall contained the music school, made up of a collection of small, acoustically separated, practice/teaching rooms, instrument stores and a larger chamber room.

The western zone became the stage, wings and 'back of stage' area. In addition this space could double up as separate, large teaching/practice space for drama or school orchestra's practice, with bi-folding doors to separate it from the hall. The northern zone was designated as the official stage 'wing' and large props and staging store. Finally the eastern zone, adjacent to the front of the hall, housed the kitchen, plant, AV/sound/control booth and main entrance space.

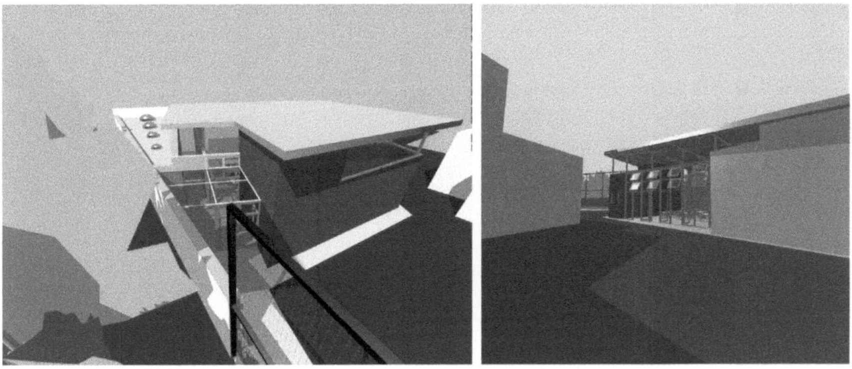

The height of the hall sloped down from two stories at the western end to a single story at the eastern end; making it ideal to locate the plant and control rooms in the higher section above the kitchen, and the proscenium arch at the opposite lower end.

A workable 'multi-purpose' facility was created with the use of low tech kit, such as curtains, manual bi-folding/sliding doors/partitions [for the stage] and a clerestory lit access corridor that doubled as an acoustic divide between the music school and main hall.

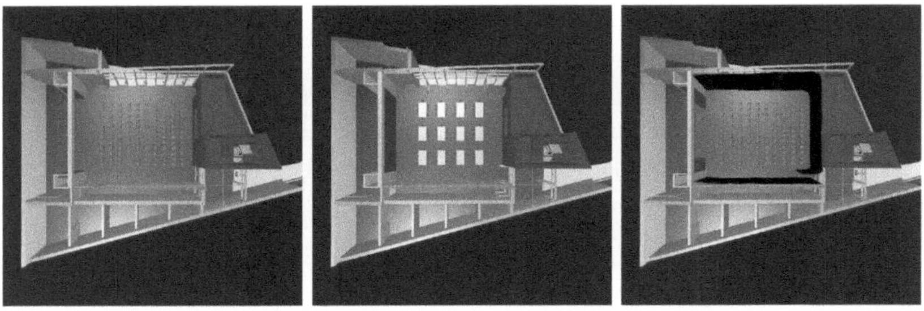

It seemed logical that the ventilation solution, easily one of the largest consumers of energy in such building types, should also follow in this vein. Proposing a building that utilized the entire site area, and was constrained on two of its three sides, left little room for courtyards or opportunity for fenestration along these boundaries. In addition, local authority planning restrictions and the brief severely restricted any form of vertical chimney or flue.

The design team turned to the only 'space' available outside the designated site: the remaining playground areas to the south and east.

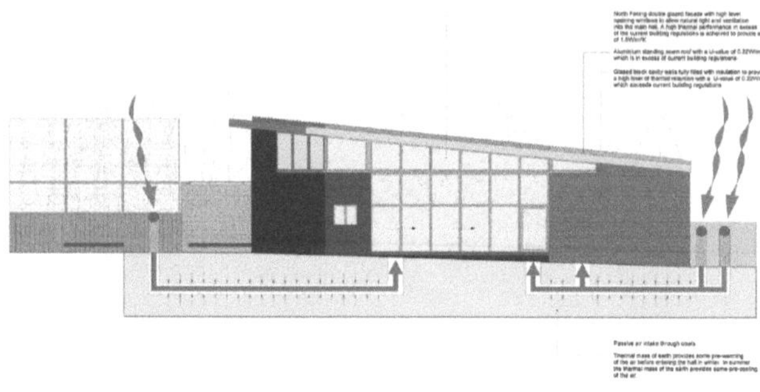

We were aware of some recent schemes that had utilized chilled beam technology to clever effect, but were conscious of the cost and limitations of such options in our case. However, ground source heating was becoming a more viable alternative, and we wondered whether an equivalent might not exist to facilitate the fresh air ventilation required for the facility, but in a passive way.

The design team was confident that other examples of passive ventilation would provide comfort to the client when adopting such an approach on their new building. The challenge was to convince the client that their particular site and circumstances would call for a re-working of the more traditional forms of passive ventilation, by proposing ground pipes. Ultimately it was just such a low tech approach, coupled with the added value of incorporating the system from the outset that won the client over.

Pioneered in various forms in 'eco' buildings as early as the nineteen sixties, the principle relies on the relatively constant, stable temperature of the ground at a depth of 1.5m; 14° C, and the difference this makes when compared to the ambient air temperature experienced at ground level [both in winter when the below ground temperature is warmer and summer when the reverse is the case].

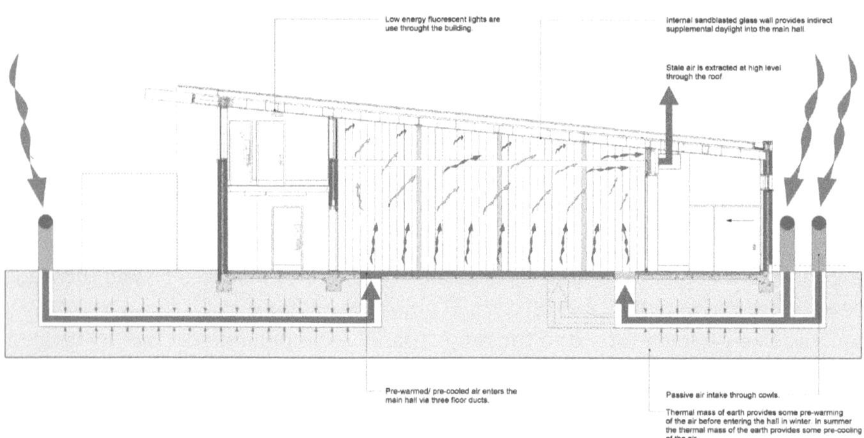

This below ground temperature constant has in recent times been more commonly exploited in modern, ground source, heat pump technology.

Utilizing such a constant temperature below the surface would require a suitable physical conduit, and in this case the design team focused on sealed pipes. Given the area of surrounding un-built playground, it was assumed there would be adequate facility for the burying of such sealed pipes. The theory held that this same constant ground temperature could be harnessed to cool or warm fresh surface air as it passed through submerged pipe runs en-route to supplying the building's ventilation needs.

For the system to be truly optimized, sufficient pressure would need to be generated, and this was proposed by way of specifying a given pipe diameter versus controlled supply/feed louvered damping to ensure that a constant flow of air supply was generated, with adequate exhaust provision to enable the warm stale air to leave the building.

This last part of the process also offered the additional option of heat recovery for re-circulation in the winter months.

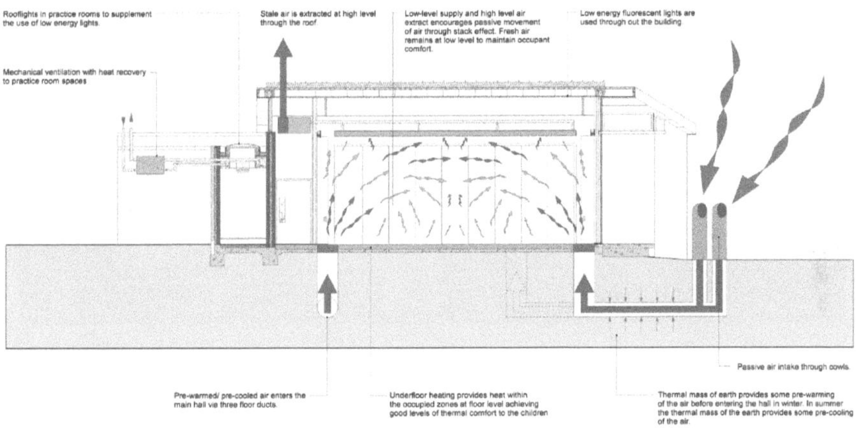

Tempering the air supply in this way meant an abundant, passive, low energy form of background cooling/heating could be easily supplied, combined with fresh air ventilation, to result in a low tech low maintenance installation.

Planning for such a system required a coordinated approach from the design team, especially as no commercially supplied, readily available 'kit' existed. Once detailed design commenced, the design team set about devising a solution that would prove both practical and 'low tech'. The system that was settled on was to comprise a series of large diameter underground pipes tasked with supplying fresh air to the central hall space.

The restricted site, and limited space available in the adjacent playgrounds, meant any subterranean system of pipes would need to be installed in such a way as to minimize the disruption to the normal functioning of the school, and this included leaving large areas of the playground cordoned off and out of bounds to the contractors; resulting in only two possible locations for the pipe trenches.

Further constraints resulted from the proposed diameter of the pipes; calculations by the Mechanical and Electrical (M & E) engineers suggested that limiting the number and length of pipe runs had resulted in an increased diameter for the supply pipes, allowing for maximum surface area to be exposed to the ambient thermal effects experienced below ground.

The final construction solution proposed the use of large, dense, concrete drainage pipes [with a diameter in excess of 500mm] placed in trenches that would in part run below the building's ground-bearing slab at a depth of at least 1.5m. In keeping with the low tech approach, these pipes were readily sourced from a general construction material supplier. Two runs were identified; the first along the south west boundary of the site to feed an area of the hall adjacent to the music school corridor; the second at the far north eastern corner of the site to feed the northern section of the hall.

Each run required a unique intake design as both were of different length, but were required to provide the same level of passive thermal cooling and heating.

The southern run intake had to be located as close to the boundary wall as possible in order to keep the playground area free, but could not extend further out from the building envelope than the extent of the glazed entrance canopy. A low, wide ground level vent was eventually proposed, carefully hidden below a seating bench running from outside into the entrance foyer.

Behind the grill, variable controlled louvers were employed to temper the in-coming fresh supply air and to provide the necessary restricted flow deemed adequate to generate sufficient pressure at the outlet end of the run inside the hall.

The north western run intake was located at the corner of the building to minimize the potential clash with an adjacent play frame and playground for primary-aged school children. Enough space existed for the intake to be more 'expressive' in form enabling the duct to provide visual cues for the school children, helping them to better understand the sustainable approach adopted for the ventilation.

The south eastern intake was subtle and only just noticeable below the entrance area bench; by contrast the north eastern intake was fully expressed, in the form of a funnel-like structure inspired by the vents employed to such iconic success in the Pompidou Center in Paris and Lloyds building in London [to name but two].

As with the south eastern vent, the diameter of the flue was dictated by the required pressure and flow rate for the supply air; resulting in a pleasing form that could be clearly expressed above the surrounding playground.

In addition to the ground pipes, a solution for the supply vents was required to provide fresh air into the building. As the brief stipulated a multi-purpose hall that could support assembly, lunch and performance; each use imposed a different load on the ventilation requirement. This was further complicated by the use of a 'low tech' approach to providing the required flexible program, by means of a curtain and bi-folding screens, all of which limited the options for the sizing locating the supply vents.

To overcome these complications, two long recessed trenches were devised to run the entire length of the hall. Located on both the northern and southern sides, these would be carefully coordinated with the retractable curtains to ensure that air-flow and circulation could not be hampered.

M&E calculations indicated that despite the sizable scale of the below ground installation, at peak times the passive air supply would require some augmenting to keep comfort levels at an acceptable level. To combat this shortfall an air handling unit was proposed that incorporated re-circulation and modest heat recovery functions. This unit could also double as the ventilation source for the music school WCs, acoustically sealed practice rooms and back of stage area. Eventually located at the rear of the stage behind the proscenium arch, the system incorporated a single, long, horizontal intake vent set into the face of the proscenium above the bi-folding partitions, AV installation and stage curtains, and also provided additional high level extract of the warm stale rising air likely to occur during peak usage times. Provision of this augmented mechanical ventilation would also act as a 'boost' for the passive supply, accelerating the flow and generating more air movement within the hall.

To meet wintertime heating demands, it was concluded that the most sustainable solution to augment the passive warm tempered ventilation was to install a low temperature background under-floor heating system throughout the main hall and principal rooms. In addition, by way of tempering the passive supply air trench 'gilled tube' type radiators were installed within the two long floor mounted supply vents.

At the commissioning stage, the M&E engineers were called upon to assess whether the desired effect of a flow of tempered naturally ventilated air was entering the hall via the ground ducts and vents as intended.

Initial testing indicated the system was functioning as required, however the client was not convinced, and to this end staff and governors were invited to place a thin sheet of paper over the internal vents to see the effect first hand.

After six months of use, a second survey of the building's use was conducted and the results indicated the following:

The school rarely turned on the under-floor heating in the winter months, as the hall temperature remained comfortably warm; even on the coldest days.

On an average warm summer day, at peak usage times, aside from opening the high level window vents the school rarely needed to open the north facing external sliding doors for additional ventilation.

Renewable and sustainable design informed a number of other facets to the scheme.

Careful consideration was given to the materials and their recyclability, durability and fitness for purpose as well as their sustainable credentials in terms of manufacture from renewable resources, and end of life recyclability.

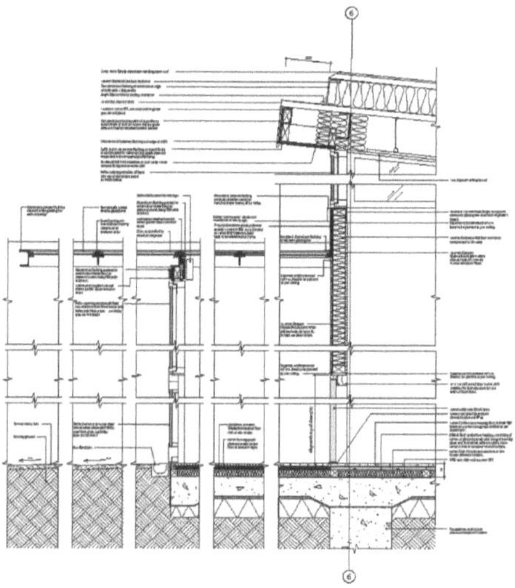

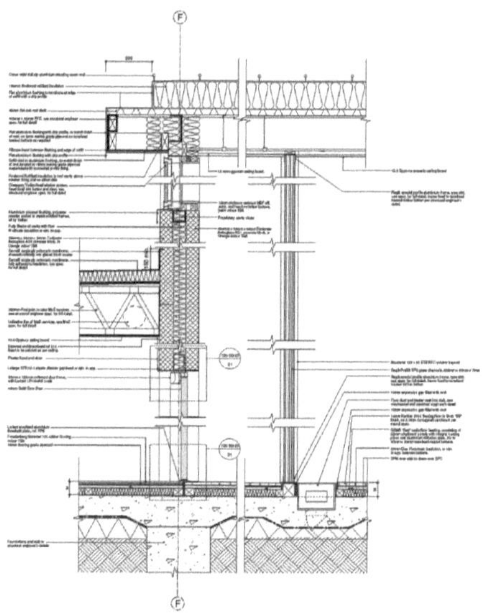

The following key materials were specified:

- A profiled, standing seam, aluminum roof – providing a long maintenance-free life span, with excellent post life recyclability, and very good solar deflection.
- Composite timber/aluminum, thermally broken, double glazed window/door units – with excellent U-Value, sound and draught performance – and constructed from renewable timber and recyclable aluminum.

- Profiled, semi-structural, full height, Georgian, wired glass panels between the hall and music school allowing for minimum framing and secondary support elements; these panels were a tough industrial grade system which were both durable and long life.
- Silicone resin dipped, concrete fair-faced blocks – incorporating a tough surface finish to a standard concrete block and offering durability, long life, and a single skin finish eliminating the need for a second applied surface finish to the exterior and interior of the hall.
- Resin impregnated, stratified, engineered, timber floor planks – these were installed throughout the main spaces of the building – utilizing timber from a certified sustainable source, the resin impregnation provided for an excellent life span and tough durable low maintenance finish.

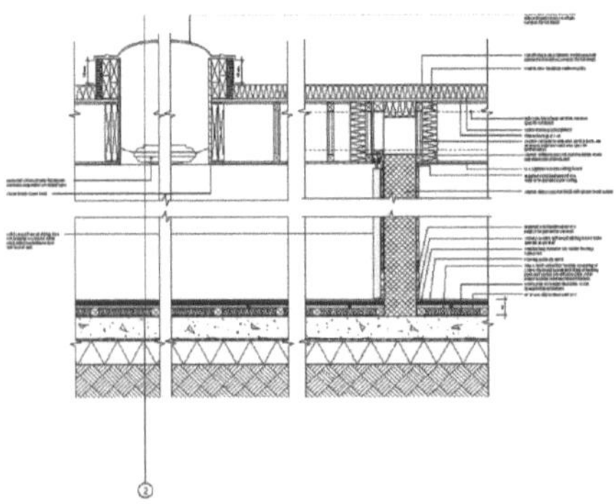

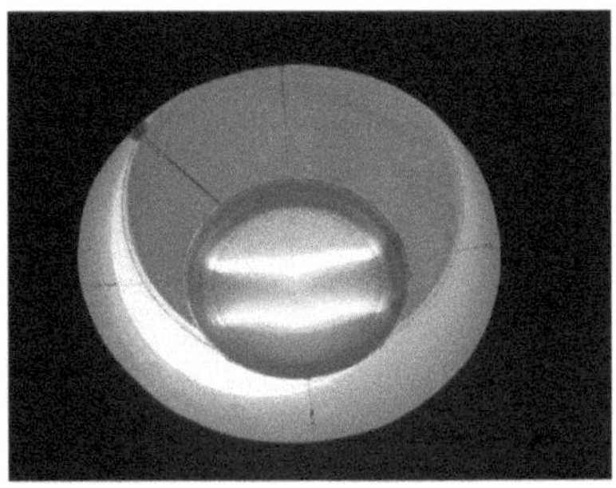

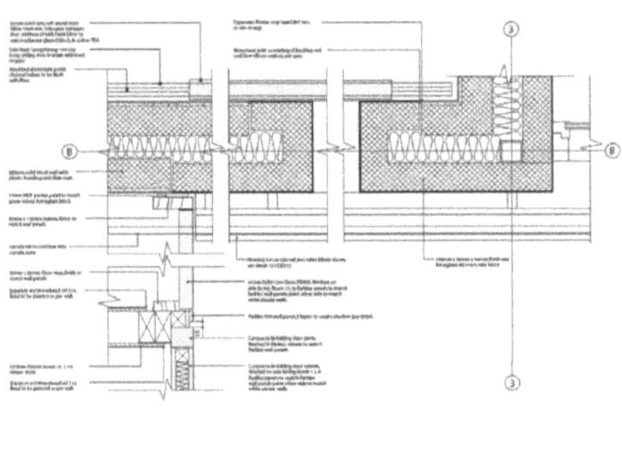

The use of natural daylight provided another area of energy saving. A large area of north facing glazing provided good levels of diffuse northern light to the main hall; the music school corridor was both top lit via facing clerestory panels and side lit via the tall vertical profiled glass panels; finally the acoustically sealed, small practice rooms gained excellent levels of natural daylight from circular domed roof lights, with clear polycarbonate circular fittings placed so as to 'float' in the center of the ceilings while minimizing the loss of natural light.

Extensive use was made of fluorescent low energy fixtures throughout the hall, including in the frosted, blown glass entrance lights, which concealed standard E27 low energy light bulbs.

2.4 CASE STUDIES 4 & 5 – OVERVIEW

The last section of this chapter focuses on two smaller schemes in which sustainable design has played a key role. Both are residential and both involved committed clients, keen on utilizing sustainable, environmentally conscious measures from the outset.

As with the larger projects discussed earlier in the chapter, both schemes adopted a low tech, low key, low cost approach to the provision of sustainable design.

However, the standout feature of both was undoubtedly the nature, type and selection of materials utilized, coupled with the unique form of construction adopted.

2.4.1 Black Diamond House, Tutti Frutti, New Islington, Manchester – Case Study 4

Designed in response to a competition held by UK regeneration company Urban Splash, Black Diamond House was to be one of a terrace of twenty-six new dwellings located in the center of Manchester, in a former industrial district.

The brief called for an exemplar house fit for the 21st Century, designed to suit the individual requirements of its respective owners, with the terrace taking inspiration from similar housing schemes in Holland. The Dutch schemes had provided prospective residents with a regular sized plot within which they could self-build their own semi-detached house, to any design and style they desired, with the only constraints being plot size and a height restriction. At Tutti Frutti three plot sizes were offered – small, medium and large – all of equal length but differing in width. A height restriction of five stories was also imposed.

Our client's brief called for a dwelling capable of housing a large family with good levels of storage and amenity space.

The local planning authority required the scheme to meet stringent housing design criteria, including 'Lifetime Homes' and for the dwellings to be fully wheelchair compatible or easily converted to accommodate wheelchair use. This was to be combined with an environmentally aware approach to the design to ensure that the scheme complied with a minimum 'Code For Sustainable Homes' score of 4 or above [with 6 being the highest].

The link between the old and new embodied in this regeneration project provided the inspiration; the old former industrial center built on the cola trade now given over to residential and service sector activities, and how this was echoed in the natural transformation that occurs between coal and its distant carbon cousin – the diamond. From rough to smooth, from raw to polished, from dark to gleaming, from solid to transparent; a sustainable design for such a location would surely need to respect the form of carbon so closely associated with such wide-scale environmental damage.

Through a process of physical modeling, a faceted rough/smooth texture was the starting point for a series of outline design studies that eventually led to the faceted features of the final scheme design. In our initial research, numerous forms of construction were considered for the primary structure of the house: from traditional block/brick cavity wall with joisted floors, to steel frame and ultimately concrete frame. However most options were quickly discarded for reasons of cost and sustainable performance/credential with the exception of two: full timber frame construction and SIPS [structurally insulated panel system] construction.

Timber frame was our initial choice, as it ticked all the right boxes in terms of renewable and sustainable design. However it was found to be problematic given the shape and form of the faceted design, and there was also some concern over the risk of differential movement and settlement of the frame/structure creating difficulty in setting out movement joints and cladding panel sizes.

We turned our attention to SIPS, which we found to be ideally suited to such a design approach, being stable, lightweight and rigid with the added bonus of flexibility as the panels could be easily re-configured to suit last-minute on-site setting out and tolerance issues. With the exception of the separating party walls, which building regulations required to be block work for fire separation and acoustic performance, it was possible to construct the entire dwelling [structure, floors, internal/external walls and roof] entirely from SIPS.

The system comprised a thick layer of highly insulating, expanded foam [low CFC] sandwiched between two sheets of orientated, stratified, timber boards [each approx 18–25mm thick]. As the scheme included a complex faceted front, rear and roof elevation, the sheets could be custom made to suit the various shapes and forms and easily jointed with a system of smaller in-fill link panels to provide a seamless external surface.

Another advantage of the SIPS approach was the relatively ‘thin’ wall thickness that was necessary to achieve a high degree of thermal insulation when compared with the equivalent traditional build approach, with the wall/insulation/structure all combined into one single building element providing further space saving efficiencies.

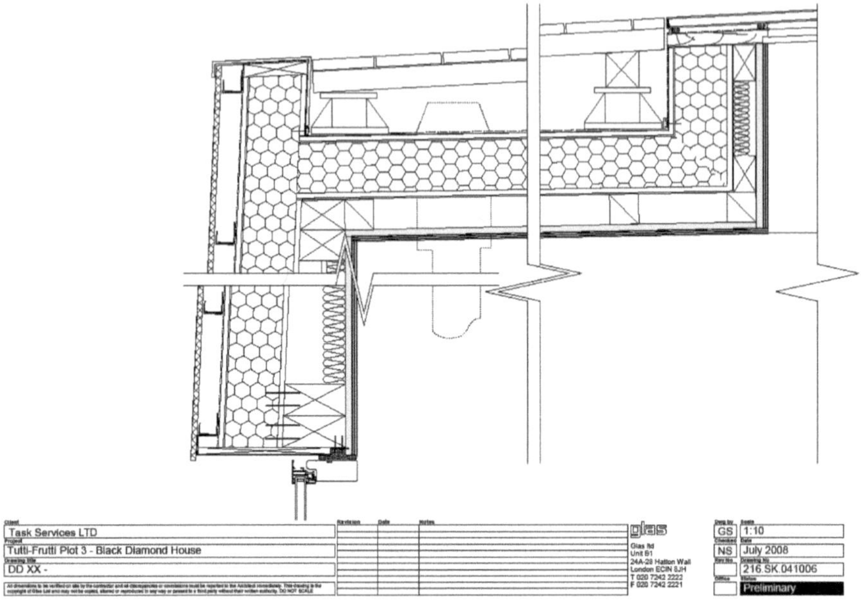
Task Services LTD
Tutti-Frutti Plot 3 - Black Diamond House
DD XX -
Glas ltd
Unit B1
24A-28 Hatton Wall
London ECIN 8JH
T 020 7242 2222
F 020 7242 2221
GS
1:10
NS
July 2008
216.SK.041006
Preliminary

Learning from our previous experience with passive ventilation at the school hall, the design team felt strongly that passive ventilation should be introduced as an integral part of the scheme design for Black Diamond House.

However, significant differences between the two projects presented some key challenges. Whereas the site for the school hall was open, and the location and orientation of the building partly decided by the design team, the available plots for this dwelling were all rectangular, and hemmed in on two sides with pre-determined north/south facing elevations. Any thought of utilizing a similar ground source tempered design were out of the question as the foundations were all to be pre-designed and pre-installed prior to the dwellings' construction.

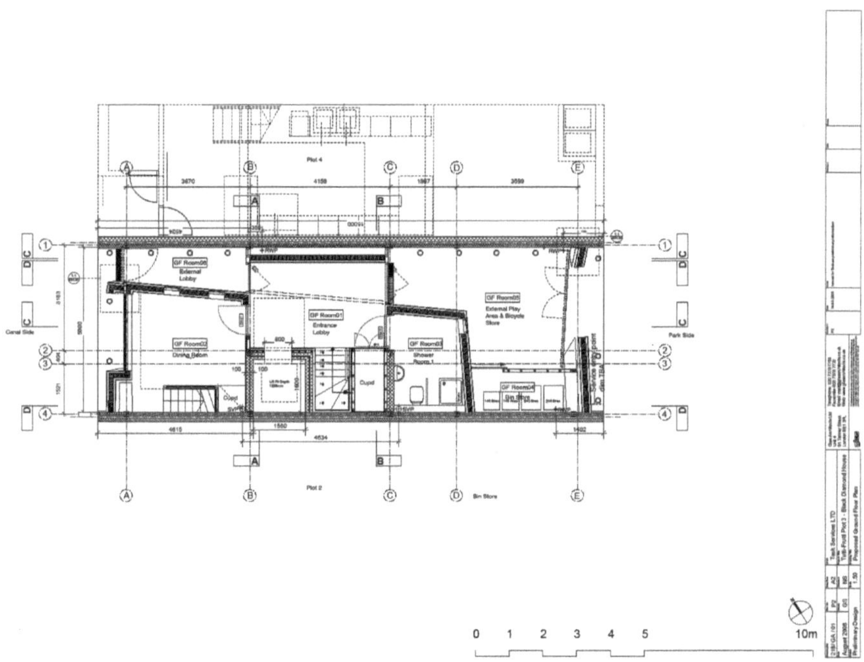

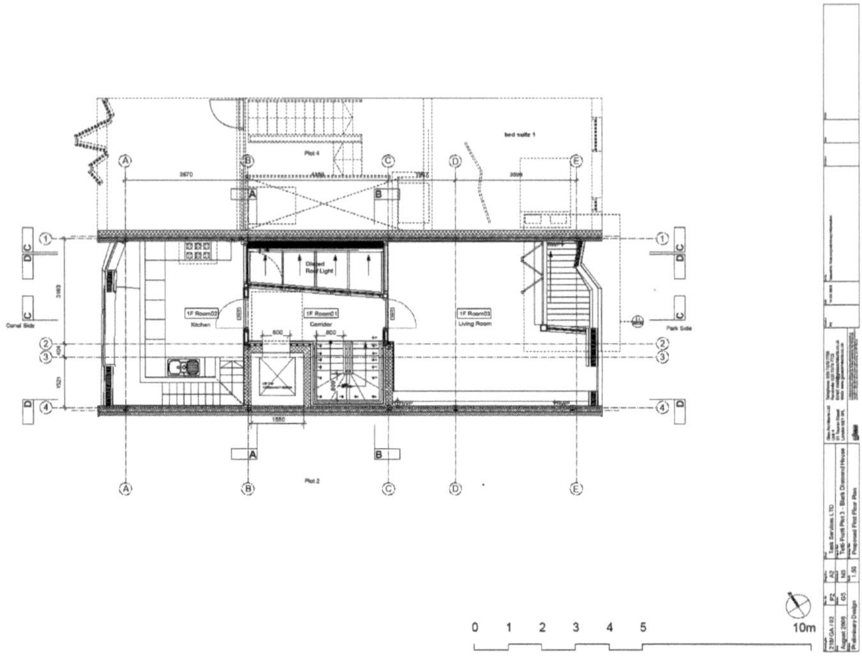
Canal Side
Park Side
1F Room02
Kitchen
1F Room01
Corridor
1F Room03
Living Room
Glazed Roof Light
Plot 4
Plot 2
0 1 2 3 4 5 10m

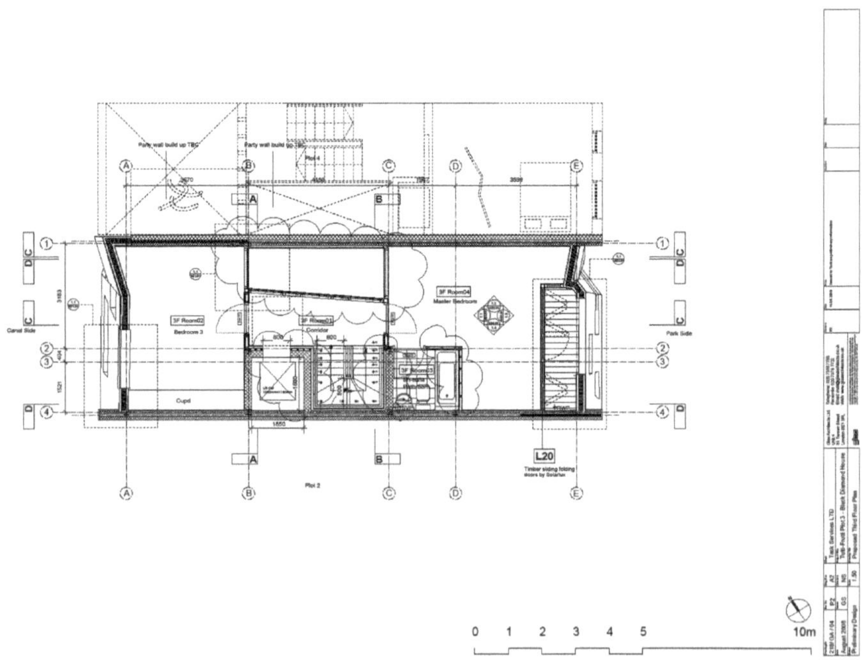
Canal Side
Park Side
3F Room02
Bedroom 3
3F Room01
Corridor
3F Room04
Master Bedroom
Plot 4
Plot 2
L20
0 1 2 3 4 5 10m

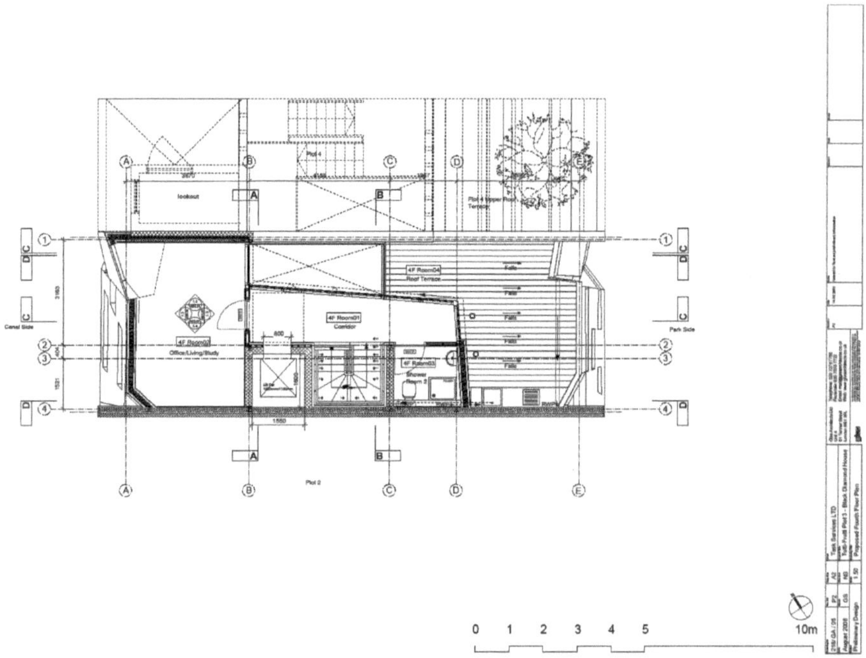
lookout
Corridor
Office/Living/Study
Plot 2
0 1 2 3 4 5 10m

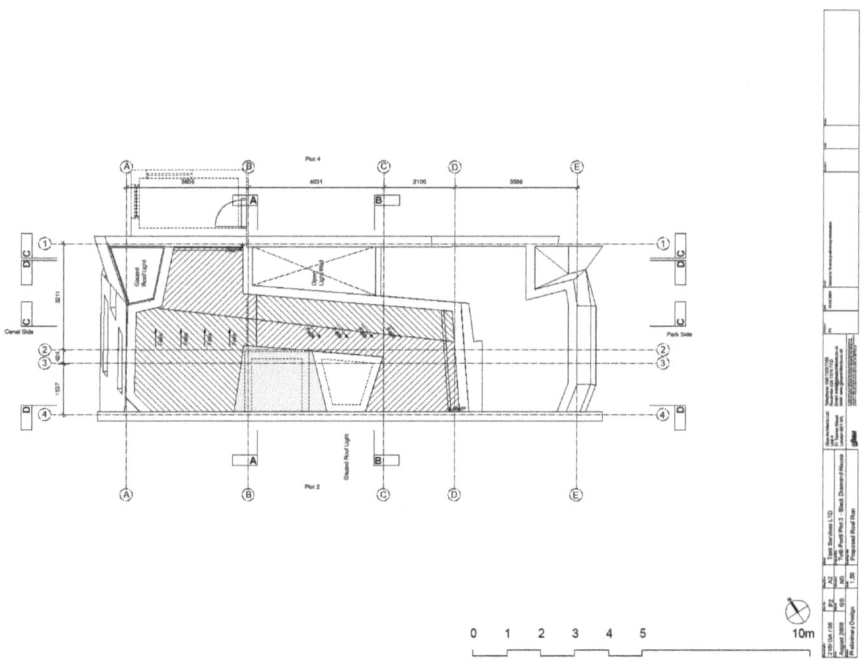
Plot 4
Plot 2
0 1 2 3 4 5 10m

The only advantage of such a constrained site was height. House designs were permitted to rise up to five/six stories, which, combined with our client's choice of medium width plot, provided just enough space to propose a vertical shaft through the center of the dwelling.

This shaft developed into a light well, providing a source of secondary daylight to the otherwise long, narrow, single aspect rooms. The shaft also allowed for the habitable rooms to benefit from passive cross ventilation, by means of louvered openings. At ground floor level, the base of the well comprised a zone of louvered shutters designed to open and close as the need for ventilation changed throughout the day and from season to season. A constant supply of fresh air was to be provided via open slatted security doors at either end of the ground floor entrance areas, both of which provided semi-enclosed, weather protected entrances and secure play areas.

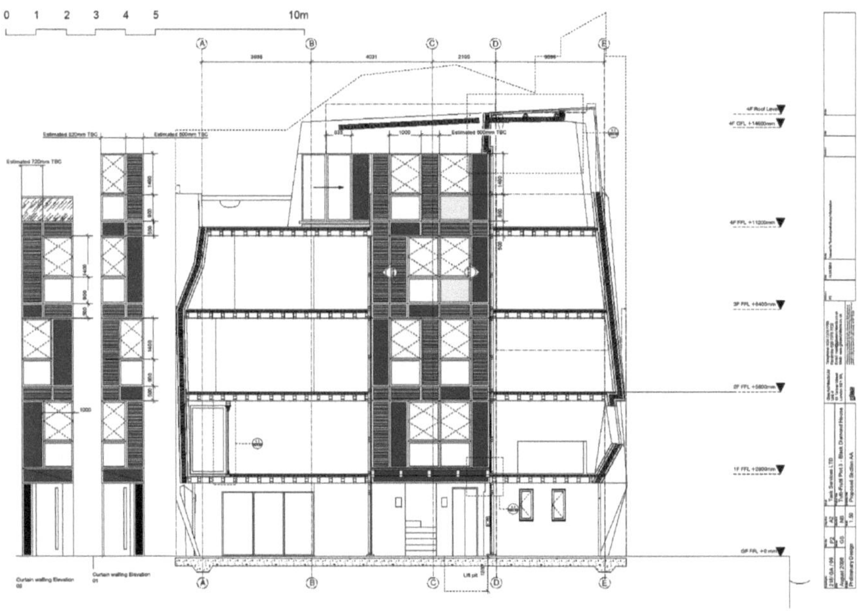

Review and first hand observation of Structurally Insulated Panel System (SIPS) panels installed at a British Research Establishment (BRE) research center flagged up a key disadvantage of such systems – lack of thermal mass. There was concern that utilizing a super insulated, lightweight structure would render it prone to overheating, especially in the summer months and when subjected to strong direct sunlight; leading to an uncomfortable internal living environment.

The solution was to introduce thermal mass via the external cladding. A material would need to be sourced that provided a reasonable degree of physical mass, while still being light and durable enough to work as an external panel, and flexible enough to suit the complex faceted form.

Our eventual choice was Fiber C – a type of fiber-reinforced concrete panel – that possessed all the above qualities, as well as matching the aesthetic qualities we were looking for.

2.4.2 Unit 2, The Light Works, Brixton, London – Case Study 5

The final but by no means least significant case study illustrates that sustainable design is not the sole preserve of larger projects, but can be effectively and attractively applied to even the smallest scheme and budget.

Our client had purchased an empty shell and core space in a former light bulb factory, and asked that we devise our scheme to deliver a live/work family dwelling. The one key proviso was that as many eco-friendly, sustainable materials be utilized as possible.

To minimize cost, and preserve as much of the internal concrete structure as possible, the existing cut outs to the floor plates were utilized to provide vertical access throughout the dwelling. Being centrally placed in the long narrow plan, clever use of the space in and around the stairs was required in order to provide the necessary accommodation.

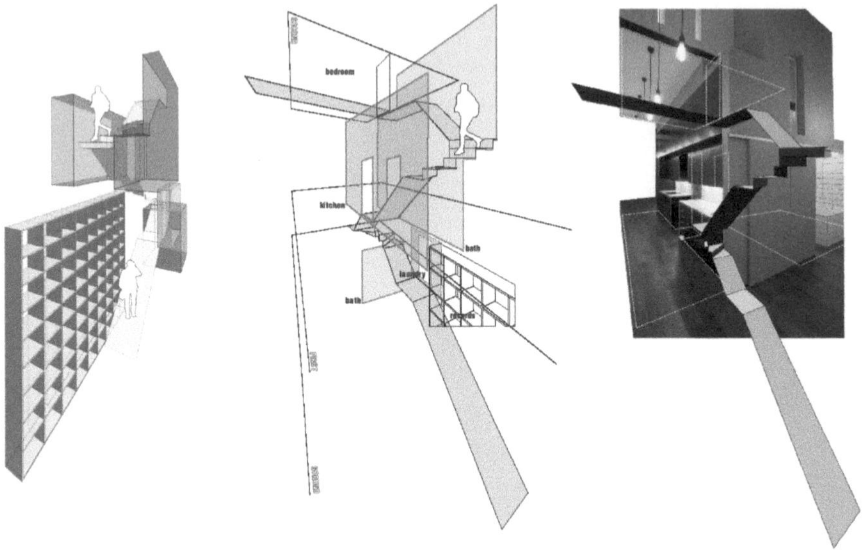

A tightly winding square spiral stair was devised that could allow various functions to be placed in and around the leftover space. The scheme was divided into ground floor work and guest area, first floor living and top floor bedroom space. At first floor level, the kitchen and bathroom slotted into the stair enclosure leaving two clear zones for living and dining, and on the second floor, built-in cupboards lined the stair core wall leaving a clear zone for the master bedroom.

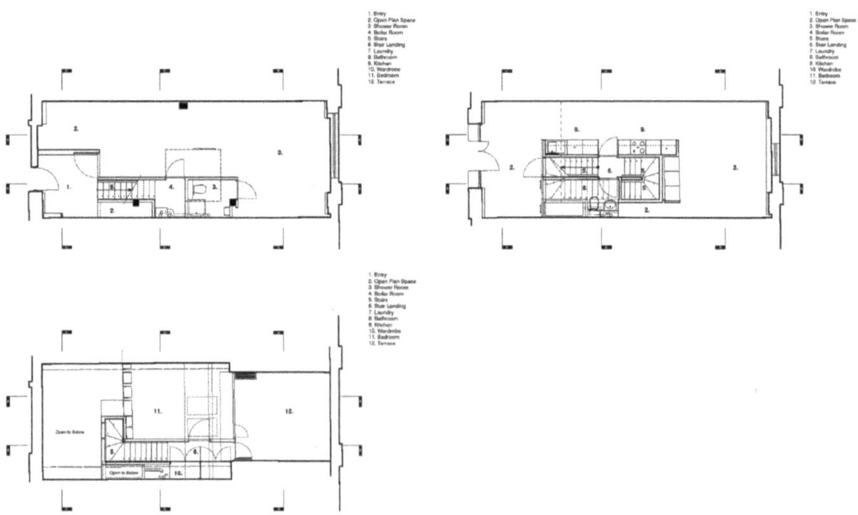

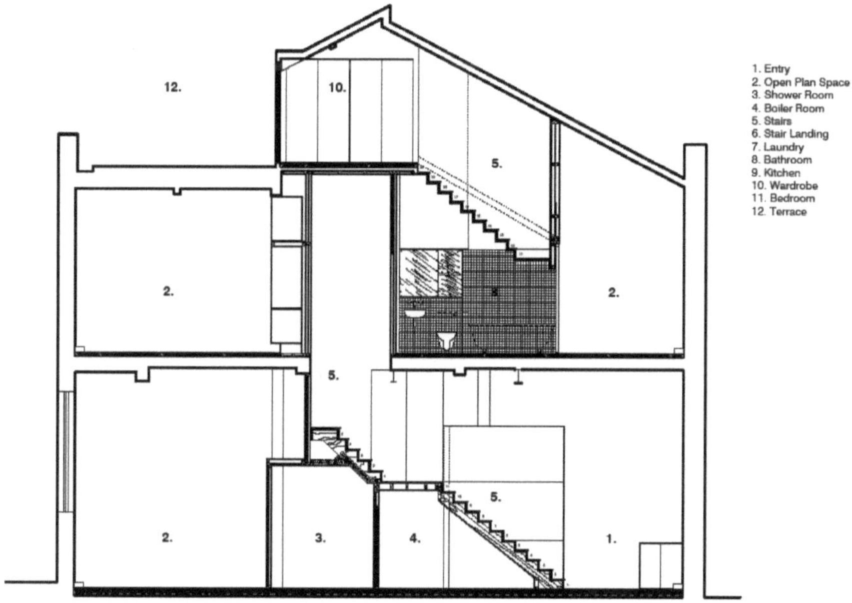
1. Entry
2. Open Plan Space
3. Shower Room
4. Boiler Room
5. Stairs
6. Stair Landing
7. Laundry
8. Bathroom
9. Kitchen
10. Wardrobe
11. Bedroom
12. Terrace

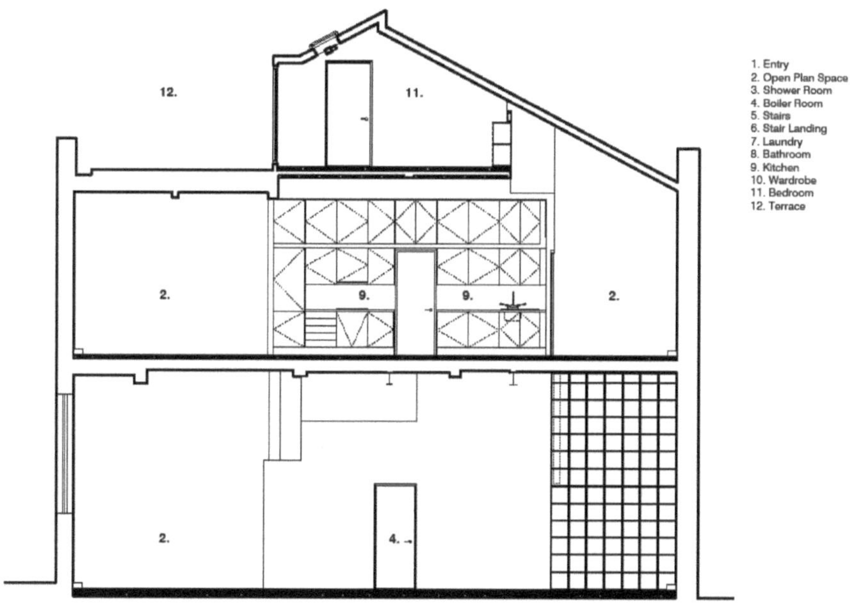
1. Entry
2. Open Plan Space
3. Shower Room
4. Boiler Room
5. Stairs
6. Stair Landing
7. Laundry
8. Bathroom
9. Kitchen
10. Wardrobe
11. Bedroom
12. Terrace

The entire new stair structure as well as the partitions and suspended ceilings were built from Forest Stewardship Council (FSC) certified timber. Phenolic shuttering ply was utilized as the finished panel for all kitchen cabinet doors, living room cupboards and storage shelves, as well as lining the ground floor staircase and for the balustrade detail. Lamb's wool insulation was utilized throughout all the partitions, and a recycled glass and resin worktop was provided in the kitchen. Organic eco paints were applied throughout, and the lighting was provided by way of heritage squirrel cage lamps [a nod to the former light bulb factory] set on a universal dimmer and run at 30% power to conserve lamp life and reduce energy use.

It could be argued that the reuse of a redundant structure was itself a highly sustainable approach to development, however as is often the case the existing fabric required a considerable degree of renovation and improvement, and this included upgrading the top floor doors, adding higher levels of insulation to the roof and below the ground floor screed, and refurbishing the existing industrial steel window frames. Low surface background under-floor heating was also introduced, powered by a highly efficient condensing boiler.

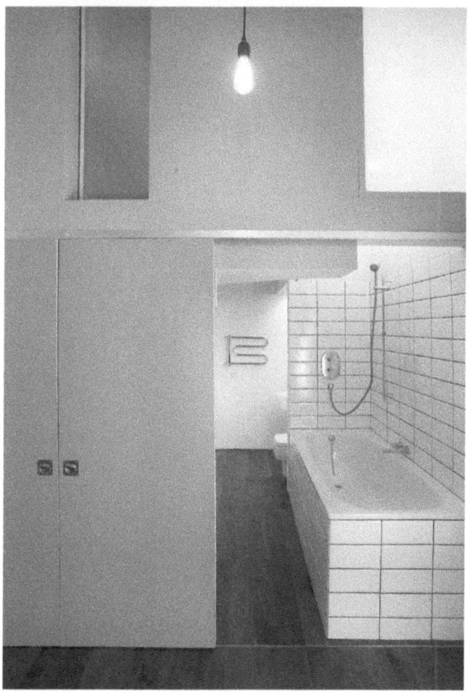

2.5 CONCLUSION

Our experience with case studies 1 and 2 illustrates that large scale projects need not rely on extensive procurement and implementation of add-on systems such as PV, wind or ground source heat technology, all of which can prove costly and ultimately end up being value engineered out.

By careful consideration of site specific criteria such as orientation, aspect and context, combined with the use of sun path mapping and intelligent space planning, a sustainable approach to building design can be made integral to a scheme.

In the case of these two schemes, careful orientation and the benefits this offered in respect of passive cross ventilation, solar thermal gain and an improved level of user amenity by way of high levels of natural daylight and sunlight played a significant role in reducing the buildings energy use.

The standout feature of case study 3 was the ground source passive ventilation system. By providing a simple, low cost, low tech approach to the provision of tempered fresh air supply to the school hall, a sustainable, low energy solution became intrinsic to the design of the foundations and site works. Once installed and attenuated, the system provided a long term, low maintenance, economic solution to the ventilation needs of the building, which also helped reduce its carbon footprint.

Size and budget need not preclude the intelligent implementation of sustainable and renewable design. Case studies 4 and 5 demonstrated that even at a small scale, careful design and specification can have an impact on a scheme's energy use and ecological footprint. Both schemes employed a passive approach to reducing their building's energy and carbon emissions; one with the use of modern highly insulated SIPs panels that acted as structure, fabric and insulation all in one panel; the other with imaginative sourcing of renewable and sustainable materials throughout, from organic paints to lamb's wool insulation.

Chapter 3

Sustainable Construction Materials

Andrew Miller and Kenneth Ip
Brighton University, Brighton, Sussex, UK

Chapter Outline

3.1 INTRODUCTION

3.1.1 World Resources

As we reach the end of the year 2011 the population of the world is approximately seven billion inhabitants, nearly four and a half times as many as there were at the beginning of the twentieth century. Over the same time period, standards of living and health care have risen and we are living longer. The consequences of these changes are that our demand for buildings to provide shelter for us to live, work and play in are ever increasing.

Sustainability, Energy and Architecture. ISBN: 9780123972699

Modern buildings consume the natural resources of the planet; requiring materials, energy, water and land. It is therefore essential that we use these resources in a responsible manner, optimizing their use and ensuring their sustainability for future generations.

Consumption of different resources are often interlinked, as the materials used for construction will require energy and water for their manufacture and transportation. There will be further impact on the land when mining or harvesting the raw materials and the development of sites for constructing the buildings themselves.

Construction materials can be considered in two categories; finite and renewable. Finite resources such as minerals and metals are mined from the earth and cannot be replaced, whereas renewable materials such as trees and plants grow within manageable life cycles so that their availability for future generations can be maintained.

Traditional construction was dependent upon local materials such as stone, timber and straw, which required little processing or transportation. Modern buildings, on the other hand, source building materials from around the world, and industrialization has enabled the production of materials with appropriately enhanced properties to meet changing demands. The geographical location of minerals and metals is often not within the regions of the developed world, resulting in heavy transportation impacts as well as environmental and social impacts on the developing countries in which they are found.

Some materials such as iron ore are considered as plentiful within the composition of the earth and there is a danger that they may be considered as an infinite resource. They are however becoming more difficult to mine, increasing costs and damage to the surrounding land and its ecology. Efficient use of these materials, facilitating re-use and recycling when the original building has come to the end of its useful life, should be targeted in order to minimize the burden on the planet. We are currently in the era when we expect technological advances to enable more sophisticated exploration techniques and the development of equipment to retrieve resources from less accessible locations. Economics are considered as the limiting factor in the mining of these resources; nevertheless they are not being replaced and are therefore truly finite.

3.2 DEMAND FOR CONSTRUCTION MATERIALS

The demand for new buildings and the upgrade of existing ones to meet modern standards of energy efficiency and quality of life for the occupants is increasing. Developing countries often have the fastest growing populations and increasing requirements for better standards of housing. They are therefore in need of increasing amounts of resources in order to reach the standards which have been enjoyed by developed countries for decades.

In the developed areas of the world, in addition to the demand for more houses, hospitals, schools and buildings of all types, the threat of climate change and of depleting fossil fuels has led to the need to upgrade the existing stock. Shifting trends in lifestyle, such as the increasing numbers of single parent families, have led to further pressures on the housing stock and the need for different types of accommodation.

Worldwide, the construction industry uses probably 40%-50% of total mass of materials consumed [1]. The current total figure for materials consumed by this sector [12] is around 60 billion tonnes per year, with a prediction that it may rise to 100 billion tonnes per year by 2030.

The construction industry in the UK is responsible for the consumption of approximately 420 million tonnes of materials each year. Not all of this material ends up as part of the completed building, as approximately 60 million tonnes is discarded as the spoils of mining the raw material, and further losses occur from wastage during construction. The overall efficiency of materials consumed in terms of additions to the building stock is only 64% [2].

3.3 MATERIAL RESOURCES

Since the demand for buildings throughout the world is increasing, so the pressure on the supply of materials is also increasing. The stock of virgin finite materials is deminishing and the efficiency of their use is being addressed through attempts to reduce waste.

Material scientists have provided the construction industry with many new synthetic and composite materials that provide alternatives to traditional materials. These materials have been developed to provide enhanced physical properties, facilitating innovative building design and simplifying the construction process. Inevitably they still require an original source material, and for many, this is oil. Use of these materials also consumes energy and water in their manufacture and transportation.

3.4 RENEWABLE MATERIALS

Renewable materials, which can be sown and harvested within manageable lifetimes, can provide for the construction industry in a sustainable way, subject to appropriate forestry and agricultural management and control. National and international organizations such as the Forest Stewardship Council (FSC) and the Program for the Endorsement of Forest Certification (PEFC) serve to validate sources of timber that are sustainably managed. The construction industry in many parts of the world is encouraging the use of timber from these sources, by undertaking the environmental impact assessment of buildings prior to construction. Such schemes include the Building Research Establishment

Environmental Assessment Method (BREEAM) award credits for materials that are sustainably sourced.

Replacement of softwood, such as spruce and pine, can be managed on a 50 year cycle whereas hardwoods growing to maturity will require 100 years or more. There are however a variety of construction materials that depend on annual rotation crops and coppice timber that can be harvested in 15 to 20 years.

Some crops may be grown specifically for use as construction materials, but often they are by-products of crops grown for food or other purposes, or sometimes the construction materials are produced from what otherwise would be considered as waste. Straw bales are made from the residue of cereal crops after the wheat or rye has been harvested. Traditionally these are used for animal bedding and mushroom compost, but they can be used in construction, because of both its insulation value and its structural properties. Rectangular bales can be used either individually, built into a load bearing wall, or as the infill within a structural insulated panel (SIP) within timber frames.

There are many examples from around the world where plant fibers are used to bind or strengthen earth, clay or cementatious materials to form different parts of the building envelope. Advantage can be taken of local crops such as pineapple leaves, coconut husks or hemp shives, so utilizing parts of the plant that may otherwise have gone to waste.

Whilst timber has an important role in the structure of a wide range of buildings, many renewable materials have tended to be used for domestic or small scale commercial buildings. However one element of building work that has a history of renewable material use, and where these are widely considered to have advantageous properties is the provision of internal finishes. Bamboo is used as a very hard wearing floor material which is made by cutting into strips and laminating into boards, whilst jute and sisal are made into floor coverings. Linseed oil is the principal constituent of the traditional linoleum floor covering which is primarily mixed with wood and cork flour and limestone powder. The majority of these components are renewable and biodegradable.

The advantages of using renewable materials can be seen in terms of both their short and long term environmental impacts. Their use must however be part of a broader context of stewardship of the planet, balancing the development of crops that are grown for materials, for fuel and for food. There is no advantage in achieving a sustainable flow of construction materials if we are giving over the land that was required to provide sufficient food to support ourselves.

3.5 RECYCLED MATERIALS

An important aspect of the efficient use of materials is to consider what happens to the individual components when the building itself gets to the end of its useful life. For many years, the economic value of aluminum, copper and steel has driven the recycling industry, as the cost of separating the materials, transporting and reprocessing was recouped from the recycled product. Many other materials are sent to landfill because the cost of separation or the value of the recycled product does not make their recovery financially viable.

The use of landfill itself has environmental implications in that in the UK. We are running out of places to dump our waste, even if we neglect the implications of the pollution to land and water caused by the sites themselves. In recent years there have been dramatic rises in landfill tax which have forced contractors to think carefully about demolition waste and to consider separation and recycling as a cost saving measure.

Building designers are aware of the resource issues surrounding the materials used in their buildings and are looking for means of recycling the materials at the end of the useful life of the building. 'Design for Deconstruction' describes a design strategy that facilitates the deconstruction rather than demolition of the building, and considers the opportunities for re-using or recycling the individual components or materials.

In some cases, the recycling industry and associated networks are not yet developed, resulting in the need to transport used materials across countries or continents to an appropriate reprocessing plant. However, as the demand grows local industry will develop to provide the required facilities which will alter the environmental balance further towards recycling.

3.6 LIFE CYCLE ANALYSIS

The selection of materials for any construction project must inevitably be driven by their physical and aesthetic properties and also by their availability and cost. Sustainable design demands that further consideration is given to the environmental impacts of the materials, making it less acceptable to use materials that place a greater burden on the natural environment. It is the designers' responsibility to balance all of these issues whilst meeting the expectations of their clients. Building regulations dictate minimum standards for properties like structural strength and thermal insulation, whilst leaving the designer to balance elements such as durability, aesthetics and costs once minimum standards have been reached.

Environmental impacts of the construction materials are generally not part of the legal requirements for building construction, but often form part of a general assessment of quality which is recognized through assessment systems such as BREEAM or the Code for Sustainable Homes (CSH). They are not legal requirements but are a priority for some clients and planning authorities.

Analysis of the environmental impact of a building requires a life cycle approach, evaluating the environmental impacts from the winning of the raw materials, which may include planting and growing renewable materials, through all the stages of its life to eventual disposal.

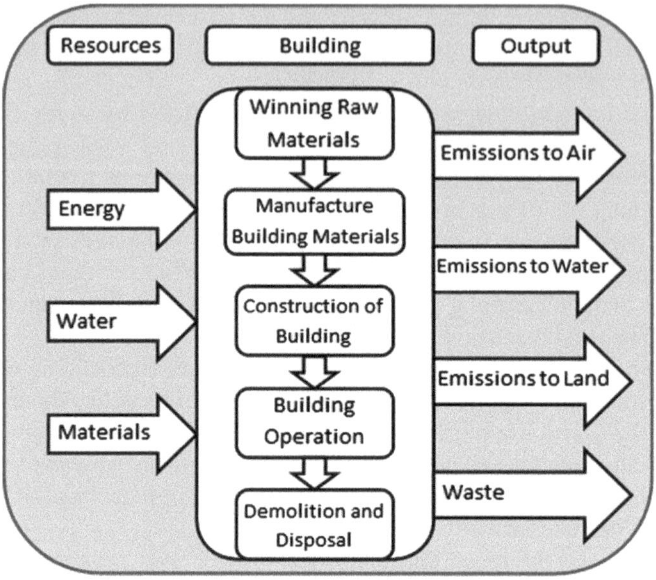

Ref: Overconsumption? Our use of the world's resources, Friends of the Earth, 2009

Schemes such as BREEAM and the CSH evaluate a wide range of environmental issues including energy, water, materials, surface water run-off, pollution, waste, health and wellbeing, management and ecology in the context of the whole building.

The life cycle approach enables the designer to consider the full environmental impact of design decisions and material selection on a like-for-like basis. However, before this can be effective, of the boundaries under consideration must be clearly defined. The expected life of the building needs to be declared, as this will determine the number of times that components will need to be replaced within the useful life of the building itself. There should also be clarity with respect to whether indirect burdens, such as the impact of making the machines that produce the construction materials, or the lorries that transport them, have been taken into account.

Assessment of individual building materials and building elements has been undertaken by the Building Research Establishment in its Environmental Profiles database. The methodology is clearly described in reference [3] and these profiles form the basis of the Green Guide to Assessment [4] which is used by both BREEAM and the CSH. The Green Guide to Specification produces a rating from A$^+$ for the materials with the least environmental impact down to E for those with the greatest.

The Environmental Profiles system involves consideration of 13 separate impacts which are then weighted into a single rating. Details of the individual issues and the weightings employed are presented in Table 3.1.

TABLE 3.1 Environmental Impact Categories, Issues Measured and Weightings for Environmental Profiles [4]

Environmental Impact Category	Environmental issue measured	Weighting %
Climate Change	Global warming or greenhouse gas emission	21.6
Water extraction	Mains, surface and groundwater consumption	11.7
Mineral resource extraction	Metal ore, mineral and aggregate consumption	9.8
Stratospheric ozone depletion	Emission of gases that destroy the ozone layer	-9.1
Human toxicity	Pollutants that are toxic to humans	8.6
Ecotoxicity to fresh water	Pollutants that are toxic to freshwater ecosystems	8.6
Nuclear waste (higher level)	High and intermediate level radioactive waste from nuclear energy industry	8.2
Ecotoxicity to land	Pollutants that are toxic to terrestrial ecosystems	8.0
Waste disposal	Material sent to landfill or incineration	7.7
Fossil fuel depletion	Depletion of oil, coal or gas reserves	3.3
Eutrophication	Water pollutants that promote algal blooms	3.0
Photochemical ozone creation	Air pollutants that react with sunlight and NO_x to produce low level ozone	0.2
Acidification	Emissions that cause acid rain	0.05

Climate change can be seen to be the impact that is given the greatest weighting within the system. It is based upon the greenhouse gases emitted in the production of the material; which is directly related to the energy consumed in those processes. Embodied energy is often used as an indicator of the impact on climate change for individual materials. However this can be misleading as the correlation between energy consumed and carbon emissions will be dependent upon the type of fuel used and the way it is generated; embodied carbon is a more direct indicator.

3.7 EMBODIED ENERGY

The embodied energy of a material refers to the amount of energy consumed in providing that material. There are however many interpretations of the term 'embodied energy' and in order to make useful comparisons between different materials it is necessary to have a clear understanding of the definition being employed.

A common definition of embodied energy is the energy consumed up to the end of the manufacturing process (cradle to gate). However it may include delivery to the construction site (cradle to site) or even the construction processes into the completed building. The latter two definitions are clearly site specific, and relate only to an individual building.

The definition is further complicated by the consideration of the scope of the analysis which can include the Gross Energy Requirement (GER) or more simply the Process Energy Requirement (PER). The inclusion of energy consumed for transportation of materials between processes is a further element requiring clarification before figures can be compared.

3.8 GROSS ENERGY REQUIREMENT

The Gross Energy Requirement is a measure of all the energy inputs to a specific material. It can be considered as the total energy consumption for which that material has been responsible. Thus it includes not only all the energy consumed in winning the raw materials, transporting them and their manufacture into building materials and components, but also the manufacture and maintenance of the plant used in winning and processing those materials, the transportation of the labor force for all of these activities and the repair of the environmental damage caused by these processes.

3.9 PROCESS ENERGY REQUIREMENT

The Process Energy Requirement is the energy consumed in the processes directly undergone by the building material or product. It will include the energy consumed by the plant and machinery throughout the processing, but not that of the second and higher generation consumptions of the plant that made the plant and the repair of damage caused by the processes.

There are, however, different approaches to the evaluation of these figures, and care is required when using published data in order to make comparisons. Common issues where clarity of assumptions and methodology is required are:

- Apportioning energy where more than one product undergoes the same process or one is a by-product of another.

- Inclusion of transportation and how return journeys of the delivery vehicle are considered.
- Apportioning transportation energy for a part load on a ship or other mode of transportation.

The most comprehensive document currently available to assist in identifying assumptions made in published figures is the Inventory of Carbon and Energy (ICE) published by Building Services Research and Information Association (BSRIA) based on the research of Hammond and Jones at Bath University in the UK [5].

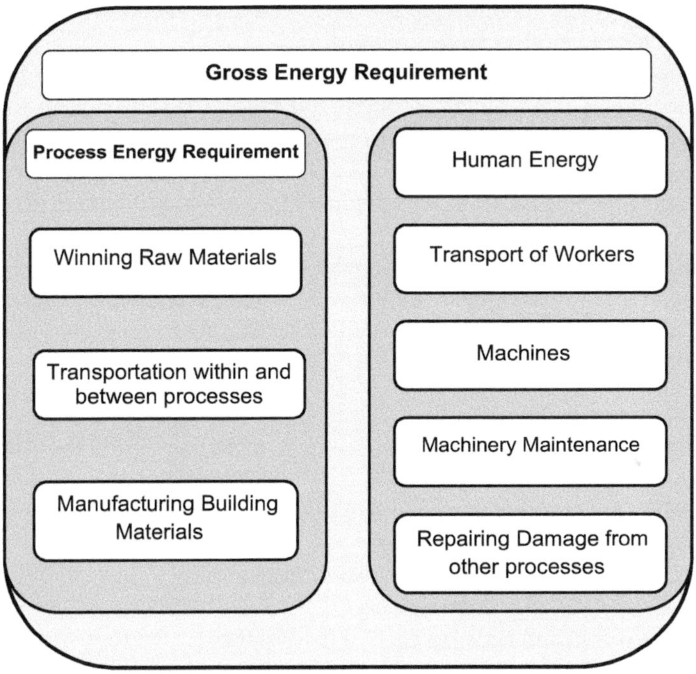

3.10 EMBODIED CARBON

Although embodied energy and embodied carbon are directly related, the impact of any material on resource depletion and on green house gas emissions may be very different. It will depend upon the primary fuel consumed and the means of generation of electricity. Consumption of renewable energy may be considered to have zero emissions provided the embodied energy of the collectors and generators are neglected. Similarly nuclear energy will also have zero carbon emissions. The embodied carbon is therefore dependent upon the fuel mix in the location where the processing takes place.

Some materials can even be considered as having negative embodied carbon when carbon sequestered during their growth has been accounted for. Trees and short term crops used for building materials sequester atmospheric carbon dioxide during their growing period, the weight of which may be greater than the emissions produced during manufacture.

3.11 NATURAL BUILDING MATERIALS

3.11.1 Renewable Construction Materials: Timber

The environmental credentials of timber as a construction material are very good. It sequesters carbon during its years of growth, it often requires little in the way of energy for manufacture and at the end of its useful life it has inherent energy which can be released by combustion.

The photosynthesis process by which plants and trees grow absorbs carbon dioxide and water from the atmosphere and emits oxygen back to the atmosphere. It is estimated that a tree absorbs 55kg of CO_2 and gives off 40kg of oxygen when growing 2kg of wood [12]. During its growth period a tree therefore has a positive impact on the environment through a reduction in greenhouse gases.

The life cycle of timber from sustainable sources can be traced through every stage from cradle to factory gate and beyond; to construction, maintenance and to disposal when appropriate. However the cradle to factory gate analysis is most common. It includes:

- Seed gathering and propagation
- Seedling planting and forest management including fertilizing, protection from animals and thinning
- Harvesting the mature trees including forestry activities
- Drying and seasoning felled timber
- Processing slab wood and rough sawn timber
- Secondary processing timber joinery
- Transport within and between each of these stages.

Through appropriate forest management and production, efficient use can be made of the material discarded from the felled trees. It can be used for composite timber products and garden mulches, and the supply of timber for the construction industry can be fully sustainable when harvested timber is replaced through re-planting.

In the UK, a high proportion of construction timber is imported, yet some regions including Sussex in the South East of England are heavily wooded. It is often quoted that the climate in England is not good for the growth of construction timber as the trees grow too quickly to achieve the required structural properties. Whilst rapid growth means faster absorption of CO_2, the resulting timber cannot be used directly for construction. However processed timber products such as glulam (glue laminated timber) provide excellent alternatives to natural timber sections and require less embodied energy than alternative structural materials.

3.11.1.1 Structural Timber

The trees used for structural timber mature over many decades. Softwood forests may be harvested after fifty years but hardwood trees may take over one hundred years to mature. Thus, the sustainable management of the forests is a long term program. The trees planted to replace those that are felled take several generations to mature to an appropriate size for construction timber.

Forest management techniques such as coppicing enable timber to be harvested in far shorter time scales, but the cross sections of the trunks are smaller and therefore they are less useful for construction without further manufacturing processes. Traditional coppicing techniques would have been carried out to supply poles for fencing, hop poles and thatching stakes as well as charcoal burning.

The technique of coppicing entails harvesting the poles after a prescribed length of time depending on the diameter of poles required. After harvesting the root system sprouts several new poles from a single stem increasing the rate of production.

Comparatively recently, in the first half of the twentieth century, manufacturing techniques were developed to utilize coppiced timber and fabricate large section structural elements by gluing together lengths of coppiced materials. The technique of finger jointing established strong connections between short lengths of material and the individual lengths could be glued together to produce structural elements of the desired strength and spanning width.

3.11.1.2 Glue Laminated Timber (Glulam)

The process of glulam manufacture involves an initial grading and milling to lengths of uniform cross section (25mm x 75mm). These lengths are then inspected, and any knots or other imperfections are cut out leaving sections of good quality timber of varying lengths which are then finger jointed together. The jointed material can then be planed and sawn to required lengths before being glued and laminated to the required dimensions. It is then possible to bend the assembled product to form arches or curved beams.

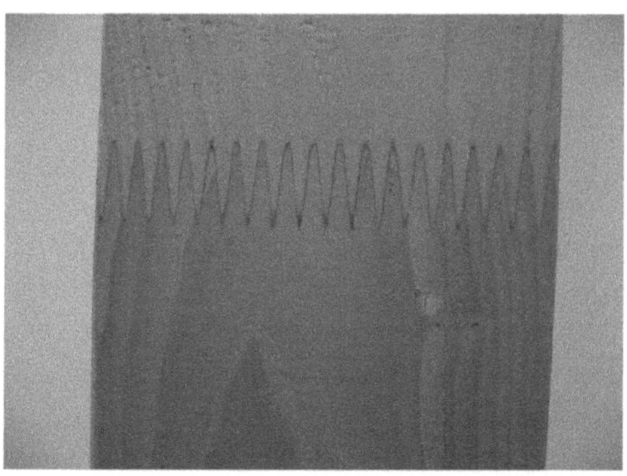

Traditional glulam processes use phenol-resorcinol-formaldehyde (PRF) adhesives that require the timber to be dried to a moisture content of 15%, however modern adhesives based on polyurethane are more tolerant to moisture and can be used with green timber. These adhesives therefore reduce the production time, reduce the embodied energy and provide the opportunity for utilizing timber that would not normally meet the quality requirements of construction timber.

The Building Research Establishment [6], have published a feasibility study for the green gluing of timber in which they conclude that it has the potential for improving the use of forest resources in the UK. Currently the vast majority of construction timber is imported and yet the UK has 2.8 million hectares of woodlands [10] and some areas, such as Sussex, are the most heavily wooded in Europe.

The advantage of green gluing and engineering timber products such as glulam enables coppice materials, forest thinning and small, poorly shaped trees to be manufactured into construction grade materials. The development of the supply chain for these products will encourage more active forestry management as it is currently not economic to harvest these low-quality materials.

3.12 SHORT ROTATION RENEWABLE MATERIALS

The use of timber for construction can be seen as a sustainable, renewable resource when appropriate forest management is practiced. The renewable nature of the material may take several human generations to complete, but this forest management has additional benefits such as enhancing biodiversity and tree growth.

There is, however, a history of using natural materials that are renewed on an annual or even shorter cycle. These materials are produced through agriculture and animal husbandry and are often byproducts of the farming cycle, hence their use reduces waste and adds value to existing processes. Where these materials can be grown locally there is a further advantage of reducing the embodied energy of transportation.

The Construction industry Research and Information Association (CIRIA) publish a handbook for crops in construction [7], in which they review the socio-economic benefits as well as the environmental impacts of natural construction materials. The handbook includes structural and insulation materials as well as finishes and floor coverings, identifying their physical properties as well as their ecological advantages and disadvantages.

Materials such as thatch and straw bale have been used in buildings for centuries, but have generally been replaced by modern materials that are quicker and often require less skill to erect. The establishment of sustainability as a key driver for development in many countries throughout the world and the re-introduction of the use of natural materials can stimulate the rural economy through new business opportunities and skilled work for local people.

3.12.1 Hemp

Hemp is a crop that grows quickly, requiring little attention or fertilizing during its growth period. It has a range of properties that are useful to the clothing, construction, food and pharmaceutical industries as well as producing biomass which can be used as a fuel. For construction the long fibers provide strength for structural elements and the shorter fibers are used in both thermal and acoustic insulation.

Hemp was traditionally grown in the UK for food and fiber. The fibers were used to make clothing and rope and the seeds were eaten raw or ground into a meal. The hemp plant is *Cannabis sativa*, which contains tetrahydrocannabinol (THC), a psychoactive substance which led to a ban on its cultivation in many countries. Industrial hemp however has been developed to have a very low percentage of THC and can be legally grown for its fiber.

The hemp plant is robust and grows vigorously not requiring herbicides or pesticides. It is the second fastest growing plant on the planet, after bamboo, and grows up to 5m tall, maturing within three to four months. The growth process requires the intake of atmospheric carbon dioxide and results in the emission of oxygen; thus locking in carbon for the lifetime of the plant and its resulting products.

Commercial products are available that are a mixture of lime and hemp, and can be used in wall construction either as a pre-cast block or poured in-situ. The material does not have great structural strength, but is used as an infill in timber frame construction. It has advantages over other infill materials in that it is lightweight and has good insulation properties. It is also breathable, hence enhancing the durability of the timber frame.

3.13 SUMMARY

Our demand for construction materials has considerable influence on sustainability and our ability to maintain an equitable and healthy existence on the planet. The demand for new and more efficient buildings is growing with increasing world population, and the minerals and metals that form a major proportion of buildings are a depleting, finite resource. Careful management of these resources is essential with emphasis given to less wastage and greater employment of recycling at the end of the useful life of the building.

Renewable materials such as trees and crops have an additional advantage in that they sequester carbon during their period of growth, which can offset the carbon emitted whilst being processed into building materials. There does however need to be an overall worldwide strategy for balancing the supply of natural products for food and fuel as well as for materials.

The building industry will continue to need both finite and renewable materials and their efficient use will be a critical component of sustainability. It is necessary to evaluate the environmental impacts of construction materials from a whole life cycle approach, assessing the impacts of winning and disposing of the materials as well as their emissions during manufacture and delivery.

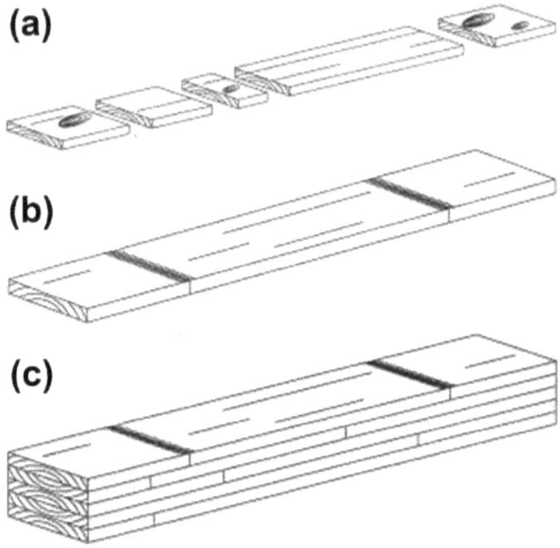

a) Remove knots and imperfections. b) Finger jointed sections. c) Laminate jointed sections.

REFERENCES

[1] CIB W115, Publication 318 Construction Materials Stewardship – The status quo in selected countries, in: B. John (Ed.), Storey CIB, 2008.

[2] R.A. Smith, J.R. Kersey, P.J. Griffiths, The Construction Industry Mass Balance: resource use, waste and emissions, Viridis, UK, 2002.

[3] N. Howard, S. Edwards, J. Anderso, BRE methodology for environmental profiles of construction materials, components and and buildings, BRE IHS, London, 1999.

[4] J. Anderson, D. Shiers, K. Steele, Green Guide to Specification, fourth ed., BRE Press, London, 2009.

[5] G. Hammond, C. Jones, Embodied Carbon The Inventory of Carbon and Energy (ICE), [s.l.], BSRIA, 2011.

[6] M. Cornwell, W. Thorpe, G. Cooper, Green gluing of timber: a feasibility study. IP 10/05 IHS BRE Press UK 2005.

[7] A. Cripps, R. Handyside, L. Dewar, J. Fovargue, Crops in Construction Handbook, CIRIA London, 2004.

[8] BS EN ISO 14040: 2006, Environmental Management – Life cycle assessment – principles and framework, Geneva, Switzerland, BSI 2006.

[9] Forestry Commission, Forestry Facts and Figures 2010: A summary of statistics about woodland and forestry, Forestry Commission, Edinburgh, 2010.

[10] N. Lazarus, BEDZED Construction Materials Report. Toolkit for Carbon Neutral Development, part 1 Bioregional Development Group, 2004.

[11] Timber Trade Federation, Consequences: The environmental impact of timber throughout the lifecycle, Timber Trade Federation, London, UK.

[12] Overconsumption? Our use of the world's resources, Friends of the Earth, 2009.

Chapter 4

The Sustainable Corporate Image and Renewables: From Technique to the Sensory Experience

Neveen Hamza
School of Architecture, Planning and Landscape, Newcastle University, Newcastle, Tyne and Wear, UK

Chapter Outline

4.1 INTRODUCTION

The corporate image is often linked to an image of environmental seclusion facilitated by advancements in building heating and cooling technologies, lighting and the advent of lifts. These technologies underpinned the possibility of spreading design philosophies and an export of the 'Modern' movement

Sustainability, Energy and Architecture. ISBN: 9780123972699

globally. The rising concern and interest in the design and construction of sustainable buildings is also dependent on know how of renewable energy generation technologies and advanced solar shading systems that if not carefully treated can only lead to an extension of a western export of a sustainability image.

In this chapter, the buildings represented as case studies are seen as opportunities for experimentation with a genuine pragmatic approach, underpinned by the aspirations of altruistic stake holders for a better world. The case studies chosen share an agenda of sustainability based on aiming for a British Research Establishment Environmental Assessment Method (BREEAM) rating of 'excellent', with varying levels of success. They highlight attempts to achieve 'experiential sustainable architecture'. The research analysis shows that these corporate aspirations moved on from buildings with 'no location' into an architecture specific to its 'genius loci' reflecting sustainability in the iconography of its facades and as a sensory and experiential experience for its occupants. Supported by scientific advancements and the environmental design rules of thumb underlying 20th century development, building designs have now moved on to use more sophisticated tools in which building performance and human comfort can be predicted at the design inception stage.

4.2 SUSTAINABLE INNOVATION, OR THE TRIED AND TESTED

Canizaro [1] warns that theorists constructing the discourse of sustainability in architecture have rarely built a connection between the past historical practices of building and sustainability.

'Like many developments in the modern era, sustainability has been seen and promoted primarily as something new, progressive, and future oriented…the result is a discourse and practice dominated by technical solutions to mostly technically framed problems'.

He advocates 'regionalists' as architects and theorists concerned with the manifestation of realness of places and people who live in them, leading to a historical thread of concern that calls for a more environmentally responsive practice. Architects as 'regionalists' are expected to deliver more environmentally responsive buildings, that consciously respond to experiential, ecological, social and cultural constructs of the place. Sustainable architecture is a revised conceptualization of architecture in response to a myriad of contemporary concerns about the effects of human activity. The label 'sustainable' is used to differentiate this conceptualization from others that do not respond to this agenda. It shifts the concept of a building from being a shelter from the environment to structures that minimize the impact of their occupants, their comfort expectations and technological demands on the environment [2]. Giddens, 1999 We started worrying less about what nature can do to us and more about what we have done to nature very recently in historical terms.

Farmer and Guy [3] stated that the reality of each of these completed buildings suggests that the nature and meaning of the technologies employed tend to be shaped by fluid, dynamic processes in which sustainable design discourses and environmental values have subsequently been pragmatically shaped by the particularities of context in a process of co-evolution. This would suggest that architects should see moral values and the processes and material practices of green architecture as intertwined and contributing to sustainability in the making.

Farmer and Guy [3] presented an extensive literature review of philosophical debates on the role of architects in conceiving sustainable buildings. Their findings are largely aligned with the arguments found by Hamza and Greenwood [4]. Alarmingly, they present a picture of an eroding role for architects in the design team for sustainable buildings, and they indicate that the prevailing techno-economic construct of sustainable buildings casts designers in a passive role. Designers are continuously attempting to resolve external societal and regulatory commands while using technology to reduce the carbon footprint. These pressures mean that they tend to use efficient construction materials by tried and tested methods, rather than aspiring to engage with these technologies in innovative ways to highlight a sensory experience.

A philosophy of sustainable building should not merely arise through efforts to conceptualize a building that operates the readymade modules of renewables, but should also create a building of civic pride as a symbol and also extend this to its occupants as a haptic experience.

Architects need to capitalize on the opportunity to utilize renewable energy technologies and passive design as an innovative architectural formal language. This should be extended to finding means to engage occupants in these design elements creating a sensory environment.

This chapter argues that designing for sustainability gives an opportunity to design buildings beyond regionalism and bio-regionalism, to a more site-specific response, capturing the 'genius loci' in an attempt to provide a sensory and experiential environment for its occupants, while responding to local climate and site, and also using renewables and complying with building regulations.

The case studies represented layer the building envelope to reflect a discourse that contextualizes connectedness to place, and to reflect sustainability as a built cultural message, and technology as a vehicle for environmental performance and responsiveness. This notion moves away from the mere concept of technique of the 'modern', or the sacred fictitious geographic boundary that determines the history and characteristic of its inhabitants, and from picturesque follies promoted by 'regionalism'. The case studies present a global aspiration for sustainability and reduction of their carbon footprint but by using a local interpretation relevant to its site and the comfort of its occupants, and hence a 'global' expression that highlights the experience facet of sustainable architecture.

4.3 THE 20TH CENTURY, THE CORPORATE IMAGE AND SUSTAINABILITY

The rise of office buildings as multi-storey icons in the late 19th century is attributed to changing corporate and business needs and a plethora of technological advancements, including the introduction of new construction methods such as steel frames, the lift, air-conditioning systems and artificial lighting. The seminal buildings that followed all offer a continuous trajectory and evolution in the principles underpinning human productivity and wellbeing, and the corporate image. It is argued in the following case studies that the realization that the corporate image is underpinned by sustainability aspirations led to a developed strong visual message which aspires to achieve a sensory experience for its occupants as well as its outside viewers.

However, it is the corporate image as reflected in the 'modern' style that brought with it a misinterpretation of many of the movement's architectural values. The Seagram building, and the Sears towers (among others) were used as icons of detachment from their environments. The Seagram building in 1958 was seen as representative of the Seagram Corporation, its role in markets and contemporary life. However these facade treatments were meant to create changing reflections from their glazed surfaces, and there was a preoccupation with expressing their technology of construction while moving around the buildings. The emphasis was placed on the external experience of the monumentality of the building, rather than a preoccupation with what happens behind the skin. Replicas of the steel structural grid on the external façade as exposing steel members were used merely to emphasize the modular construction of the curtain wall and had no function in structure or for shading. This was a practice that under today's sustainability lens can be argued against, as being a decorative and wasteful use of resources.

With the need to export this building function to other localities came the need to transfer the image. However, the case studies discussed here will demonstrate how corporate agendas are pushing sustainability, with its three pillars, to the fore while emphasizing the sensory and changing experiences indoors for its occupants.

The practices attached to the 'modern' movement consciously disregarded its protagonists' philosophies to respect the local environment, the 'genius loci' and cultural values. The role of the building envelope as an environmental moderator or a cultural message (or both) has been contested by the Modern Movement. The flexibility of construction behind curtain walls, the 'technique', flexibility of large open spaces and the economic rental values superseded the original intentions and philosophies. Hitchcock and Johnson [5] in promoting 'modern' views stated that:

'in the 1920s it was maintained that modern architecture should follow the same principles regardless of 'location' or 'region.'

A notion that Walter Gropius and Siegfried Gideon were eager to dispel. Gropius stated that:

'architecture should not be conceived as a mere practical product but has to deliver 'aesthetic satisfaction to the human soul.' [6]

Gideon [7] goes further in calling on the building envelope to deliver a 'new monumentality' in which it:

'springs from the eternal need of people to create symbols for their activities and for their fate or destiny, for their religious belief and for their social convictions [demanded], respect for the 'way of life' to be studied with 'reverence'.'

However, the 'international style' is seen as lacking stimuli, being over-dependent on a technological expression of construction technology, and becoming a commodity reflecting an imported corporate image, as in the blind transportation of the image of glass curtain walls into hot regions such as the Arabian Gulf.

The works of Team 10 in the post-war era reflects a preoccupation with the expression of the curtain wall to the outside environment, its modularity and offsite construction. The Battle Bridge, London, by Alison and Peter Smithson (1972–74) [8], follows a statement that:

'the building's position on the basin's edge procures a smooth continuum of wall screen into mirror image in the water, the building responds to this calmness of untroubled repose by presenting a single skin of stainless steel and glass; the layered dimension of the sky and the buildings opposite are ever changing, responding to season, weather and time.'

It is argued that the occupants' experience and how the building skin moderates rather than separates the external environment is still unrealized, although there is an attempt to look into the 'genius loci' of the site.

In the early seventies, Dutch Structuralism Movement hoped to overcome the reductive aspects of 'functionalism'. Herzberger's well-cited 'CentraalBeheer' in Holland in 1974 presented an exploration that focused on human interaction and workflow indoors to create a built form, while ignoring the local outdoor context. Daylight is introduced from the top level of an atrium like central space creating an introverted environment. The company had to put up signs for people to find the entrance. Herzberger described the building as a 'bunker like labyrinth'. Although cited as one of the great architectural buildings for exploring the human relationship between office workers and the flexibility of the organization [9] it can be argued that the detachment from site and context is opposed to the philosophy of sustainable buildings.

Norberg-Schulz [10] continues from there, accusing:

'those who got stuck with the early images of a green city and standardized form, were the epigones and vulgarizers of modern architecture.'

He alluded to the function of the building envelope as a visual message and a layer in which dealing with the external environment takes a distinctive character. He laments:

'The character of the present day environments is usually distinguished by monotony…the 'presence' of new buildings is very weak, very often 'curtain walls' are used which have an unsubstantial and abstract character. Most modern buildings exist in a 'nowhere'; but live their life in an abstract life in a kind of mathematical-technological space.'

He goes on to warn of an 'environmental crisis' in which buildings do not offer any meaningful or indeed intentional variation in engagement with their environment. The quest for an architecture that addresses all the senses still carries on in Pallasmaa's writings [11]. He pursues an architecture that engages all human senses, and that seems to warn against the 'globalization of bland environments' to address:

'Qualities of space matter and scale which are measured equally by the eye, ear, nose, skin, tongue, skeleton and muscle.'

Palsamaa extends his call for buildings to denounce 'fragile architecture' and to also include those with sustainable aspirations:

'Flatness of surfaces and materials, uniformity of illumination, as well as the elimination of micro-climatic differences, further reinforce the tiresome and soporific uniformity of experience. All in all, the tendency of technological culture to standardize environmental conditions and make the environment entirely predictable is causing a serious sensory impoverishment. Our buildings have lost their opacity and depth, sensory invitation and discovery, mystery and shadow.'

Why should our engagement with sustainable technologies not be included in this discourse then?

However, the question of the aspirations for 'sustainable architecture', with its use of advanced building services, knowhow and renewable technologies still remains unanswered; will architects' responses to a sustainability agenda follow those of 'functionalist' and 'globalized' architecture? Will architecture present renewables as an alternative to the esthetic of the 'module and prefabrication' reflecting a commitment to technique? Will sustainable architecture present itself externally as technology laden shelters while providing sensory deprived environments to its occupants?

4.4 THE TECHNO-CENTRIC SUSTAINABLE BUILDING IN THE 21ST CENTURY

The building envelope, and particularly for public buildings, is where a visual statement of commitment to place and people is exhibited. Therefore it is critical to move forward from renewables being considered as an add-on afterthought to an integral visual engagement and a design language that is by its nature very specific to the 'genius loci'. It endangers the public acceptance of these technologies for them to be projected as techno-centric and 'sustainability bling'. Sustainable buildings have to reflect a deeper expression of an engaging and memorable building experience that manifests the climate-specific context of the building, the clients' sustainability aspiration and a sensory message that enhances the wellbeing of occupants. It is acknowledged that delivering a sustainable building is a holistic concept that integrates the building and its services. The building's envelope has a longer life cycle than its supporting mechanical systems and based on the plan depth can contribute to about 30–40% of its energy demand [12]. This chapter presents case studies with an underlying commonality; they have all achieved an environmental performance on a 'core and shell' principle achieving a BREEAM rating of 'very good' or 'excellent'. This means that it is generally expected that, once these buildings are occupied by different users with different technology demands, the energy performance of the building will vary and exceed the predicted performance. Apart from the first case study, these buildings seem to share a rational anticipation of circumstantial contingencies changing internal use and layout, but still aspiring to convey a message of responsiveness and responsibility towards the environment and a sensory engagement for its occupants

4.5 THE SUSTAINABLE WORKING SHED, LION HOUSE, ALNWICK, NORTHUMBERLAND, UK

The Department of Environment, Food and Rural Affairs (DEFRA)'s Lion House 2009, Alnwick, UK, as designed by Gibberd architects, presents a common expected aesthetic for a building with a plethora of renewable add-ons. It has received many awards for its 'predicted' and 'in use' performance and also as a small building within budget. This building presents a common case of a shed with techno-centric aspirations and interiors deprived of sensory experience, in which occupants are informed that their building has a sustainability agenda by information on a monitor screen near to its entrance.

The design architect, Raymond Gill of Gibberd, stated [13].

FIGURE 4.1 The DEFRA's Lion House 2009, Alnwick, UK, A BREEAM 'excellent' building. *Photo courtesy: the author.*

'We have designed a modest building, whose esthetic is derived from our aspirations to make it environmentally sound. We have maximized its passive sustainable potential and integrated active measures, like the PVs, to make the very best use of them in multiple ways.'

The building is celebrated on the BREEAM web site (www.breeam.org, retrieved 10th of September 2011) as a flagship ultra-low emissions office, designed to achieve exemplary standards of sustainability and environmental performance, and which demonstrates DEFRA leadership in embracing and delivering sustainability. Lion House was able to achieve the maximum amount of credits for the reduction of CO_2 associated with the running of the building due to the specification of the building's fabric and services. Net true carbon dioxide emissions amount to −6904 kg CO_2 per annum because the building and surrounding technologies feed energy back into the national grid.

The passive environmental measures involved are:

- Orientation of the PV cells on the south facing facade. These are also used as a fixed shading device.
- Natural cross ventilation can be achieved through the 13.5 m deep plan.
- Minimization of window areas on the east and west elevation to avoid solar overheating.

The low-carbon technologies in place are:

- Sub-metering of substantial energy uses, high energy load installations and tenancy area allows monitoring of the tenants' energy consumption. This is displayed on a screen at the building entrance, giving users an indication of the real-time performance of the building's energy and water consumption.
- The installation of Low or Zero Carbon Technologies (LZCs) resulted in a reduction in emissions of 52.2% or 43,452 kg CO_2/annum. Some of the technologies used were:
  - A solar thermal panel producing hot water, hence reducing the demand on the boilers.

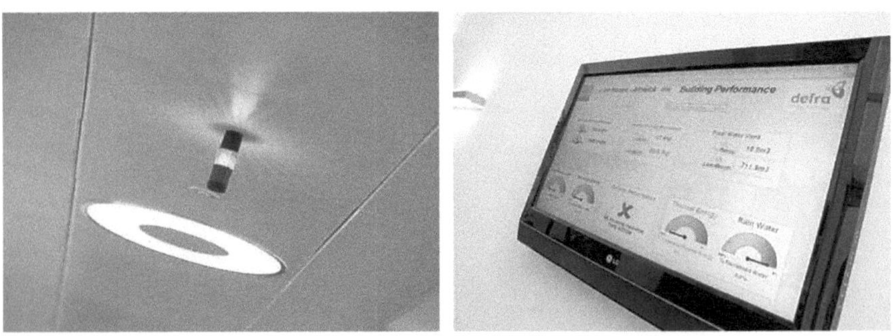

FIGURE 4.2 Traffic light system for natural ventilation (Left) and the energy performance dashboard (right). *Courtesy http://www.kier.co.uk/uploaded/fileupload/Defra%20Alnwick%20 Case%20Study.*

   - Photovoltaic cells for supplementary electricity, reducing the need for grid-supplied electricity.
   - Wind turbines to supply the office with its electricity needs, as well as providing electricity back to the national grid.
   - The specification of lifts that includes a stand-by mode, variable frequency control of drives, energy efficient car lighting and a regenerative unit to reduce the consumption of electricity.
   - Cyclist facilities were provided for building users to encourage staff to cycle to work rather than drive, in order to reduce transport-related CO_2 emissions.
   - An annual water consumption of 1.42m^3/person/year has been achieved by reducing the use of potable water in sanitary fittings. This is also supplemented by rainwater harvesting for flushing WCs and urinals.
   - Sanitary supply shut off is provided by a solenoid valve attached to a passive infrared sensor to prevent leaks in the toilet facilities.
- All materials were responsibly sourced to reduce their environmental impact.
- Construction site waste management of all materials ensured that hazardous waste from the site was no greater than 4.03 tonnes per 100m^2.

The esthetic appearance gives an expected message of many inclinations, to present sustainable buildings as energy-saving machines with as many renewables as possible added to their building envelope. While it is true that the photovoltaic cells are also a visual message of the building's orientation to the south, the danger here is falling into the trap of indicating that sustainability is merely a set of PV cells and a wind turbine added onto a functional core and shell building that could presumably be constructed anywhere! This is the same message that even the pioneers of the 'modernist' movement warned against.

4.5.1 Sustainable Architecture, An Experiential Sensory Approach

Sustainable architecture, performative architecture, and low-carbon design are all terms used to describe an intention to integrate current construction and environmental technologies in designing the building. Building performance simulation tools are increasingly used to predict the performance of a building, not only in terms of energy use reduction but also its impact on occupants and its site. Aided by a thrust in the development of building performance regulations, 'Building Performance Simulation Tools' found an increasing role in the design phase of buildings to demonstrate regulatory compliance. This led to a tangible collaboration between architects and their consultants at the very early stages of a project's inception [4]. However, Leatherbarrow [14] rejected the notion that the development of new instruments and methods of predicting a building's structural or environmental behavior will radically redefine architectural practices or theories. But that attention to performance will contribute to a new understanding of the ways buildings are imagined, made and experienced. Thus, a holistic human and technical interpretation of performance to avoid an inadequate reductive and an uncritical reaffirmation of pure functionalist ideology is called for.

FIGURE 4.3 **The Integrated PV array on the South Façade of the Doxford Solar Building.** *Photo courtesy: the author.*

4.6 EXPERIENCING RENEWABLES IN BUILDING SKINS

An integrated PV array designed by Studio-E in 1998 for the Doxford Solar Park, Sunderland the UK, reflected a strong message of commitment to sustainability (Figure 4.3).

The intention to present the renewable aspect of the facade was influenced by the environmental simulation of indoor environments. This lead to the facade being extended higher than the building to deal with the heat stratification behind the facade produced by the glazed areas and the heat generated from the transparent 532 m^2 PV arrays (Figure 4.4). The PV array produces 75 kWp, with the surplus being fed into the national grid. The building is estimated to generate a quarter to a third of its electricity needs [15]. The first speculatively constructed office building to incorporate building-integrated photovoltaics (PV). The building consists of a 66 m long, south facing, PV facade covering 532 m^2. This was the largest PV facade installed in an office building in Europe to date. The facade is the main feature of the building and is inclined at an angle of 60° to the horizontal for optimum capture of sunlight [16].

At the time of writing this chapter a visit to the site revealed that the photovoltaic array was not operational due to technical reasons and was technical refurbishment work was being undertaken to reinstate its function as an energy generator.

The building is well insulated and uses a passive solar design to maximize the use of natural daylight and to minimize space heating and air conditioning needs. It is also designed for natural ventilation and night-time cooling.

The brief required the design of a speculative office building that fully meets the requirements of the commercial market. It also required the building to be designed using 'best practice' low energy, environmentally sound principles. The client supported initiatives taken by the architect to include a worthwhile PV

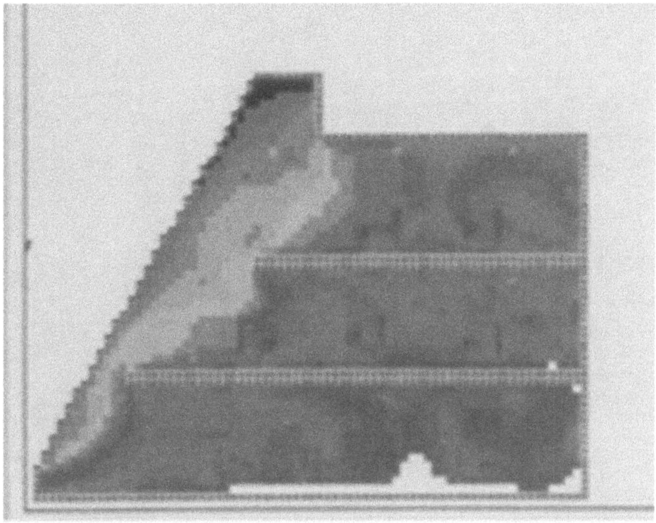

FIGURE 4.4 **Computational Fluid Dynamics (CFD) simulation of heat stratification behind the PV array and in atrium space.**

installation, but would only agree to include it in the brief if it was wholly funded from outside sources and it did not extend the design and construction program.

Accordingly the 4,600 m^2 three storey building was constructed to a 'shell and core' specification. It was fitted out to suit the specific requirements of the occupying tenant. The tenant was encouraged to operate the building in its low energy 'passive solar' mode, but chose to augment this strategy by utilizing the provision made for comfort cooling due to high project incidental cooling loads. The building can, if necessary, be divided into up to six separate tenancies.

It is noticeable how the choice of the PV system and the use of the atrium as a buffer created a distinctive architectural language and internal indoor environments (Figure 4.5). To reduce infiltration into the atrium there are no windows on the PV integrated facade, and all ventilation is supplied from windows into the office space which face other orientations. The costs of the PV array were covered by a European Union funding, as this alone would have exceeded the building's budget.

Here the building reflected a commitment to the performance aspect responding to the client's brief and the architect's vision of translating 'sustainability' in the building brief. As an early example, the exterior is inclined towards the 'technical' although the interior is an accidental 'experiential'.

4.7 THE RESPONSIVE SKIN AND CORPORATE IMAGE

During the 1980s, the greenhouse effect was linked to the increase in CO_2 emissions from the built environment. In office buildings, the heavy utilization of energy to provide comfortable indoor environments was found to be at a profligate

FIGURE 4.5 The interior of the building, with its varied shadows and light from PV arrays and transparent areas.

level of five to six times higher in a conventionally sealed envelope office environment than in one that was naturally ventilated and lit [16].

The following case studies are a broader translation to clients requiring buildings to reflect their sustainability policies, to provide a sensory and experiential experience through responsive and dynamic building skins integrated with holistic performance.

The Devonshire building was established as flexible lab and research accommodation in the University of Newcastle Upon Tyne by Dewjoc architects in 2004 (currently Devereaux architects) – see Figure 4.6. Newcastle University (as the client) was aiming for a BREEAM standard of 'excellent' to reflect the university's commitment to sustainability in its research and societal responsibility. This was achieved by designing for the inclusion of many renewable systems, including a PV array on the roof, an atrium with roof lights to introduce natural daylight, rainwater harvesting systems for toilet flushing and an automatic motorized solar shading system 'Brise Soleil' to the south elevation incorporating horizontal aerofoil blades. The blades are controlled by an integral

FIGURE 4.6 **The Devonshire Building completely closed in response to a sunny day.**

sensor and automated control system. This intelligent program allows the blades to track the sun's azimuth so as to eliminate direct solar glare. A high quality internal environment, with all internal spaces designed as mixed mode systems to take advantage of natural ventilation with additional mechanical cooling and heating during summer and winter is incorporated into the design.

The PV array with its 184 m^2 and estimated output of 25KW creates a visual message of commitment to sustainability. No real data regarding its contribution to the building's electricity consumption could be released and unfortunately the manufacturers had to change the whole array in 2010, as the PV cells became delaminated. The attempt to create an experiential, visually engaging environment by the automated louvers, which close completely on a sunny day to prevent glare, actually just prevents a view out on the few sunny days that occur in North East England. Automatically controlled natural ventilation was disengaged to allow for more control by occupants on a floor-by-floor level. The curved facade was created to improve pedestrian air movement and a visual vista to avoid a claustrophobic visual experience from the imposing buildings surrounding the site.

However, it is important that these buildings are seen as successful experiments that underpin the development in current knowledge and future thinking about improvements in building performance. The building facade provides a strong message of commitment to building sustainability, and corresponds to the corporate image of a university that is positioned as a leader in research on sustainability in technology and social research.

In another example, the client (ThyssenKrupp AG) wanted their new headquarters (Q1) building (Figure 4.7) to reflect their commitment as a company to: 'innovations, sustainability, openness and knowledge sharing'[13]. It was to be constructed

FIGURE 4.7 Changing daylight levels in Q1 (top), Facades of Q1 (bottom).

on a 200-year-old steel manufacturing site in Essen, Germany. JSWD Architecten and Chain & Morel et Assoils won the competition in 2008. The building was occupied in June 2010. The primary energy consumption of the new buildings in the complex is estimated to be 20 to 30% lower than statutory requirements. There are currently only very few buildings of this standard in Germany.

A 1,000 m^2 geothermal field on the site contains loops which extend to a depth of 100 m^2 below the earth's surface. The geothermal ground loops utilize the heat and cold stored in the earth, and are used to reduce the buildings' energy demand for heating, while passive measures using natural ventilation and solar shading systems are used to further reduce demand from air-conditioning systems. A 10 storey high atrium introduces natural ventilation and has neither heating nor cooling. Throughout the Quarter, rainwater will be collected from the roofs of the buildings – an area of 25,000 m^2. After removal of impurities, the water will be fed into the lake in Krupp Park.

FIGURE 4.8 The Leicester John Lewis facade, the inspiring original pattern (top left) morning facade (left) and night (bottom left) interior-building. *Photos courtesy Dr. Shimaa Hosni.*

The facade design is not only about providing a stainless steel mesh that traces the sun movement, but also serves to give the building a distinctive skin that changes its performance in response to natural daylight levels. The electrical lighting systems are connected to sensors that dim the lights according to availability of daylight from the facade.

In both the cases discussed in this section, the interactive building skin has shading systems that are considered part of the intelligent facade systems and treatment. Their performance is measured by their ability to respond to unforeseen and changeable external climatic conditions, in order to ameliorate the adverse effects of the latter on the indoor environment and its occupants. The richness of expression is mostly created by the variable movement of sunlight on these systems, which creates variability in opacity, reflectivity, transparency and color. These variations create engagement and appreciation of the systems, which in most cases are an accidental delightful surprise encountered after the building is constructed.

However, the extensive use of automated climatic responsive facade systems has to provide a robust detail. The inherent vulnerability of these systems needs to be minimized as they are in continuous motion to respond to both the outdoor climate and occupants' needs. This leads to various parts of the automated facade configuration to experience varying levels of stresses requiring maintenance and replacement of some of the parts.

Apart from these technical reservations, the experiential quality of the buildings and their reflections of their 'genius loci' is a promising precedent.

4.8 INCREASING FACADE LAYERS: DOUBLE SKIN FACADES AS A PASSIVE MEASURE AND A CULTURAL MESSAGE

Double skin facades are increasingly deployed in architectural applications, offering a passive climatic buffer zone for the building, which can be utilized effectively to introduce natural ventilation indoors for higher floors while reducing noise propagation. Such facades also have the potential to reduce mechanical ventilation loads even in hot arid climates [17]. Natural buoyancy drives out heat stratified in the gap between the two facade layers.

Double skin facades have been deployed in the Willis, Faber and Dumas building (Foster and Partners, 1975), the Occidental Chemical Center in New York (Canon Design 1980), Commerz bank (Fosters and Partners, 1991–1997), Strador in Dusseldorf (Petzinka, Pink and Partners), Swiss Re Building in London (Fosters and Partners, 2002) and the Leicester John Lewis Store (Foreign Office Architects, 2008).

It is the latter that points to a newly emerging trend of romanticism in reflecting the sustainability agenda of the corporate image and its intention to respect local history.

John Lewis Partnership's 'sustainable construction policy' aims to reduce its carbon footprint, and states that:

'Our current target is to reduce CO_2 emissions as a percentage of our sales by 10 per cent by 2010 (against a 2001/02 baseline) and to improve energy efficiency by 10 per cent by 2013 (against a 2003/04 baseline). An expanding business cannot avoid rising energy consumption. Our sales have risen by 28% over the last five years, but we have managed to contain the Partnership's absolute CO_2 emissions to 19% over the same period' (sustainable construction framework) [18].

The High Cross complex in Leicester includes a multi-screen cinema and a four storey John Lewis store. The store's facade uses a fabric analogy expressed as a series of pleats and swirls. The pattern was taken from an archived piece of fabric in John Lewis, and pays tribute to Leicester's textile manufacturing history. However, the use of a double skin facade here might provide a bonus environmental performance rather than its integrated intentional performance. The two layers have the printed pattern directly aligned and are lit at night, producing an engaging changing facade. But as the depth of the shop floor reduces the effect of both daylight and thermal transmittance from the double skin facade, it performs as a decorative drape rather than an environmental moderator. It transmits considerable daylight levels while reducing the direct solar penetration (which may discolor fabrics) as the external fabric is covered with a reflecting layer. The exact alliance of the two patterns on the glass reflects direct solar radiation and permits direct vision to the interior as it is viewed tangentially. The facade does not provide any natural ventilation to the indoors, which is completely air-conditioned. The facade is structurally suspended from the top of the building to reduce the supporting structural members and a distracting appearance behind the double skin.

To cover the structural system, the increase in its height above the building's roof allowed the stratified hot air to be kept away from the top floor of the building. However, all this appears to be an accidental bonus rather than a planned-for integration. This building was awarded a BREEAM grading of 'very good' and can be seen as a new landmark to lead the way into thinking of layering a cultural message within a performative framework.

4.9 SUSTAINABILITY AS HAPTIC EXPERIENCE

Away from the ocular messages of sustainability through the building's facade intelligent design, extending this appreciation of the building's sustainability seldom finds its way into the sterile and white washed interiors. In the UK Display ENergy Certificates are mandatory displays in buildings to inform occupants of their energy consumption. It is good practice to display energy generated from renewables by the use of screens as in the case of the Devonshire building and the Lion House. However, there is a real dichotomy between occupants and their sustainable buildings. The monitors are situated where people will only glance at them quickly on their way in to a busy working schedule. Furthermore, the electricity generated by PV panels is not used on site but rather exported to the national grid, and wind turbines malfunction and increase skepticism of these technologies. Hot water from solar heaters and recycled water on site do benefit occupants directly, but it has been reported on several occasions that the recycled water in toilets is not without odor problems. While it should be noted that these technologies are all in their experimental phase, other haptic and more engaging experiences should be sought in sustainable building (i.e. Devonshire building).

CONCLUSIONS

This chapter argues that reflecting sustainability of the building as an experiential and sensory experience had its roots in ancient civilizations, and its delivery is still an ongoing aspiration.

Although the first generation of the 'modern' movement advocated site specific architecture, the misinterpretations of the early thinking underpin a global, widespread, corporate image of buildings that are misconceived environmentally.

The fully glazed and sometimes bland facade of a building, which forms a curtain wall acting as a climatic separator, reflects a preoccupation with technique rather than the experiential quality of the occupants inside the building. The simulation tools available for architects today (illustrated here by the Doxford solar building) is becoming common practice and creates an opportunity to predict with a high level of accuracy the indoor environments at the design stage.

As new corporate sustainability agendas develop, so does the realization that a sensory and experiential sustainable building improves employee's wellbeing and increases productivity. This will move sustainability aspirations from a reductive performative notion, which treats buildings as a mere optimized machine, to looking deeper into human experience of housing experiences rather than housing functions.

The corporate image seeks to find a surface treatment that reflects its commitment to its location and climate, but also which offers an engaging urban and indoor sensory and experiential statement about its commitment to sustainability. All case studies presented, whether using integrated renewable energy, moveable responsive facades or double skins, are uneconomical solutions compared to traditional single skin configurations. These investments reflect a willingness to achieve a higher perception of sustainability, and to use it as means of engaging with the site and its environment.

This chapter does not attempt to discuss how these facade technologies lead to real reductions in the carbon footprint of a building, as the original assessments using BREEAM were based on a shell and core principle which means that the building, with its changing technologies and occupancies, will naturally lead to variations in the targeted reductions achieved by both passive and active measures.

REFERENCES

[1] V.B. Canizaro, Regionalism, Place, Specificity, and Sustainable Design, in: S.S. Moore (Ed.), Pragmatic Sustainability: Theoretical and Practical Tools, Routledge, New York, 2010, pp. 150–167.

[2] T. Williamson, A. Radford, H. Bennetts, Understanding sustainable architecture, Spon Press, London, 2003.

[3] G. Farmer, S. Guy, Making moralitity: sustainable architecture and the pragmatic imagination, Building Research and Information 38 (4) (2010) 368–378.

[4] N. Hamza, D. Greenwood, N. Hamza, D. Greenwood, Energy conservation regulations: Impacts on design and procurement of low energy buildings, Building and Environment 44 (5) (2009) 929–936.

[5] H.R. Hitchcock, P. Johnson, The international Style, W.W. Norton&company Inc, New York, 1932.

[6] W. Gropius, The New Architecture and the Bauhaus, MIT Press, Cambridge Massachusetts, 1965, p.18.

[7] Geidion, Architecture You and Me, Harvard University Press, Cambridge Massachusetts, 1958.

[8] A. Smithson, P. Smithson, The Charged Void: Architecture, The Monacelli Press, New York, 2001.

[9] M. Emanuel (Ed.), Contemporary Architects, St. Martin's Press, New York, 1980.

[10] C. Norberg-Schulz, Genius Loci, Towards a Phenomenology of Architecture, Rizzoli, New York, 1980, pp. 194–195.
[11] J. Pallasmaa, Notes on Fragile Architecture, Architectural Review, May 2000.
[12] N. Hamza, S. Dudek, H. Elkadi, Impacts of changing face configurations on office building energy consumption, proceedings of CLIMA 2000, on CD, Naples, Italy, 2001.
[13] R. Gill, Defra's green office: better than BREEAM excellent? Building Magazine, 17th of March 2009. Available on line www.building.co.uk. Accessed 20.5.2013.
[14] D. Leatherbarrow, Architecture's unscripted performance, in: B. Kolarevic, A. Malkawi (Eds.), Performative Architecture, Spon Press, New York, 2005.
[15] B. Evans, Solar Power Gets Serious, Architects' Journal 205 (24) (1997) 44–45.
[16] D. Jones, Architecture and the Environment, Bioclimatic Building Design, The Overlook Press, New York, 1998.
[17] N. Hamza, Double versus single skin facades in hot arid areas, Energy and Buildings 40 (3) (2008) 240–248.
[18] John Lewis Partnership, Sustainable construction Framework. Available online at http://www.Johnlewispartnership.co.uk, 2007 Accessed 10.11.2010.

Chapter 5

Residential Deep Energy Retrofits in Cold Climates

Shawna Henderson
Bfreehomes Design Ltd, Halifax, NS, Canada

Chapter Outline

5.1 INTRODUCTION

While new housing can, with some careful planning and analysis, incorporate many sustainable aspects, in Canada, this represents less than 2 % of the housing stock.[1] A good portion of Canadian housing is lightweight construction, built prior to energy efficient methods, advanced materials and solid building

1. 'New housing', in this context, is defined as being built within the last 10 years, according to Canada Mortgage and Housing Corporation, text from 2004 request for proposal (RFP) for 'Approaching Net Zero Energy In Existing Housing'.

Sustainability, Energy and Architecture. ISBN: 9780123972699

science. Regional issues such as horizontal driving rain and repetitive freeze/thaw cycles (Atlantic Canada), high flood zones and expansive clays (Southern Manitoba), permafrost (North of 60°) and seismic loading (Coastal British Columbia) have led to specific regional approaches to structural and safety concerns. However, with inexpensive hydro and natural gas in the highly populated areas that hug the Canada–US border and a wealth of fresh water and forests, conservation of resources and sustainability has not been the driving force behind the evolution of the Canadian home renovation industry.

This is ironic, given that Canada has a strong history over the past 40 years of research and innovation in energy efficiency and building science, especially in the new housing sector. The R-2000 program[2] has been in existence since 1982, and the Model National Energy Code of Canada for Buildings (MNECB) was developed in the late 1990s. The National Building Code of Canada 2012 includes energy efficiency requirements for the first time. The updates to the 2012 edition can be found at http://oee.nrcan.gc.ca/residential/new-homes/r-2000/standard/16118. But some regional and municipal building codes have included energy efficiency requirements in the past decade. Most of these energy efficiency requirements bring new housing to within range of the pre-2012 R-2000 standard.[3] While the industry has the capacity to achieve the R-2000 standard in new housing, renovations and retrofits that meet this standard are few and far between.

Deep energy retrofits look at reducing the energy loads associated with space conditioning and water heating. Then, instead of requiring a heating system that can provide heat to the occupants from October through April, the house now requires small 'shots' of heat to maintain a reasonable comfort level over a six to eight week period of very low temperatures. Lighting, appliance and miscellaneous electricity (LAME) use is completely dependent on occupant lifestyle and energy awareness, and is difficult to predict or dictate. While they cannot be disregarded, the analysis of LAME loads is fraught with variables.

The primary goal for energy efficiency in existing houses is to decrease the overall amount of energy required to keep the occupants comfortable. The starting premise is Amory Lovins' adage from the 1970s: 'It's cheaper to save a Watt than make a Watt'.[4] At the utility level, this is demand side management. At the homeowner level, this is reducing energy used in operating the house: electricity, natural gas, fuel oil or wood. In terms of a deep energy retrofit, this is achieved via envelope improvements.

Deep energy retrofits shift the relationships between purposes and energy use, as noted in the chart below. The 'as is' house started out with space and

---

2. R-2000 is a voluntary technical performance standard administered by Natural Resources Canada (NRCan), first introduced in 1982. It requires builders to meet targeted reductions in energy use while maintaining indoor air quality and comfort levels. See www.oee.nrcan.gc.ca/residential/personal/new-homes/r-2000/background.cfm?attr=4 for more details.

3. The original R-2000 standard was in place from the inception of the program. The first update of the standard was adopted in 2012.

4. Note that Lovins was referring specifically to electricity at the utility scale when he coined the term 'negawatts'.

water heating accounting for 82% of the overall energy use. A conventional, energy efficient renovation would have dropped this to 78%. The deep energy retrofit proposed for this house dropped the space and water heating requirements to 41%; half of the original load. As the space heating requirements drop, meeting the domestic hot water load becomes the bigger challenge, along with reducing the lighting and appliance loads.

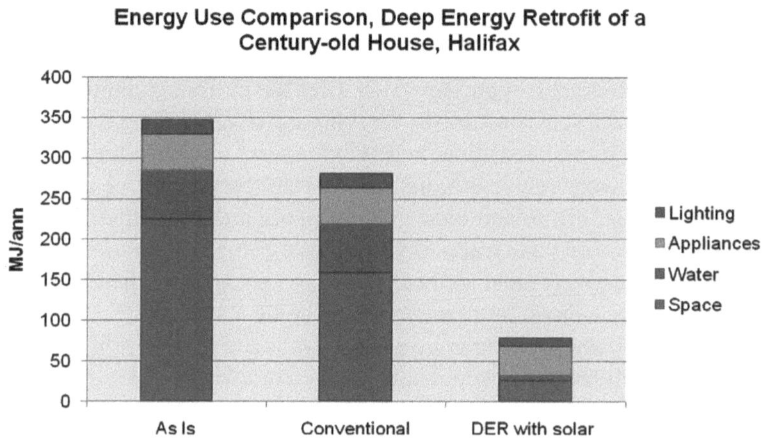

The following table is a summary of a deep energy retrofit of a bungalow in Vancouver. It shows how the energy required to operate a house can be reduced in distinct phases that address the envelope first, the mechanical system efficiencies second, and then add in renewables, once the total load associated with the house has been optimized.

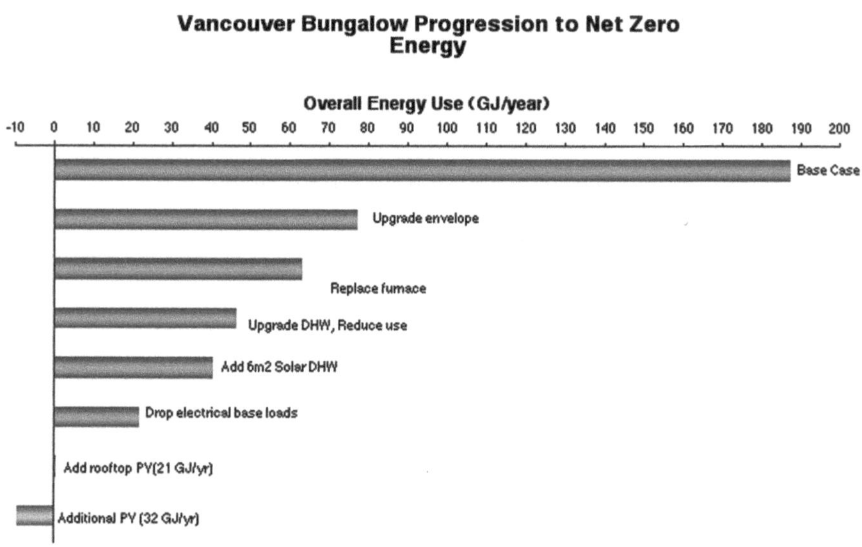

Let's be clear: a building is not energy efficient if it has a high performance heating system installed and no envelope improvements have been carried out. It is then a building with a high performance heating system. Improving the envelope is always a better long-term solution both at the homeowner and at the community/municipal scale. However, homeowners often end up with a high-cost, high-efficiency heating system, because technology is sexy and thus easy to sell. Insulation is just not that sexy, and air sealing is downright homely.

Changing out a space heating system is a reasonably straightforward, fairly tidy operation that can happen in a day or two, perhaps longer for a geothermal system, but one contractor typically looks after the system change out and the project has a distinct start and finish. Working up a viable deep energy retrofit that includes insulation, air sealing, new windows, interior or exterior finishing, and/or excavation work requires more time, several contractors or a larger crew and a lot more dust, grime and crud. Not to mention the fact that it's tough to show off the insides of your walls to your guests and neighbors to 'prove' that your house is energy efficient. A heat pump is visible, it makes noise, it has switches and lights – much more amenable to being shown off.

Then again, any energy efficiency measure runs a high risk of being trumped when a budget collides with a super-sexy kitchen renovation. Faced with a decision based on budgetary constraints, a homeowner is likely going to snort in derision and say: 'You want me to put $30,000 into my house and all I get is fat walls?'[5] The long-term invisible benefits of a deep energy retrofit are a difficult sell against the immediate visual and tactile enticements of a new kitchen with (hopefully) improved functionality.

This scenario is set on endless repeat mode by low energy costs. Canada has a lot of low-cost natural gas. From British Columbia to the Ontario–Quebec border, gas-fired forced air systems are the ubiquitous space heating system, and stand-alone gas-fired water heaters are typically leased to householders. Other parts of the country don't have access to natural gas. Where there is also no access to low-cost hydroelectricity, the choice is oil, either on-site feeding a boiler or furnace, or oil/coal feeding an electricity generation plant. Both of these latter two options result in higher costs per unit of energy delivered.

Deep energy retrofits are also confounded by the fact that existing housing, and the fabric into which it is woven, is undervalued. Currently, builders and renovators often dismiss a renovation in favor of demolition and a re-build, or a new build on a green field property when the cost of a proposed renovation exceeds 30% of the estimated value of the existing house and property.[6] It is often the case that the builder perceives his job would be easier if the client started anew rather than to tackle a major retrofit project.

---

5. In discussion with Greg Pedrick, Project Manager Buildings R&D at NYSERDA, May 2010.
6. Experience of the author with individual contractors as well as discussion in focus groups led by the author.

The builder is, rightly, focused on getting the job, and delivering a level of service in a specific time frame, and without the anticipated hidden challenges (rotted sills, pest infestations, asbestos, lead paint, sub-par do-it-yourself additions, etc.) that increase the cost of renovations.

The builder's point of view (cost-effective business management) does not take into consideration what is driving the householder to consider a retrofit. While the bottom line delineates the scope of any renovation project, it is not the primary driver. Several studies have shown comfort, occupant health and safety, personal goals and community leadership are more likely to be at the forefront of homeowners' minds when they are making decisions in regard to energy efficient renovations.[7] A broader economic analysis would take into consideration the environmental cost or embodied energy of all new materials, and/or those same costs as they relate to green field development. This will become apparent as the cost of developing new land on the outskirts of sprawling cities and obtaining new materials and resources becomes prohibitive.

Architects are not often involved in single family renovations in Canada, but they are involved in planning neighborhood redevelopment/improvement. While builders and contractors are typically on-the-ground with individual householders, they are not the ones who have the opportunity to weave elements of sustainability into neighborhoods and recognize the value of the fabric that is already there. Climate-sensitive retrofits combined with neighborhood energy system options need to be addressed and implemented at a broad community level, while the importance of energy security needs to be addressed and implemented at individual household level.

5.2 BUILDING MATERIALS AND ASSEMBLIES

One of the more typical approaches to a deep energy retrofit is to improve the thermal envelope dramatically, from the exterior or from the interior, or a combination of both approaches. Adding insulation to the exterior of the building shell can work in many situations, especially when it is incorporated into a project where the siding or cladding of a house is being replaced. Issues that arise in this approach primarily revolve around required setbacks in zoning by-laws and fire separation between buildings.

In typical terms of energy efficiency, there is a 'law of diminishing returns' in play when increasing insulation levels. While the physics don't change, the first layer (25mm/1inch) of insulation is always going to be more cost-effective than the tenth. However, the way that the law of diminishing returns has been

7. Zero and low energy homes in New Zealand: The value of non-energy benefits and their use in attracting homeowners, Stoecklien, A & Skumatz, L. ECEEE Summer Session 2007 • Saving Energy – Just do It! Why Comprehensive Residential Energy Efficiency Retrofits are Undervalued, Knight, R. ACEEE Summer Session 2006, Session 7, paper #726

analyzed is lacking. A conventional analysis of the value of insulation involves the relationship discussed next.

5.2.1 The Cost of Insulation vs. the Cost of Fuel

Under this scenario the 'financial return' on increased insulation stops at a certain point. This financial return may be below current code requirements. As it fluctuates dramatically with the cost of fuel, it makes insulation levels dependent on current fuel prices, yet the householder will be subject to escalating heating costs into the future. It is clearly not the best analysis. The analysis used for optimal insulation levels in the Model National Energy Code (MNEC) for Canada uses a 'square root' approach that includes annual heating degree days, present worth factor and internal heat gain factor weighed against the cost of insulation. This level of analysis is often used by energy efficiency incentive programs to derive their insulation-increase-to-available-rebate formula.

Looking just a tad more closely at the relationship between energy use and insulation and occupant comfort, we see that insulation does not exist in isolation. The insulation and the heating system work in tandem to provide comfort. Thus, the cost-effective evaluation of a deep energy retrofit should consider the total cost of the thermal comfort system including the heating system, and not just the cost of the insulation. In a very simple sense, it's simply shifting figures on budget lines: $20,000 is being spent, for example, on a $15,000 central space heating system and $5,000 is being spent on insulation. Or, $5,000 is spent on point source space heating units and $15,000 is spent on insulation. Recall that this central heating system costs the householder every time it runs from October through May.

Below is a list of insulation levels used in the Canada Mortgage and Housing Corporation's (CMHC) Equilibrium House Initiative projects, a set of new demonstration homes (and one retrofit) that aimed for Net Zero Energy standing. These are similar to other high-efficiency houses built in the northern US and Canada.

RSI (R)	Ceiling	Main Walls	Exposed Floors	Below Grade Walls	Slab
Vancouver	10.6 (60)	7.0 (40)	7.0 (40)	7.0 (40)	1.8 (10)
Calgary	14.4 (80)	10.6 (60)	10.6 (60)	7.0 (40)	1.8(10)
Toronto	10.6 (60)	7.0 (40)	7.0 (40)	7.0 (40)	1.8 (10)
Montreal	14.4 (80)	10.6 (60)	10.6 (60)	7.0 (40)	1.8 (10)
Halifax	10.6 (60)	7.0 (40)	7.0 (40)	7.0 (40)	1.8 (10)
Whitehorse	14.4 (80)	10.6 (60)	10.6 (60)	7.0 (40)	1.8 (10)

There is ongoing debate about what insulation materials are 'greenest'. Recycled jeans, newsprint, wool, straw; these materials are seen to have lower embodied energy than other materials, especially those made from petroleum products. All currently available fibrous insulations offer about RSI 0.22 to 0.026/mm (R-2.9

to R-3.7 per inch). Rigid foams and spray foams – all made using petroleum products, offer RSI 0.042/mm (R-4 to R-6+ per inch). Fibrous insulations offer very little in terms of air sealing qualities, which may be seen as a good option in some schools of thought, but most building science points to an airtight envelope with mechanical ventilation as the best option for reducing energy use effectively. Foam insulations of various manufacture offer air sealing qualities as well as insulating qualities, and some foams are also well-suited to damp conditions.

Currently, the existing housing stock uses such a surplus of gas and oil in the form of heat that the amount of petroleum product being used as insulation is a mere drop in the bucket in comparison. The simple fact that huge amounts of gas and oil are burned as fuel, effective only in the moment they are burned, and then impact the globe as greenhouse gas emissions after combustion, is of bigger import than foam insulation – which remains relatively inert for its lifespan, effectively reduces the need for combustible heating fuels and in some forms, can be re-used. That doesn't equate to petroleum-based insulation being used with impunity.

As an example, spray foam insulations can be superb materials for air sealing, they can work well in areas where moisture issues may not be resolved completely, and high-density foams have better insulation value per unit of thickness than other materials. The drawback of any spray foam is that it is difficult to remove from the wood or other material that surrounds it. This means 'mining' old houses for useable wood and other materials at the end of the useful life is a complicated matter – materials needing to be pried apart, due to the adhesive qualities of the foams, and the spray foam needing to be forcibly scraped off any surface it came in contact. Where rigid board or fibrous insulations are used, the re-use of the insulation material itself as well as that of the salvageable structure around it is much easier. Using these types of materials in areas where their outstanding qualities are best used makes sense: full-depth in tight spaces, 'flash' coats in attics to eliminate thermal bypasses, at rim joist/header areas, in damp foundations. While deep energy retrofits, done properly, can extend the life of a house by several decades, we still need to consider ways that the resources tied up in the house can be extracted in the future.

5.3 VENTILATION AND AIR MOVEMENT

Controlled (mechanical) ventilation is required in cold climates as building envelopes get tighter. The rate of mechanical ventilation – and how it is calculated – varies somewhat, but in Canada and the US we are typically looking at the rates as laid out in CSA F-326 or ASHRAE 62-89. This level of ventilation minimizes the occurrence of problems associated with condensation and moisture build-up within the living space and the envelope components. In houses where a forced air system is present, there are several options that can tie into the existing ductwork. Happily, this is one of the most common heating systems in most of Canada, the notable exceptions being Quebec, with inexpensive hydroelectricity making electric furnaces and strip electric the heating systems of choice, and the Maritime region, where low-cost natural gas is not available and the traditional choice of heating system has been an oil-fired boiler or strip electric.

Ventilation requirements can be met by bathroom and range hood exhaust fans, however, to optimize the indoor environment, balanced ventilation systems, such as heat recovery ventilators (HRVs) using low-volume ductwork are preferred. These can be costly to install in houses without existing ductwork when only a modicum of energy efficiency work is being carried out. But, if the goal is to reduce energy use required for space heating by 70 or 90%, the cost of a high-efficiency HRV, dedicated ductwork, and a heating coil in the supply duct could be offset by the savings possible, and would supply both ventilation and space heating. Other approaches that lay outside of current building code requirements in Canada include laminar flow windows.

The most challenging obstacle to overcome in discussing deep energy retrofits with householders and contractor/builders in cold climates is the approach to space heating. Instead of requiring a standard central heating system, a retrofit house may need only point source space heating. Traditionally, inexpensive strip electric heating is considered an expensive operating solution, and a poor use of electricity generation. More electric-based space heating systems equate to more generation capacity requirements. When space heating needs are dropped by 70 to 90%, this no longer holds true.

Here are three quick case studies looking at variations in approach to deep energy retrofits. All of these are real-world examples. Motivation for improving the buildings ranged from 'the right thing to do' to the business case for improving/expanding a rental property and maintaining long-term tenants via low energy charges that could be bundled into the rent.

5.4 CASE STUDIES

5.4.1 Case Study: A Larsen Truss 'Chainsaw Retrofit', Regina, Saskatchewan

In the annals of deep energy retrofits, the exterior application of insulation above (and below) grade using a Larsen Truss[8] is classic. One of the first was done in Saskatchewan in the 1970s, and was dubbed the 'chainsaw' retrofit, after the way in which the soffits were hacked off to allow for higher levels of roof insulation.[9] In all seriousness, the best approach to retrofitting a house is to give it a new skin – a continuous layer of insulation from top to bottom. This eliminates all thermal breaks, takes care of all air leakage bypasses within the envelope and provides a continuous drainage layer so bulk water is shed and drained away from the house.

In this house, the improvements to the building envelope dropped the size of the required space heating system by 60%.

8. A Larsen Truss is a building method that allows large amounts of insulation to be added to the exterior of a house. Light weight, ladder-like trusses are set around the perimeter, and fastened to the outside of the structural wall.

9. See: A Major Energy Conservation Retrofit of a Bungalow, by HW Orr and RS Dumont, available from the National Research Council Canada, Institute for Research in Construction, Internal Report No. 540.

THE CHAINSAW RETROFIT THERMAL ENVELOPE UPGRADE

	Pre-Retrofit	Post-Retrofit
Attic Insulation	RSI 3.5 (R20)	RSI 10.7 (R60)
Above Grade Wall Insulation	RSI 1.2 (R7)	RSI 8.4 (R47)
Below Grade Wall Insulation	RSI 1.2 (R7)	RSI 8.4 (R47)
Basement Floor Insulation	RSI 0 (R0)	RSI 3.5 (R20)
Windows	RSI 0.35 (R2)	RSI 0.53 (R3)
Air Tightness (AC/H at 50Pa)	2.95	.29
Peak Heat Loss (kW)	13.1	5.5

Notes: all insulation values in table are nominal and related to insulation materials added, not U-value of overall assembly

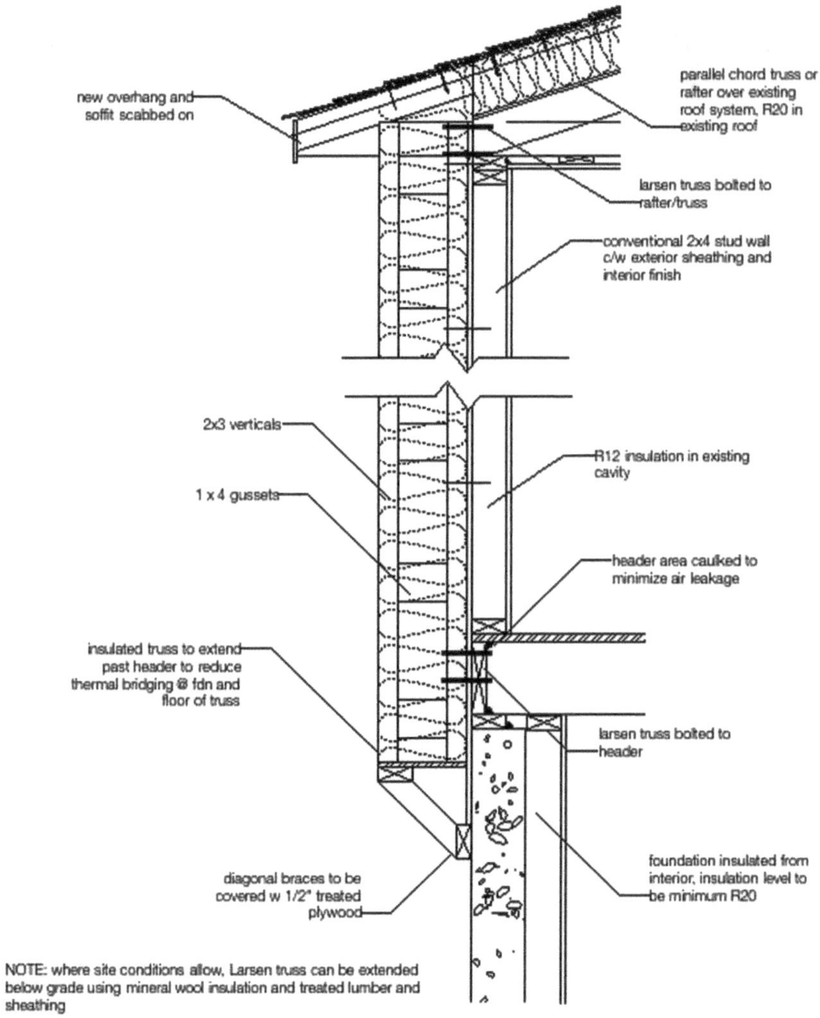

Chainsaw Retrofit, Larsen Truss Construction, interior foundation insulation.

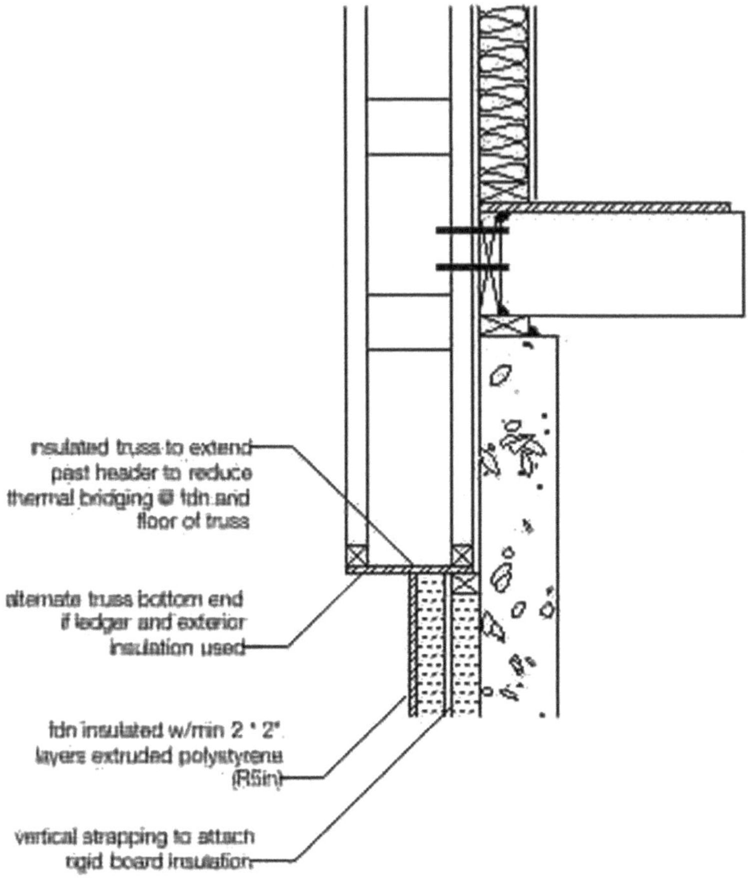

Chainsaw Retrofit, Larsen Truss Construction, exterior foundation insulation.

In another, more recent example, the house was excavated down to the footings, where serious water leakage was causing severe damage to the building below grade. Repairing the foundation remedied the leak. The house was built as a quadplex, but was virtually uninhabitable in the lower two units because of water leakage. The owner of the building was not only able to salvage the whole building, but, because of the major improvements to the building, was able to rent all of the suites out at a higher market value. The business case was very easy to make in this situation.

Larsen Truss construction can be limited in some areas because of side yard restrictions. It can also be the least expensive way to gain vast improvements to the thermal performance of a building, as the low-cost fibrous insulation (fiberglass) can be used above grade. Below grade, a free draining material such as mineral (rock) wool is required.

5.4.2 Case Study: Interior Above and Below Grade Insulation, Halifax, Nova Scotia

This project has a dual focus. First, it is a deep energy retrofit of an old, not insulated house with a rubble basement. Second, it is a retrofit of a two storey property that doubles the living space, thus increasing the potential density of a neighborhood of similar vintage and style.

A 100 year old, two storey house with four bedrooms was gutted, and the living space was doubled by digging out the basement and replacing the roof. The end result is a house with two four bedroom flats. In addition, the modeled heating energy per unit of area for the finished house was less than half that of the original house.

Insulation and air sealing was carried out on the interior of the building, as the tight urban lots in this particular area don't allow for encroachment. While the rubble foundation had some spilling, the rim joist/header area was dry and well above the finished grade. The floor was partially dirt, with a thin, cracked concrete slab over the remainder. This was a major source of moisture in the foundation. Water and moisture issues were dealt with prior to any insulation and air sealing work, and water-hating materials such as high-density polyurethane spray foam were used in the basement renovation.

THICK WALL ENVELOPE UPGRADE ON A PRE-WORLD WAR II CITY HOUSE IN HALIFAX

	Pre-Retrofit 176.6 m^2 (1900 s.f.)	**Post-Retrofit** 353.2 m^2 (3800 s.f.)
Attic Insulation	RSI 1.2 (R7)	RSI 10.7 (R60)
Above Grade Wall Insulation	RSI 0 (R0)	RSI 6.3 (R36)
Below Grade Wall Insulation	RSI 0 (R0)	RSI 6.3 (R36)
Basement Floor Insulation	RSI 0 (R0)	RSI 1.75 (R10)
Windows	R 0.35 (R2)	RSI 0.88 (R5)
Air Tightness (AC/H at 50Pa)	+15	Target: 2.0
Peak Heat Loss (kW)	19.4	12.4
Solar Thermal Capacity (kW_{th})	0	± 9

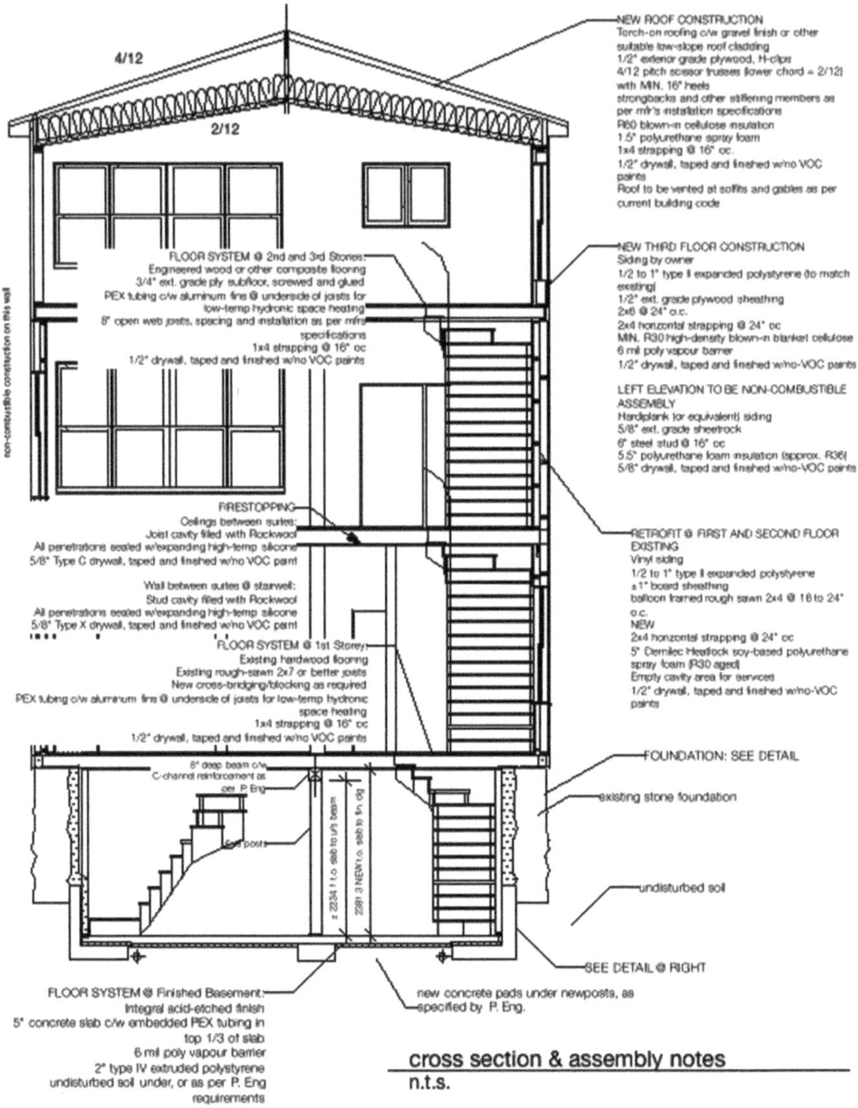

X-section of Retrofit.

A portion of the foundation sits on shale, which was a challenge to dig out without damaging the foundation. A sturdy interior footing with appropriate drainage was engineered to protect the original foundation wall and allow for a full-height living area in the basement. The new living area walls were insulated with 125mm (5 inch) high-density polyurethane spray foam, while the new slab was insulated from below with 100 mm (4 inch) of Type IV extruded polystyrene with free-draining gravel under.

The above grade wall insulation was done in two steps. The rough-sawn 100mm (4 inch) studs cavity was filled with high-density polyurethane foam, with attention paid to the header area to ensure a good seal. Then, the walls were strapped out with 89mm (2x4) framing members and an interior layer of dense-pack cellulose was installed. This meant a loss of useable floor space in the existing living space and required detailed planning around the stairwells. Windows and exterior doors were replaced and drywall returns used instead of extra wood for casings and trim on the interior of the thickened walls. The windows specified were high performance triple-pane units, however, the installed units were double-pane, with one low-e coating, argon fill and insulating spacers.

Cross bracing and interior insulation.

Interior insulation with stand-off wall in basement.

The new third storey was insulated in the same manner as the existing walls, and the low-sloped roof trusses were filled with blown cellulose.

The cast-iron radiators were taken out and low-temperature hydronics were used throughout – in the basement slab and as 'underfloor' with extruded aluminum fins elsewhere. Originally, space heating was delivered via two large, mid-efficiency, oil-fired boilers, and domestic hot water supplied by two electric tanks. All of the equipment was located in the unheated, uninsulated, rubble foundation area. The house now features an integrated solar thermal space and water heating system, with a high-efficiency natural gas boiler as auxiliary supply. Two high-efficiency HRVs were installed as well as low-sone bathroom fans and range hoods to ensure high indoor air quality.

5.4.3 Case Study: Exterior Insulation Above Grade/Interior Insulation Below Grade, Halifax, Nova Scotia and Utica, New York

While the best approach to insulating the whole house is to create a continuous thermal break/drainage plain, this is simply impossible to carry out in many situations. Many urban lots are unsuitable for exterior insulation on the foundation, primarily due to the proximity of the neighboring building. Suburban and rural lots may be unsuitable for exterior foundation insulation because of landscaping or decks, or the high cost of machinery. If there are drainage problems with the foundation, then excavating the exterior of the foundation could be warranted.

Several urban retrofits in Nova Scotia and upstate New York[10] have used thickened above grade walls matched with interior foundation insulation. Some jurisdictions allow thickened walls if there is no 'structure' added to the building. This means that an all-exterior retrofit that uses a Larsen Truss system, like the Saskatchewan example, would be unacceptable, as the Larsen Truss is deemed to be structure. However, rigid board insulation under new siding is not considered to be structural. Under this interpretation of the zoning bylaw, the distance from the outside edge of the foundation (and/or the outside edge of the stud or sheathing) is considered to be the perimeter of the house and thus, is the point of measurement to the property line at the sideyard and/or front yard setback. Not all municipal units interpret this the same way, and fire safety issues (i.e., materials used, openings to the adjacent building) may need to be addressed in more detail in some jurisdictions than others.

10. Upstate New York projects were carried out by the New York State Energy Research and Development Agency (NYSERDA), detailed information on that project can be found at http://www.nyserda.ny.gov/BusinessAreas/Energy-Innovation-and-Business-Development/Research-and-Development/Buildings-Research/Advanced-Residential-Buildings/Deep-Retrofit.aspx?sc_database=web (accessed 26 Sep 2011).

MINIMAL THICKNESS ENVELOPE UPGRADE ON A PRE-WORLD WAR II CITY HOUSE IN HALIFAX

	Pre-Retrofit	Post-Retrofit
Attic Insulation	RSI 3.5 (R20)	RSI 10.7 (R60)
Above Grade Wall Insulation	RSI 0 (R0)	RSI 7.0 (R40)
Below Grade Wall Insulation	RSI 0 (R0)	RSI 3.5 (R20)
Basement Floor Insulation	RSI 0 (R0)	RSI 3.5 (R20)
Windows	RSI 0.35 (R2)	RSI 1.1 (R6)
Air Tightness (AC/H at 50Pa)	5.5	Target: 1.4
Peak Heat Loss (kW)	18.7	6

Exterior above grade insulation.

Interior foundation insulation.

Adding exterior insulation is a cost-effective process when coupled with the replacement of siding or cladding. However, most contractors would consider 25 mm or 50 mm (1 inch or 2 inch) sufficient. In fact, in most of Canada and the upper tier of the US – where the heating season dominates – more exterior insulation is required, especially where wall cavities are insulated and there is no interior air barrier. To ensure that moisture does not become a problem within the wall structure, 75mm to 100 mm (3–4 inches) of rigid board is required. This lifts the dewpoint out of the cavity, which is most commonly filled with fibrous, moisture absorbing material such as fiberglass or cellulose. It also allows the exterior board to be taped and sealed at joints and edges to create an exterior air barrier, effectively reducing air leakage.

Interior foundation insulation applications vary, according to a range of factors. Basement areas can already be finished, but still be experiencing moisture-related problems, unfinished basements can be dry, damp or wet, drainage can be rectified to the exterior or not, foundation area can be planned to become living space, or used for storage, or is outside of the building envelope completely, etc. One approach to insulating basements with unresolvable moisture problems from the interior involves installing rigid board insulation with a grooved surface on one side to allow for drainage. This is actually an exterior-grade product. The grooved side is placed against the foundation wall to allow bulk water to drain down the wall, and the top of the rigid board is sealed to the top of the foundation and the rim joist/header area with an impermeable product such as high-density polyurethane. The foundation floor, sloped to a covered sump pump, is fitted with a dimpled exterior waterproofing membrane, and this is covered with a layer of rigid board. Essentially, the idea is to mimic a pool liner, using hygroscopic materials and drainage spaces to keep water from infiltrating the useable space of the foundation.

5.5 VISION: DEEP ENERGY RETROFITS AND NEIGHBORHOOD ENERGY SYSTEMS

One-off renovation projects are costly and labor intensive. There is little that can be done to remedy the fact that every house has been customized by past and present owners. However, at the point where high energy costs, high material costs and high development costs collide, the situation will have to change, as householders will find it more and more difficult to afford to heat older homes and municipalities will potentially be faced with a less stable tax base. It is feasible that groups of householders and/or ratepayers associations could take on deep energy retrofits under municipal sustainability plans, which can embrace deep energy retrofits and neighborhood energy systems.

As an example, some Halifax householders have been discussing a shared geothermal system, which would be jointly owned under a condominium arrangement. This system would be sized to accommodate the space and water heating systems of four houses after they have gone through a deep energy retrofit. With expected reductions of 70 to 90% in space heating requirements, the overall load for this shared system would be slightly larger than a system dedicated to only one house prior to the deep energy retrofit.

While this scenario is rather rosy, and assumes much for the sake of a simplistic illustration, there is great benefit in moving towards micro-scale neighborhood energy systems that could fit under several schemes for financing through municipal tax rolls and/or utility-based demand side management (DSM) programs. Energy service companies (ESCos) could offer long-term leases of equipment that roll the energy efficiency measures carried out on the houses into one package, taking all issues of renovation management and ongoing maintenance out of the realm of the householders. Essentially, this approach addresses the 'point source' of energy use in communities (i.e., the individual dwelling), by reducing it and then amalgamates a number of smaller energy loads into a similar or slightly larger point source. With some type of diurnal storage, utility concerns about peak loading during cold, clear nights would be assuaged.

In this scenario outlined above, neighborhood systems start out delivering space and water heating via geothermal systems or air-to-water heat pumps. The infrastructure for solar thermal and/or on-site energy generation (photovoltaic or micro-wind) would be allowed for at the initial retrofit phase so that the heating system can be expanded to an energy system, with feed-in to the grid. Where decent feed-in tariffs are in place and careful energy management is an agreed-upon goal, householders could see a small profit that could be used to cover their condominium fee. In an ESCo model, the company providing the equipment could benefit from the feed-in tariff.

5.6 WHAT CAN WE DO TO IMPROVE THE TEACHING OF ARCHITECTS?

While architects are not involved in a large portion of single family new construction and retrofit projects in Canada, they are often employed by regional social housing authorities, developers and large-scale improvement projects. This means they have the opportunity to improve the fabric of existing neighborhoods and communities. 'Improving' neighborhoods and communities often relates to the day-to-day livability of an area – pedestrian friendly areas, lighting, parking issues, density for infill projects, amenities. All of these add to the inherent value of existing homes.

Existing houses have multiple levels of resources twined together. Physically, there are the materials that make up the house – in many older homes, this includes structural members of old-growth wood that is simply unavailable in today's market. In good condition, this type of wood has an inherent value that goes beyond the cost of replacing it. That value is not included in the analysis of the value of an existing house, with a few notable exceptions.

In addition to valuing the resources tied up in existing houses, architects must bring an awareness of the importance of energy security into the mix when they are working on neighborhood-scale projects that include existing housing. Lovely livable neighborhoods include affordability – at purchase and ongoing.

The architect must create a quantifiable matrix of value that includes existing resources and energy security as well as the typical amenities. This is the area where employing sustainable building practices pays off exponentially and the architect needs to be able to make the business case clearly and concisely to all players: funders, municipal and regional units, housing authorities and householders.

Chapter 6

Sustainable Building for a Green and an Efficient Built Environment: New and Existing Case Studies in Dubai

Mohsen Aboulnaga
University of Dubai, Dubai, UAE

Chapter Outline

This chapter is organized into nine sections. Each section deals with specific issues addressing the following:

1. Climate change: cities and buildings
2. Importance of sustainable buildings
3. How enacted regulations and laws on sustainability contribute to the reduction of carbon emissions?

Sustainability, Energy and Architecture. ISBN: 9780123972699

4. Taxonomy of a sustainable building
5. Selection criteria of Dubai's case studies
6. Features and elements of energy efficiency
7. Daylighting and lighting techniques
8. Assessment and lessons learned
9. Conclusions.

The new green economy would provide a new engine of growth, putting the world on the road to prosperity again. This is about growing the world economy in a more intelligent, sustainable way.

Achim Steiner, 2008

6.1. INTRODUCTION

Amid the economic recession and the global financial crises of 2008—which despite the short-term bailout packages, symptoms are still apparent in United States and in Europe, and consequently impact the rest of the world—more emphasis should be directed toward green growth and sustainability in all avenues of life. Such a move necessitates the answering of many questions as well as rigorously and transparently identifying current and future challenges. The most important question is whether we are addressing this crisis in the context of sustainable development. Do buildings and cities, in the way they are currently built, contribute to green economy and growth? To address such multiple-faceted questions, it is vital to understand the economic, environmental, and social impact of buildings—which we, as architects, engineers, researchers and students, create and build through the entire process of education, research, design and construction, and operation including retrofitting. It is equally vital to understand the policies and regulations that govern this sector of the economy—on green growth in the context of sustainable development.

The world economy is currently experiencing a lingering and an imbalanced pattern of growth. When viewed against this backdrop, the concept of sustainable development is not just an option for how humanity as a whole and each community should develop; it is a set of opinions about the model of such a society to ensure its sustainable existence. Sustainable development seeks to balance the economic, environmental and social dimensions of development in a long-term and global perspective. It entails a broad view of human welfare, a long-term perspective about the consequences of today's activities, and the full involvement of civil society to reach viable solutions [1]. Green growth, according to the definition of the Organization for Economic Co-operation and Development (OECD), can be described as:

a way to pursue economic growth and development while preventing environmental degradation, biodiversity loss, and unsustainable natural resources use [2,3].

However, the United Nations Environment Programme (UNEP) defines Green Economy as:

one that results in improved human well-being and social equity, while significantly reducing environmental risks and ecological scarcities [4].

The key elements of green growth strategy, which all stakeholders—be they architects or engineers—should be aware of, include four key environmental challenges: climate change, unsustainable use of natural resources, loss of biodiversity and ecosystem services, and unsustainable materials management [1]. Obviously, buildings and cities, directly and indirectly, contribute to these challenges in the design and construction phase or during the operation phase and demolition. There is an interlinked relationship between sustainable consumption and production when we design and erect buildings and cities; the processes involved need to be carefully examined and rigorously evaluated.

A green economy is a tool for achieving sustainable development. The approaches and methods to bring about this may be diverse [5]. Architects, engineers, scholars and researchers, as well as regulators and policy makers, have a major role to play in promoting and making green economies. To this effect, the UNEP published its Green Economy report in February 2011, which designates "Buildings Investing" in energy and resource efficiency, as a major section to achieve a greener economy [6]. It also specifies that the building sector is the single largest contributor to global greenhouse gas (GHG) emissions, with approximately one-third of global energy end-use taking place within buildings. Looking at the construction sector, it is responsible for more than a third of global resource consumption, including 12% of all freshwater use, and significantly contributes to the generation of solid waste, estimated at 40% of the total volume. These facts illustrate the significant hand of the building sector in increasing GHG emissions globally, and consequently influencing climate change. By and large, green building is among the major indicators to measure the sustainability of cities and their adherence to sustainable development in general. So, promotion and construction of green buildings and improvements of existing buildings would positively contribute to the reduction of the carbon emissions.

The use and incorporation of renewable energy in buildings, and the promotion of energy-efficient buildings, are considered among the best means of tackling the global problem of climate change, and of eliminating carbon dioxide (CO_2) emissions [7]. The zero-emission REC Conference Center in Szentendre, Hungary holistically manifests such means, and is considered as one of Europe's sustainable buildings. It aims at achieving zero emissions using renewable energy and efficient building elements (Figure 6.1). The building has 140 photovoltaic (PV) panels on the roof which harness 19% of the sun's energy (a very high proportion), and produces up to 29 kW of electricity during the summer months, generating a surplus that feeds into the local grid. Such surplus would correspond as a rebate from the local operator during winter. The PV panels have had a maintenance-free run from the very first day they went into operation. In addition, the building is free from light switches. Lighting in the building is automatically adjusted, with natural light penetrating it, when required [8].

High efficiency solar panels | Light shelves | Window's shutters | General view—REC-CC in green

FIGURE 6.1 The zero-emission REC Conference Center, Szentendre, north of Budapest, Hungary. *Photo Credit: Author.* (For color version of this figure, the reader is referred to the online version of this book.)

6.1.1. The Dubai Story

In Dubai—and in the United Arab emirates (UAE), in general—green building is high on the agenda. Excessive energy consumption and water use in buildings pose a great challenge for planners and government. As remedial measures, the period from 2007 to 2012 has witnessed a set of developed strategies, policies and enacted regulations concerning sustainable buildings and transport. Dubai has taken many initiatives to achieve sustainability and sustainable development. The announcement of the Dubai Government Strategic Green Building Policy in 2007 is a manifestation of the long-term government plans to curb high energy consumption and water use in the emirate. The driver for such policies was to conserve and manage natural resources, mainly energy and water demands and supply (Figure 6.2) [9].

The following sections of this chapter portray and argue about the way sustainable buildings can contribute to increase the efficiency of the sustainable built environment with the intent to cutting carbon emissions. They also present an overview on how sustainability regulations and laws contribute to carbon emission reduction. The objective of this chapter is to discuss and showcase sustainable buildings (new and existing) in Dubai and to present the energy-efficiency principles adopted in them. The case studies represent three new buildings and three refurbished buildings. The case studies portray Dubai's first carbon-neutral warehouse (dnata) and the first LEED-certified existing office building in the Middle East. Additionally, this chapter discusses some of the energy-efficiency features and water-recycling elements in the highest tower in the world, Burj Khalifa, in an effort to address encountered challenges of global climate change, particularly CO_2 emissions.

(a)

Year	GW
2005	3.6
2008	5
2011	8
2015	16

Electrical energy demands in GW (2005–2015), *Data source: Ref. [9]*

Burj Khalifa, world's Tallest skyscraper

(b)

Year	$10^6 \times$ gallon/day
2005	200
2008	250
2011	310
2015	780

Water demands in Million imperial gallons per day (2005–2015), *Data source: Ref. [9]*

FIGURE 6.2 Electrical energy and water demands—Dubai. *Photo Credit: Author.* (For color version of this figure, the reader is referred to the online version of this book.)

6.2. CLIMATE CHANGE: CITIES AND BUILDINGS

The need for green cities and sustainable buildings!

The impact of climate change on our cities has been clearly manifested in 2001–2010. Cities not only consume most natural resources but also produce air pollution and generate great amounts of waste and large volumes of wastewater. Over the twentieth century, cities and towns, using various urban systems, have been developed and built in inefficient ways based on conventional methods and life styles. Such models engendered wasteful production and consumption of resources.

Urban sprawl is, indeed, a persistent problem as the world's natural resources are becoming scarce and are pressured by human population increase and settlements. These pressures, particularly those related to carbon emissions in urban areas, involve power generation, transport, and waste. Thus, the act of creating buildings and of developing entire cities is one of the most complex and sophisticated.

It is imperative to notice that roofs and pavements form almost 60% of a typical city's surfaces, as shown in Figure 6.3. Subsequently, the way in which these surfaces are designed and in which the materials they constitute are chosen—environmentally friendly materials, for instance, gain less heat and reflect the majority of solar radiation impinging on these surfaces—significantly contributes to energy savings and to lowering carbon emissions.

Urban development is clearly central to most of the concerns. Cities pressure natural resources and consume large amounts of energy and water as well as emitting air pollutants and generating waste. All these activities pose potential risks to our economic, health, and environmental wellbeing, and the prime global concern at the present time is the effect of human activities upon

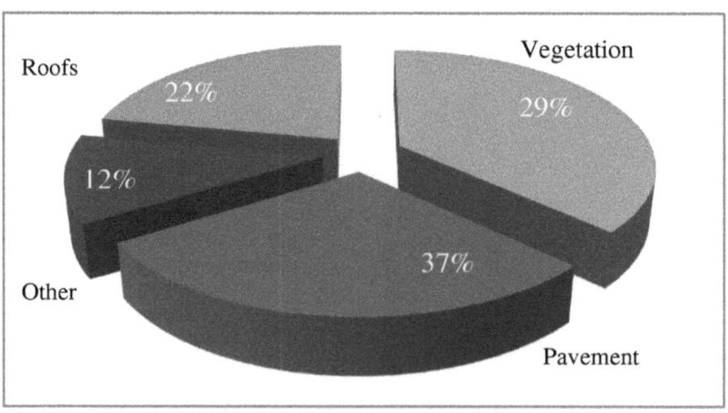

Average for 4 large cities in the USA (Sacramento, Chicago, Salt Lake City, and Houston)
Note: Elements 75% is building related in urban areas, 25% roofs, 25% pavements, 25% sites formation and 25% vegetation and water

FIGURE 6.3 City urban components (fabrics) to combat Heat Island Impact phenomenon. (For color version of this figure, the reader is referred to the online version of this book.)

climate change particularly related to urban areas. Several studies published by the United Nations Habitat showed that in 2007–2008 a major shift occurred where more than half the world's population was classified as urban dwellers [10,11,12]. Globally, the number of people living in towns and cities at the end of 2008 has been estimated at 3.3 billion and it is expected to increase to 5 billion by 2030 [10,13]. The main challenges to such increases are the pressures on natural resources and utilities, such as energy and water, as well as waste management and mobility. For instance, the UNEP data on global energy demand suggested that it will increase by 45% by 2030 [14]. So, the solution to all these challenges and facts is to make new and existing buildings and cities green.

For architects and urban planners, it is vital to efficiently plan cities to be sustainable by developing strategies and policies to promote sustainability. Significantly, capturing sustainable development indicators at local as well as global levels is relevant since climate change and pollution are among the major world problems. Also, emphasis should be stressed when it comes to cities' consumption patterns and their impact on other neighboring regions and ecosystems. Additionally, the accountability and responsibility for all stakeholders to assist in monitoring consumption patterns and addressing the requirements for a sustainable city is needed. Moreover, adaptive policies to reduce, recycle, and reuse consumer goods are important strategies for cities to achieve sustainability pathways [15].

To plan our cities to be sustainable, we need to develop bold strategies including key features to address important issues that on the long term, achieve sustainability pathway. Figure 6.4 illustrates these lines in detail. According to the latest report by the Sustainable Cities Index (SCI), sustainable city indicators mainly in US, Canada and UK centered on many indicators, namely, clean technology, sustainable planning and management, environmental quality and green mobility, quality of life, as well as green building, as core issues as shown in Figure 6.5 [16].

There are several methods which can be used to achieve sustainability in buildings, but we should discuss the importance of sustainable buildings.

FIGURE 6.4 Suggested strategy lines to achieve sustainability pathway. (For color version of this figure, the reader is referred to the online version of this book.)

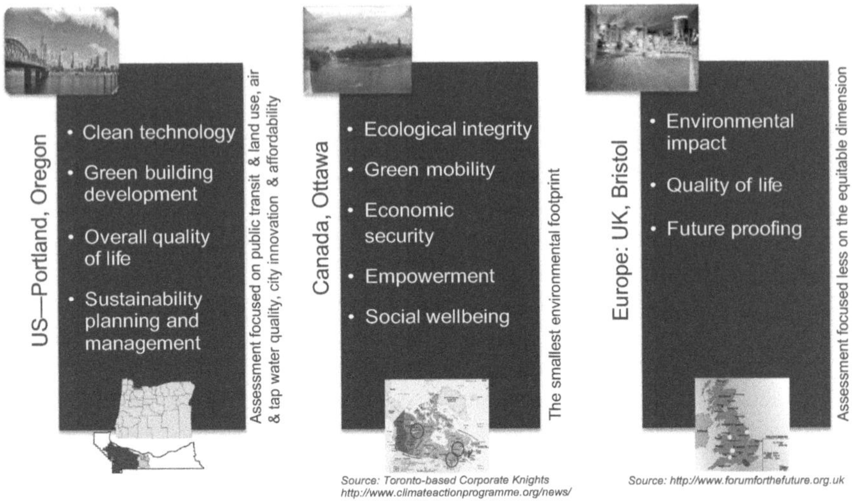

FIGURE 6.5 Indicators to measure the sustainability of cities (USA, Canada, and the UK). (For color version of this figure, the reader is referred to the online version of this book.)

6.3. IMPORTANCE OF SUSTAINABLE/GREEN BUILDING

Energy, water, efficient buildings for the future!

Talking, practicing, and achieving sustainability is an issue of paramount importance for every society and nation, primarily in the building sector. Three facts on material resources, energy use, and waste produced from the building sector in Europe should inspire the change in our current behavior and promote sustainable building [17]:

1. Fifty percent of material resources taken from Nature are building related.
2. Forty percent of the energy used in Europe is building related.
3. Over 50% of the national waste production comes from the building sector.

Making buildings (homes, schools, and universities, offices, and hospitals, etc.) healthy and green is an imperative task for all architects, engineers and, most importantly, regulators and policy makers. Students should be careful in the selection of the materials and finishes of the spaces they design and eventually build as professionals. We design and construct buildings in a region that is characterized as hot-arid/hot-humid climate. So, it is worth paraphrasing the Modern Movement's remarks on Form following Function: *Form does not follow climate: it responds to it, or Climate does not determine Form: it influences it* [18]. Thus, making buildings green yet sustainable should be the main focus in the present and the future.

Another fact worth highlighting is the percentage of energy, water, and electricity consumption in buildings and, consequently, carbon emissions [19].

Residential, commercial, and government buildings are responsible for one-fifth of Australia's GHG emissions, whereas buildings in the United States account in their use for the total energy use, total water consumption, and total electricity consumption of 39%, 12%, and 68%, respectively. These total rates of consumption equate almost 40% of carbon emissions.

Energy efficiency is a good way to use less energy, lower demand on energy resources and lessen GHG emissions. Reducing the amount of energy use in cities by transport and buildings is widely believed to be the quickest, simplest, and most cost-effective way to reduce global GHG emissions. To this effect, the International Energy Agency suggests that energy efficiency could deliver 65% of all the global GHG abatement needed to reach a target of 450 parts per million of CO_2. In that context, green building ought to be accepted as a necessity in view of the fact that the buildings of today consume a high rate of electricity, energy, and water and generate a large amount of waste and pollute the air, which in turn, causes health problems. Speaking of which, while energy efficiency is a key attribute, so is human health, from a building occupancy point of view and not just from an outdoor pollution point of view. In fact, indoor pollution is of grave concern.

In general, people spend about 90% or more (330 days out of 365 days) of their time indoors. Hence, the quality of these indoor spaces is crucial to their health and impacts them with long-term exposure. It is known that indoor levels of pollutants in airtight buildings may be two to five times higher and, occasionally, more than 100 times higher than outdoor levels [20]. Developing future efficient buildings and enhancing the performance of existing buildings to address the climate change challenges, especially amid the catastrophic nuclear disaster in Japan, which brought the nuclear power plants in forefront risk to generate electricity, is of a top priority.

Furthermore, health, livability, and productivity improvements are attributed to green building. They increase productivity of inhabitants, which can result in savings higher than those achieved from energy-efficient buildings [6]. Another factor that is a major cause of critical illness and premature death in residential buildings in many developing countries is indoor pollution resulting from gas cookers or poor combusted solid fuels (e.g. coal or biomass) coupled with poor ventilation. Lower respiratory infections, such as pneumonia and tuberculosis, linked to indoor pollution are estimated to cause about 11% of human deaths globally each year. Owing to their daily exposure, women and children tend to be most at risk [6].

In all, it was initially perceived that benefits of "green building" efforts go only to the environment, but this is no longer the case. Environmentalists, researchers, and public supporters of ecofriendly architecture took and fostered a holistic approach to the concept of environmental health, including human wellbeing, in their approach and calculations. To this end, a list of green building programs was developed and published by US Environmental Protection Agency (EPA) to increase efficiency in the use of water, energy, etc.; protect the health and increase the productivity of the building's residents; and reduce pollution and waste [19]. According to the EPA, the natural environment, human health, and the economy are, to a greater extent, influenced by the built environment.

Generally, both economic and environmental performances can be maximized when green building strategies are adopted. Integrating green construction methods into buildings is not limited to the design stage, but can be done at the construction, renovation, and deconstruction phases, as well. Nonetheless, the most significant benefits can be obtained if the design and construction teams consider an integrated approach from the earliest stages of a building project. Figure 6.6 illustrates the potential benefits of green building. These can be outlined in three main domains: environmental, economic, and social benefits [19].

Designers of buildings play a pivotal role within the building sector, particularly when it comes to energy consumption. The decisions they take have a significant impact on the building and the occupants. It is vital that architects and engineers learn and digest the extent to which materials and components they select and specify have low environmental impacts. Mike Sinclair et al. developed a full guide on green specifications: an environmental profiling system for building materials and components to aid the complex, time-consuming, and expensive process of the full life-cycle assessment. The relative environmental performance of over 250 materials and components have been assessed in this guide, using carefully researched, quantitative data derived from the British Research Establishment Environmental Database [21]. To this effect, developing and constructing green and sustainable buildings are vital to save our planet in terms of resource management and conservation.

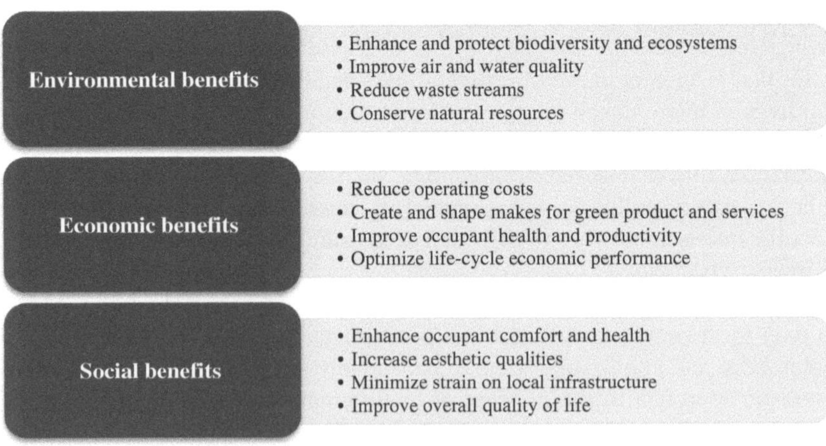

FIGURE 6.6 General benefits of green building [19]. (For color version of this figure, the reader is referred to the online version of this book.)

6.4. SUSTAINABILITY REGULATIONS AND LAWS CONTRIBUTING TO CARBON EMISSIONS REDUCTION

Regulations to promote green building for better living!

The role of regulations and laws addressing sustainable development in a boarder spectrum and sustainable built environment, particularly in urban areas and buildings, is crucial. Stringent regulations and adoptive policies in the European Union are classified among the highest in the world, in terms of environmental standards [22]. These regulations and laws play an important role in controlling energy use in buildings and, hence, in lowering carbon emissions. Policies and regulations are needed to curb energy use as buildings alone account for approximately 30% of all energy consumption globally and a significant share of GHG emissions. According to Country Report on Building Energy Codes in Australia, commissioned by the US Department of Energy (DoE), codes and regulations related to building energy help in ensuring that new buildings are energy efficient and, subsequently, result in reducing building energy use by 50% or more compared to those designed without energy efficiency in mind [23].

6.4.1. How does Dubai Measure Up?

In the UAE, the Dubai Strategic Plan (DSP) 2015, published in 2007, emphasized on sustainable development: economic development, social issues, and the built environment. It was revisited in 2009 to address the impact of the world financial crisis on development and growth. The DSP was centered on six main sectors:

1. Public sector
2. Economic Development
3. Social Development
4. Infrastructure, Land, and Environment
5. Security, Justice, and Safety
6. Government Excellence [24].

It was established to provide a universal comprehension of Dubai's vision among various government bodies, and ensure a ground framework for the operation of these entities to meet the aims set by the Government of Dubai. In the case of Dubai, the green thrust came from the very top. HH Sheikh Mohammed bin Rashid Al Maktoum, Vice President and Prime Minister of UAE, and the Ruler of Dubai, in keeping with his vision to make Dubai a green city, issued a directive to implement green building standards in the Emirate, meaning that all owners and developers of residential and commercial buildings and properties in Dubai must comply with internationally recognized, environmentally friendly specifications [25]. In other words, pursuing a green goal was not an option but mandatory. In that sense, the directive was a watershed development—in a global context. While many governments the world over—the UK is a case-in point—stopped short of insisting on green and of issuing strictures to that effect, Dubai crossed the threshold.

6.4.2. Europe

It is understood that counterbalancing climate change, preserving biodiversity, reducing health problems resulting from pollution, and using natural resources effectively are the key priorities. These objectives aim to protect the environment while encouraging economic growth. However, that the process of tackling these key priorities should look at current challenges, primarily energy demands in cities and in buildings. The first challenge test facing Europe includes rising energy prices and high consumption rates. Obliquely, the "Energy 2020 strategy" contains a sound framework for energy policy that aims to target 20% energy savings by 2020. The core objective for 2020 is energy efficiency through reducing emissions in all sectors, improving energy security, lowering consumption, and decreasing overall energy consumption by 20% by relaying on renewable resources, mainly solar roofs [22].

6.4.3. The United Kingdom

The energy used for heating rooms and water in the nation's buildings generates about 33% of the UK's carbon emissions and these must be cut by 20% if the government is to meet its legally binding emissions targets [26]. In the UK, commercial buildings contribute 17% of the nation's emissions, yet have no obligation to improve efficiency [27]. For example, the breakdown of the overall energy consumption in a typical building in the UK is 65%, 12%, 10%, 10%, and 3% for heating, lighting, hot water, appliances, and other uses, respectively. Nevertheless, the Energy Manufacturer Model, "Green labeling" and the "A–G rating" in the UK, known as the Display Energy Certificate, has an accompanying report advising on how to improve performance. The A–G rating was introduced in 2008 and was extended to become the A–E rating in March 2011 [27]. The purpose of the system is to evaluate energy performance in buildings. Although it is a good mandatory tool to assess the building performance in terms of its energy use, the course of action to rectify less efficient buildings is optional. For example, the House of Parliament (Westminster Palace) in London was rated G, but no mandatory action has been applied to improve its performance (Figure 6.7).

The UK Government's Committee on Climate Change introduced a bill known as "green deal" that intended to vastly improve home energy efficiency. This is one of the three key tests alongside delivering low-carbon energy and implementing ambitious targets for emissions cuts, which seems to entitle this government the "greenest government ever" in 2013. According to the

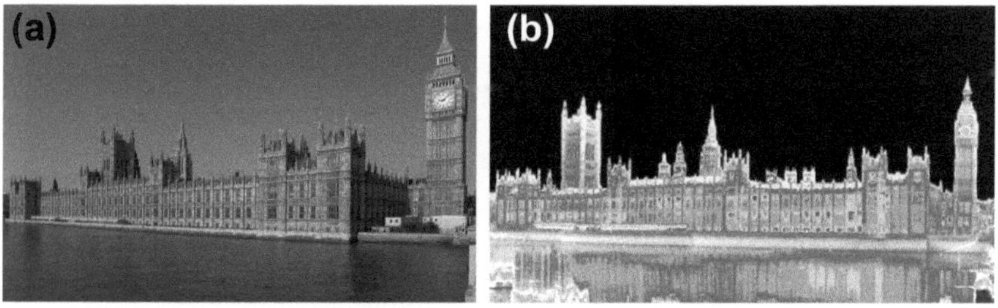

General view of the building An infrared image of heat loss

FIGURE 6.7 The House of Parliament (Westminster Palace) in London, UK, a G-ranked building. *Photos Credit: Irt Energy/PA; Source: The Guardian, UK). Note: Energy Efficiency and Environmental (CO_2) Impact Rating: A (92–100); B (81–91); C (69–80); D (55–68); E (39–54); F (21–38); and G (1–20).* (For color version of this figure, the reader is referred to the online version of this book.)

Department for Energy and Climate Change (DECC), this "golden rule" will apply to the loans made for energy-efficiency measures, which dictate that the cost of the loan repayments will always be lower than the fuel bill savings delivered. It is hoped that the green deal will allow householders, unable to pay for funding energy-efficiency measures, to gain access to upfront capital, but with a cap of about £6000 [27]. This would mean that someone thinking about moving home would not be deterred from taking up the scheme by the prospect of not benefiting from the savings delivered by the work [27]. This is seen as an excellent way forward to improve energy efficiency and cut emissions, though it may get the power to force landlords to install energy-efficiency measures after 2015.

6.4.4. Australia

In Australia, the story is different. The Ministerial Council on Energy initiated the National Framework for Energy Efficiency (NFEE) in August 2004. This is a major national step forward to promote energy efficiency, productivity, and the environment [23]. The NFEE's first phase created and interlinked nine policy packages addressing the following types of buildings and sectors:

1. Residential buildings
2. Commercial buildings
3. Commercial/industrial energy efficiency
4. Government energy efficiency
5. Appliance and equipment energy efficiency
6. Trade and professional training and accreditation
7. Commercial/industrial sector capacity building
8. General consumer awareness
9. Finance sector awareness [28].

Also, the Australian Government and State and Territory Energy Ministers launched Stage Two of the NFEE (December 2007). This phase included a package of five new energy-efficiency measures:

1. Expanding and enhancing the Minimum Energy Performance Standards (MEPS) programme.
2. Creating strategies for high-efficiency heating, ventilation and air conditioning (AC) systems.
3. Phasing out of inefficient incandescent lighting.
4. Emphasizing government leadership though green leases.
5. Developing measures for a national hot-water strategy, for later consideration [28].

In terms of building energy codes, the Australian Government developed and is administrating the Australian Building Codes Board—a Council of Australian Government, which has the following key objectives:

1. Maintaining and updating the Building Code of Australia (BCA).
2. Providing regulations to aid the design, construction, and use of buildings throughout Australia.
3. Supporting the governmental agenda such as on issues relating to climate change [29].

The BCA is responsible for technical provisions for the design and construction of buildings and other structures throughout Australia, according to various climatic zones. It covers structure, fire resistance, services, equipment, energy efficiency, and certain aspects of health and amenities. It is also worth mentioning that all the provisions for the building envelope apply to most building types, including commercial buildings and residential houses. Energy-efficiency measures were introduced into the BCA in January 2003. For energy efficiency, the code allows for either a performance-based approach to compliance (compared to a reference building) or a prescriptive approach based on requirements for specific building components.

In 2000, the Australian Government launched a "two-pronged approach" to combat GHG emissions from buildings:

1. Mandatory minimum energy performance requirements through the BCA.
2. Voluntary initiatives in industry [30].

Additionally, the minimum improvement in energy efficiency of products and enable consumers to choose their products that use less energy. Energy performance standards (MEPS) and labeling have been enacted. The MEPS and Energy Rating Labels help in Australian Government white paper announced several initiatives (December 13, 2007 and updated in 2009), including the expansion of the MEPS to cover buildings and more appliances [31,32]. This means that importers, manufacturers, or sellers of products that are regulated for energy efficiency must meet MEPS and/or energy rating label requirements [23].

General view   External facades of Geyser office building

FIGURE 6.8 Geyser building in Parnell, Auckland, New Zealand's first World Leadership Green Building. *Photos Credit: (a) SAMSON Corporation & (b) Jeff Brass, ARCHITECTURENOW.* (For color version of this figure, the reader is referred to the online version of this book.)

6.4.5. New Zealand

In New Zealand, the first World Leadership Green Building was built. The Geyser building in Auckland (5040 m^2 of primarily office space) was rated a six-star green building (Figure 6.8), and was designed to

1. save 73% of energy (utilizing 27% of the energy of a typical retail/office building of its size);
2. decrease the artificial lighting by 50% (half of that required of a typical retail/office building, 6 W/m^2);
3. conserve 50% of water (only half of that used in a typical retail/office building of its size);
4. supply 100% fresh air to occupants (compared with the average of 25% for air-conditioned buildings in New Zealand);
5. recycle 70% of the building's generated waste [33]. All these are achieved during the building use and operation.

Green building features were installed in the Geyser building [33]. In this new standard of green building (Figure 6.8), the floors are divided into five individual subbuildings grouped around a system of atriums with pedestrian connections. Such design maximizes natural daylight penetration into, and exterior views out from, in the majority of the spaces. Each "sub building" is wrapped in a dynamic three-dimensional semireflective white, twin-walled façade. The application of such technology creates natural heating and cooling ventilation through thermal currents rising up through the void spaces between the two walls. It naturally heats each building by trapping warm air between the two walls in winter, significantly minimizing heat losses overnight and "prewarming" outside air during the day. In the hottest months, the whole outer skin opens electronically to fully ventilate the cavity by creating cross-convection currents from the cooler southern elevations of the buildings. Also, a rainwater harvesting system was incorporated to collect rainwater to store and supply low-flush toilets, irrigation systems, and low-flow sanitary fixtures, to save water use form the grid.

Furthermore, tenants and visitors were encouraged to walk or to cycle, and 25% of the car parking spaces in the automated stacker car park are nominated only for small cars, the building was located very near a train station and public transport network [33]. This excellent example showcases the state-of-the-art in design and the appropriate technologies to create a green and sustainable building. *(Note: Most buildings exclude garage, storage, or laboratory buildings that do not have a conditioned space, or an atrium or solarium that is not a conditioned space and is separated from the remainder of the building by an envelope.)*

6.4.6. Dubai Green Building Policy

In 2008, Dubai Government Strategic Green Building Policy (DGSGBP) was implemented by Dubai Electricity and Water Authority and Dubai Municipality. Also, Phase 1 and 2 of Dubai Green Building regulations were announced in 2010. The framework covers such areas as ecology and planning; building vitality; resources effectiveness (energy); and recourses effectiveness (water and materials) [34]. Figure 6.9 depicts this framework in details. This move substantially contributed toward achieving DSP in terms of economic and social development, infrastructure, land, and environmental development. The driving force behind developing DGSGBP was the resource challenges (energy and water demands). Figure 6.10 exhibits a comparison between energy use in a typical building in the UK and Dubai, whereas Figure 6.11 shows electricity and water demands by sector in different building types in Dubai. To reduce the high current rate of demand, the Dubai Supreme Energy Council of the Dubai Government has introduced electricity and water slabs tariff based on monthly rate of electricity (kWh) and water use (IG) for residential, commercial, and industrial categories, (e.g. G: 0.0–2000, Y: 2001–4000, O: 4001–6000 kWh, R: 6001, and above). This tariff was revisited in 2010 and a new tariff, including fuel surcharge, was introduced effective January 2011 to encourage users to reduce their consumption rates [35].

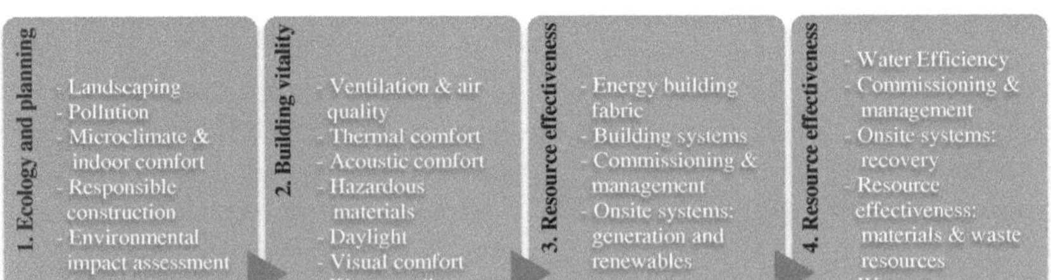

FIGURE 6.9 Framework of Dubai Green Building regulations. (Most buildings exclude garage, storage, or laboratory buildings that do not have a conditioned space, or an atrium or solarium that is not a conditioned space and is separated from the remainder of the building by an envelope). (For color version of this figure, the reader is referred to the online version of this book.)

From the above, it can be seen that enacting regulations and laws related to sustainability and diligently implementing them will certainly play a part in cutting carbon emissions. Also, awareness and training programs are vital tools in educating public and private sectors, including all stakeholders, which in turn, contribute to gearing the city's inhabitants to change their behavior regarding energy and water consumption in buildings and toward recycling.

6.4.7. Dubai's Iconic Building—Burj Khalifa

Condensation water for irrigating the site's landscape—an environmental innovation

Dubai has embarked on a unique journey—to aim for high environmental standards, but without taking the foot off the pedal in its ambition to raise extraordinary structures. Perhaps nothing typifies this better than its iconic building, *Burj Khalifa*, which is the tallest freestanding skyscraper in the world to-date. The building developed by Emaar Properties and designed by Skidmore, Owings,

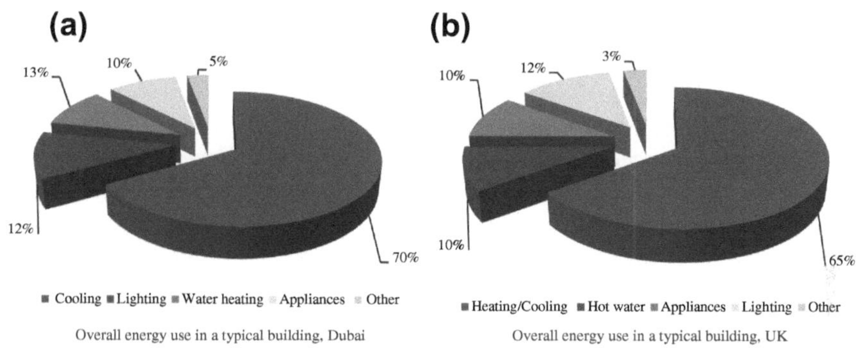

FIGURE 6.10 Comparison of typical energy use in the UK and in Dubai buildings. (For color version of this figure, the reader is referred to the online version of this book.)

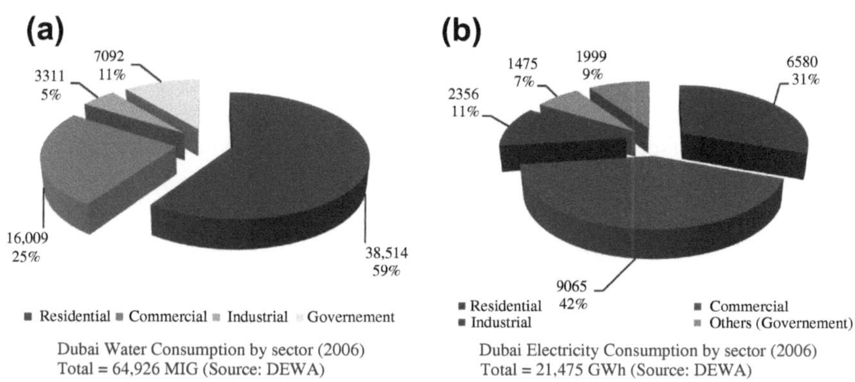

FIGURE 6.11 Electricity and water demands by sector in different building types, Dubai. (For color version of this figure, the reader is referred to the online version of this book.)

A view showing the massive area of facades glazing

FIGURE 6.12 Burj Khalifa, the tallest skyscraper in the world. *Photos Credit: Author.* (For color version of this figure, the reader is referred to the online version of this book.)

and Miller (SOM), rises to a height of over 828 m (2716.5 ft) and contains more than 160 storeys. The detailed data on energy performance and energy-efficient features are not available; however, the use of condensate water, occurring on the external leaf of the glazing surface in the hot and humid summer months, for irrigating the landscape on the site is worth studying. The skeleton of the building is covered by glass, a dominant element in the design and construction of the tower, which has a total of more than 24,000 cladding panels over a total curtain wall area of 132,000 m^2 (Figure 6.12). Its façades were designed to help minimize heat transmission and save energy [36]. Due to such a colossal area of glazing and the water condensation on the outer skin of the building during summer months in Dubai, it was decided to exploit the fact effectively to save condensation water for irrigation of the site's landscape. The AC system in the skyscraper draws air from the upper floors, where the air is cooler and cleaner than on the ground [37]. At peak cooling times, the tower's cooling is equivalent to that provided by 13,000 t (29,000,000 lb) of melting ice in one day [38]. The condensation water collection system, which utilizes the hot and humid outside air, coupled with the cooling requirements of the tower, results in a significant amount of condensation of moisture from the air. The condensed water is, then, collected and drained into a holding tank, which is located in the basement; this water is, then, subsequently pumped into the site irrigation system for use on the park [39]. The use of condensate water, accumulated on the skyscraper facades, for irrigation is as good as any environmental innovation. In the case of Burj Khalifa, it has offset the large volume of water that would have been required, had conventional thinking been applied.

6.5. TAXONOMY OF A SUSTAINABLE BUILDING

Elements and features to make a building green!

A green building can be defined as the one that has minimum adverse impacts on the built and natural environments, in terms of the building, its immediate surroundings, and the broader regional and global settings. Thus, the rational use

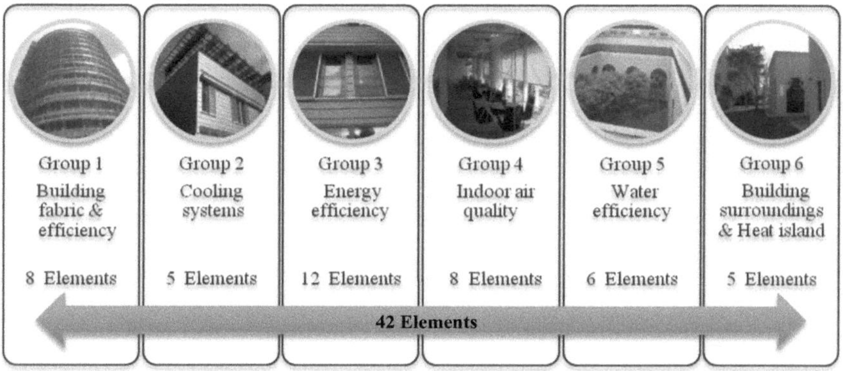

FIGURE 6.13 Groups, features, and related elements of green building. *Photos Credit: Author.* (For color version of this figure, the reader is referred to the online version of this book.)

of natural resources and appropriate management of the building stock will contribute to saving scarce resources, reducing energy consumption, and improving environmental quality. Although this definition is straightforward, it is yet too general to be applied in practice to a specific situation. Therefore, it is important to have measurable criteria to judge whether a building is green or not.

It is vital to understand the classification of a sustainable building to define its features and the elements that will entitle such a building to be green. In a nutshell, there are six main relevant features:

1. Site and surroundings
2. Energy efficiency and renewable energy use
3. Water consumption
4. Indoor environmental quality
5. Materials use and management
6. Integrated design approach.

In order to ensure a building is green or sustainable, architects and engineers should pay attention to the following groups, features, and their related elements. This should be carried out in a diligent and a rigorous examination and a comprehensive evaluation and judgment throughout the building life cycle (predesign phase, design and build phase, and operation phase). The elements of green building have been grouped by the author to enable us to feature the selected elements for the design analysis of the Dubai Government green building policy (Figure 6.13). For simplicity and comprehension, these elements were bundled in six groups (Figure 6.14). For the policy design, out of 100 green elements, 33 were assessed and then 12 were picked out for the analysis. The 12 elements are water recycling and irrigation, low-e glazing, daylighting, shading, water reduction, water fixtures, cooling systems, renewable energy, efficient bulbs and lighting fixtures, CO_2 sensors, motion sensors and control, as well as low-surface temperature roofs (cool roofs), as shown in Tables 6.1–6.4 [9].

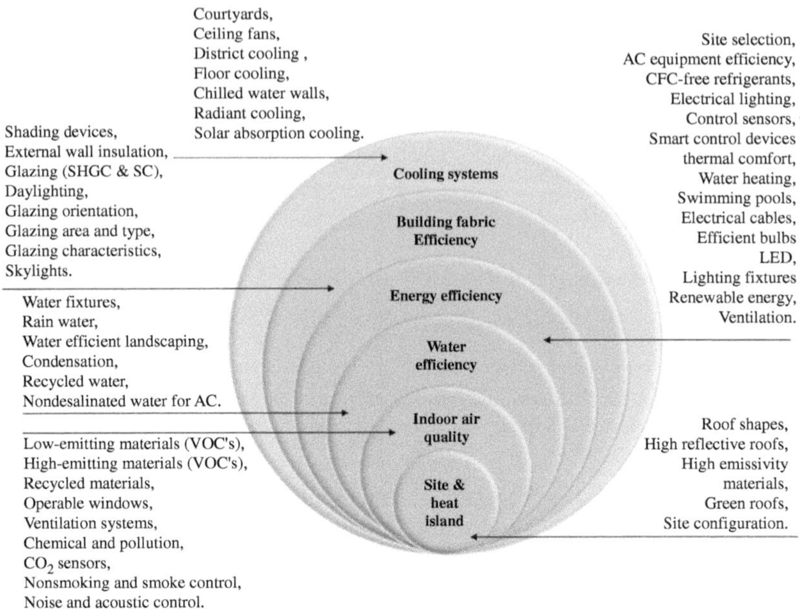

FIGURE 6.14 Classification of green building's features and elements. *Chart Credit: Author.* (For color version of this figure, the reader is referred to the online version of this book.)

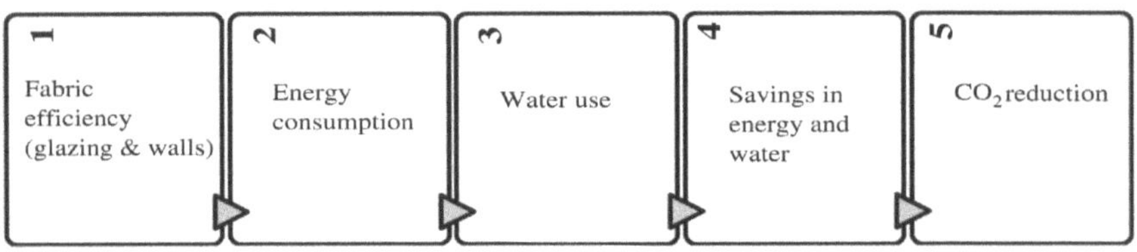

FIGURE 6.15 Key performance indicators to measure the extent of green building case studies. (For color version of this figure, the reader is referred to the online version of this book.)

In the design process of creating and developing a green building, one has to apply performance indicators (PIs) for building under design and operation, or retrofitting, to ensure meeting the target requirements. Figure 6.15 illustrates such key PIs. Five PIs of the three new case studies were also established to examine the extent of the fabric efficiency (glazing and wall), energy and water use, savings in water and energy, and footprint in terms of CO_2 reduction, as shown in Tables 6.3 and 6.4 [9].

6.6. GREEN BUILDINGS IN DUBAI, UAE

6.6.1. Case Studies—New Buildings

Dubai launched a bold initiative toward green and sustainable building in 2007. Many green buildings were designed and built. In 2011, the total green build area reached 430,000 m^2 compared to 35,501 m^2 in 2009 [40]. The following three

TABLE 6.1 Dubai Green Building Case Studies (New Buildings)

Case Study 1	Case Study 2	Case Study 3
(a)  Aviation College's southern façade *Photo credit: Author*	(b)  Pacific Controls' western façade *Photo Credit: Author*	(c)  METITO headquarters' main façade *Photo credit: Author*
Location: Al Garhoud	Techno Park, Jebel Ali	Techno Park, Jebel Ali
Type: educational	Corporate office	Corporate office
Number of floors: G+6 floors	G+4 floors	Ground floor
Total built-up area (in sq ft): 513,852	100,000	140,587.43
Initial cost[1](in US $): 53,424,657.57	14,246,575.34	9,315,069.00
Actual cost[1] (in US $): 47,945,205.48	16,438,356.16	10,958,904.00
Cost (in US $/sq ft): 93.3[1]	164.38[2]	84.54

Notes:
- The average cost of construction/sq ft of the three case studies is about US $114.00 (AED416) current exchange rate.
- The construction cost/sq ft of a quality office building in Dubai varies from US $115.10 to 164.40 (AED420 to 600).
- The average construction cost/sq ft of a conventional building is estimated as US $139.75 (AED510).

[1]*Cost does not include that of the indoor furniture and interior decoration.*
[2]*Cost was 30% more than that of a conventional building, but its operational cost is kept at 50% less over 20 years.*

TABLE 6.2 Dubai Retrofitted Case Studies (Existing Buildings)

Case Study 4	Case Study 5	Case Study 6
(d)  **Dubai Chamber's western façade** ***Photo Credit: Author***	(e)  **dnata Freight Gate-5 warehouse** ***Photo Credit: Author***	(f)  **Emirates Academy's western façade** ***Photo credit: Author***
Location: Dubai Creek—Deira	Airport Free Zone, Deira	Community 366, St. 10c, No 69, Umm Suqueim 3, Jumeirah
Type: mixed-use	Warehouse–logistics services	Educational
Number of floors: 18 floors (B+G+16)	5 floors (G+4)	2 floors (G+1)
Total retrofitted area (in sq ft): 80,000[1]	161,459[2]	130,000.00[4]
Initial cost (US $): 45,000,000	20,500,000[3]	15,000,000
Actual cost (US $): 45,000,000	21,750,000[3]	15,000,000
Cost of retrofitting (US $): 13,860,000	334,674.40	32,000[5]
Cost (US $/sq ft): 173[6]	642.073	115

Notes:
- The average cost of construction/sq ft of the three case studies is about US $123.00 (AED453) current exchange rate.
- The construction cost/sq ft of a quality office building in Dubai varies from US $115.10 to 164.40 (AED420 to 600).
- The average construction cost/sq ft of a conventional cargo building is estimated as US $81.75 (AED300.85).
- The average construction cost/sq ft of a conventional office building is estimated as US $139.75 (AED510.10).

[1]*The total area of the Dubai Chamber building is 200,000 sq ft.*
[2]*Cost includes only the warehouse portion of the building and excludes the automated freight handling equipment and office blocks.*
[3]*The total area of dnata Freight Gate-5 is 269,100 sq ft, but only 161,459 sq ft were retrofitted.*
[4]*Total built areas include two university buildings (50,000 sq ft) and three student accommodation buildings (80,000 sq ft).*
[5]*The retrofitting cost of EAHM buildings includes all new fitted equipment (solar water heaters) and fixtures to improve the building performance, as per Table 6.4.*
[6]*Includes furniture and fixtures (FF&E).*

TABLE 6.3 Performance Indicators (PIs) of the Three Sustainable Buildings in Dubai—New Buildings

Group	PI	Case Study 1	Case Study 2	Case Study 3
1. Fabric efficiency	Glazing PI 1 Shading coefficient SC PI 2 Solar heat gain coefficient SHGC	 SC=0.35 SHGC=0.305	 North=0.33 SHGC=0.29 Non-north=0.41 SHGC=0.36	 North=0.19 SHGC=0.33 Non-north=0.19 SHGC=0.165
	PI 3 *U*-value $W/m^2 K$ $Btu/h ft^2 °F$	 *R*=1.80 *U*-value=0.16 Btu	 R-19 insulated *U*-value=0.052 Btu	 *R*-value=9.5 h ft^2/Btu *U*-value=0.047 Btu
2. Water consumption	PI 4 Water use Yearly use Daily use	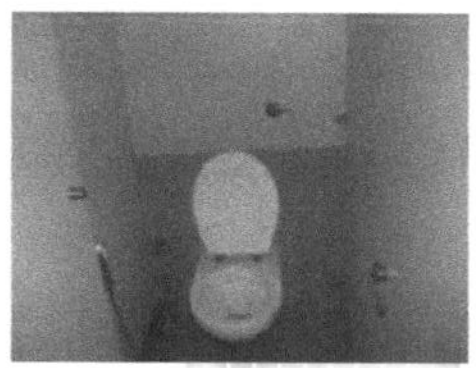 833,199.00 gallons 4 gallons/person	 3,148,316 gallons 21 gallons/person	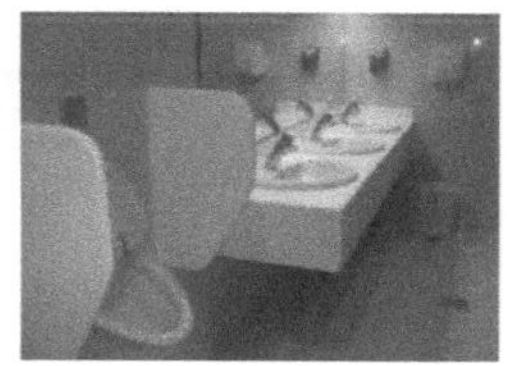 N/A N/A

Continued

TABLE 6.3 Performance Indicators (PIs) of the Three Sustainable Buildings in Dubai—New Buildings—cont'd

Group	PI	Case Study 1	Case Study 2	Case Study 3
3. Energy use	PI 5 Energy use: Annual Electrical use: Space cooling, lighting & fans kW (yearly) kW/day[1] kW/sqft (yearly)	 5,427,647.8 14,870.3 10.56	 321,517.00 1014.13 3.21	 5,083,581.9 8018.27 1.32
4. Savings	PI 6 energy Water	N/A	40–50% N/A	33.4% 74.1% (Calculated)
5. Footprint	PI 7 emitted CO_2	440 kg CO_2	≤380 kg CO_2	474 kg CO_2

[1]*Assuming the building is operated for 12 h/day.*

TABLE 6.4 Performance Indicators (PIs) of the Three Retrofitted Sustainable Buildings in Dubai

Group	PI	Case Study 4	Case Study 5	Case Study 6
1. Fabric efficiency	Glazing PI 1 Shading coefficient SC PI 2 Solar heat gain coefficient SHGC	 SC=0.30 SHGC=0.165	 North=0.41 SHGC=0.36 Non-north=0.41 SHGC=0.36	 North=0.35 SHGC=0.33 Non-north=0.30 SHGC=0.165
	PI 3 *U*-value $W/m^2 K$ $Btu/h ft^2 °F$	 $R=0.455$ *U*-value=$2.2 W/m^2$ K	 N/A[1] *U*-value N/A[1]	 *R*-value=N/A[1] *U*-value=N/A[1]
2. Water consumption	PI 4 Yearly use Daily use	 1,744,411 gallons 17.3 gallons/person	**X** (Not included in the retrofitting process)	 2,432,999 gallons 21.7 gallons/person

Continued

TABLE 6.4 Performance Indicators (PIs) of the Three Retrofitted Sustainable Buildings in Dubai—cont'd

Group	PI	Case Study 4	Case Study 5	Case Study 6
3. Energy use	PI 5 Energy use: Annual Electrical use: Space cooling, lighting, fans kW (yearly) kW/day[2] kW/sq ft/(yearly)	 3,958,118 10,844 19.8	 3,146,386 8620 19.5	 2.062,618 360 15.0
4. Savings	PI 6 energy Water	40% 70%	20% N/A	29.25% 33%
5. Footprint	PI 7 reduction in CO_2 emissions	176,000[3] kg CO_2 29,333 kg CO_2/month	167,000 kg CO_2 (year) 13,917 kg CO_2/month	109,477 kg CO_2 9123 kg CO_2/month

X not included in the retrofitting process.
[1]*N/A, not available.*
[2]*Assuming the building is operated for 12 h/day.*
[3]*CO_2 emissions over 6 months.*

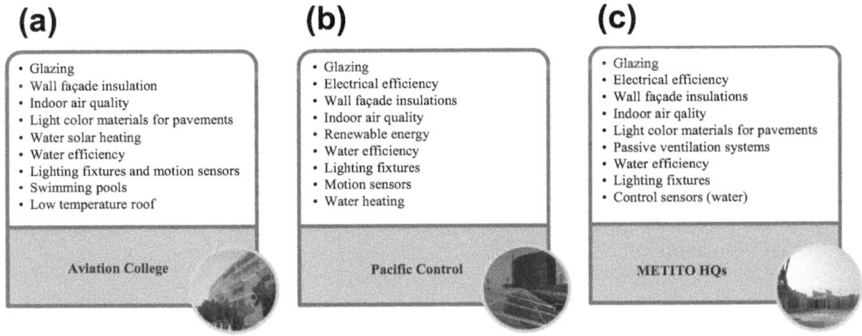

FIGURE 6.16 Green building elements incorporated in the three case studies (1, 2, and 3). (a) Four out of 14 elements were not incorporated in the design and construction: electrical efficiency, cascaded roof, renewable energy, and water heating. (b) Five out of 14 elements were not incorporated in the building: low-temperature roof, cascaded roof, light-color material for pavements, swimming pools, and passive ventilation system. (c) Five out of 14 elements were not incorporated in the building: low-temperature roof, cascaded roof, renewable energy, swimming pools, and water heating. *Image Credit: Author.* (For color version of this figure, the reader is referred to the online version of this book.)

buildings, located in Dubai (Latitude 25.27°N and Longitude 55.33°E), were selected to showcase the accomplishments: (1) Emirates Aviation College—Crew and Training building, Emirates Group, in Al Garhoud area; (2) Pacific Controls headquarters; and (3) METITO headquarters. The last two buildings are in Techno Park, in Jebel Ali Free Zone (Table 6.1). The first case study is of a building that houses the academic arm of Emirates Group to train the organization's students on aviation-related matters (Image A), whereas the second and third buildings are the corporate headquarters of Pacific Controls and METITO (Table 6.1, Images B and C).

Figure 6.16 shows each case study (new building) and the green building elements that were incorporated in the design and the facilities. Case studies (1, 2, and 3) were part of a study carried out in 2008 as part of and to inform Dubai Government Strategic Green Building Policy, pursued between 2007 and 2009. The study was presented in a research paper, which was published in the proceedings of The World Renewable Energy Congress (WREC, 2008), Glasgow, Scotland, UK [9].

6.6.2. Case Studies—Existing Buildings

Three retrofitted buildings located in Dubai were selected as existing buildings that went through the process of refurbishment to meet the criteria of green/sustainable building. These retrofitted buildings are as follows:

Dubai Chamber—Dubai's first existing-rated retrofitted green building in the Middle East, located on the side of Dubai Creek, Deira. The building, which is a mixed-use type, including offices, customer services, auditorium, and conference halls, with associate banquet facilities, was occupied in February 1995 (Table 6.2, Image D).

Box 6.1 Green Elements Applied in the Retrofitting Process of the Three Case Studies (4, 5, and 6)

Energy efficient techniques	Healthy environment and comfort
• Daylighting and light fittings	• Lighting level
• Automatic daylight dimming	• Noise level
• Occupancy controls sensors	• Water quality
• Efficient light bulb	• Air quality control by CO_2 sensors
• Passive techniques	• Mold and dust measurements
• Solar water heating	• Safety
• Sunlight ducts (light wells)	• Waste segregation and recycling
• Lifts capacities	
• Cool roofs	

Dnata Freight Gate-5 warehouse—Dubai's first carbon-neutral warehouse—known as the "green warehouse". It is located at the northern edge of Dubai airport apron and within the Airport Free Zone, Deira (Table 6.2, Image E). The refurbishment process was carried out by the Department of Economic Development, Dubai, dnata and DHL Neutral Services, part of The Neutral Group.

Emirates Academy of Hospitality Management, EAHM—this is an academy, which runs in association with Ecôle hôtelière de Lausanne, Switzerland. It is located in Jumeirah (Table 6.2, Image F).

The first case study is of a mixed-use commercial building. The second and the third case studies are of logistics, service-type buildings sheltering some of dnata's air cargo handling operations and an educational building, which is part of the Jumeirah Group, to educate students on international hospitality and tourism management, respectively (Table 6.2).

The purpose of the retrofitting process of the three case studies was primarily to reduce energy consumption, improve performance, lower carbon emissions, and make these buildings energy-efficient and sustainable. In the retrofitting process, the focus was on addressing the following elements, as shown in Box 6.1.

6.6.2.1. Case Study 4—Dubai Chamber Building

- The original building (occupied in 1995) underwent a complete internal refurbishment process in 2008, which considered all elements of the building apart from the external building envelope (glazing, thermal insulation and roof), besides grouping all meeting rooms in one floor to promote energy savings (a special lift, which is assigned for the building's visitors and customers, is available for this purpose, so the load on the main lifts operating for the 15-storey building was reduced, hence reduce energy use). Also, the cleaning routine of all offices and spaces was changed from evening to be from 7 to 9 a.m. and from 4 to 6 p.m.

This was to facilitate cleaning work during daylight. Additionally, the AC system was switched off after office hours to further reduce electricity consumption. By this exercise, indoor air temperature never rose above 3–3.5 °C above the comfort indoor temperature even in peak summer. The chillers and air-handling units were started a few hours earlier in June (from 5 to 7 a.m.). This approach brought indoor air to a comfortable temperature by 7 a.m.

- All these actions resulted in 40% saving of electricity consumption from the 1996 baseline. From 2008 to 2011, an additional 20% of the Dubai Chamber energy use was saved as a result of the retrofitting process of the building.
- Obliquely, 70% of water use was saved, plus all leakages in water closets were eliminated from the baseline consumption of 1996 until the retrofitting process, started in 2008. Also, collected water from AC systems was at an initial cost of US $1400. Furthermore, 2 months payback on summer time (May–October), estimated at 1500 imperial gallons per day (8 million liters per year), was saved as a result of eliminating almost all leakages in toilet flushed water compared to water use of 28.1 million liters per year in a convectional building, a saving of 19.5 million liters per year.
- In 2010, energy-efficient lighting fixtures were applied. Also, an EIB lighting control system, with presence detection, was used to efficiently manage and program the lighting requirements. This technology, which includes passive infrared sensors with presence detectors to check body heat, was applied to eliminate human carelessness and errors, yet to control electricity consumption when rooms are not in use. So, the control system combines motion, body heat, and illumination level sensors (in Lux), which control the lighting based on the ambient light level.
- Finally, air-quality sensors, mainly to detect the concentration of CO_2 in Dubai Chamber retrofitted spaces, were installed to ensure that the maximum concentration level does not exceed 750 ppm, which is better than that recommended by the American Society of Heating, Refrigerating and Air-Conditioning Engineers (ASHRAE) Indoor Air Quality standards of 1000 ppm.

6.6.2.2. Case Study 5—dnata Freight Gate-5 Warehouse

- The retrofitting process for dnata Freight Gate-5 warehouse was mainly to reduce energy consumption, which consequently lessens the utilities cost and lowers CO_2 emission. It perfectly aligns with dnata's environmental objectives that are driven by the requirements of dnata's ISO 14001 accreditation.
- In dnata Freight Gate-5 warehouse, all inefficient lighting fixtures were replaced by more energy-conserving light bulbs, mainly in the main freight handling hall. Solar-powered lighting fixtures were also installed in the parking zone to save energy.

- In addition, the very dark area in the second level of the warehouse was fitted with a sunlight duct to increase the natural light. In other words, no artificial lighting fixtures are on while it is sunny.

6.6.2.3. Case Study 6—Emirates Academy of Hospitality Management Building

- It is a self-designed Phase II building that was a designed-and-built project for Jumeirah Group in Dubai. The concept of the project started back in 2006 and was completed in 2007. The building includes two reception halls, a library, academic staff rooms and 10 classrooms, two lecture theatres, and an auditorium, as well as administrative offices and support facilities. The Academy building Phase II is 2-storey high, and the design concept was centered on a traditional style concept to match the tradition and climate of Dubai. The campus also includes dormitory facilities encompassing 270 studios, a social area, a gymnastic hall, restaurants, and sport facilities, as well as open spaces and green areas.
- In this building, passive design elements and energy-saving techniques were observed, such as courtyards and adjacent corridors that maximize daylighting, light-efficient bulbs, and occupant-motions sensors to control energy consumption. These courtyards were created to regulate air temperature and, subsequently, reduce dependence on AC, yet form transient spaces between air-conditioned spaces and other rooms, thus saving energy.
- For EAHM buildings, the retrofitting objective was primarily to decrease the energy consumption of 2008 from 212 to 150 kW/m^2 in 2011, i.e. almost 30% of the 2008 energy consumption. To this effect, the annual energy consumption was 2,062,618 kW/h in 2010 (a higher rate compared to 1,565,021 kW/h (original campus) in 2008). This was due to the large area of the campus, including the new extension, Phase II building, and the increase in the number of users.
- Renewable sources, mainly solar water heating systems, comprised three panels of a capacity of 900-l tanks (each is 300 l in capacity), were installed on the buildings' roof.
- Water use was (268 gallons/m^2 in 2008—original building); it was reduced to 179 gallons/m^2 in 2010 (a decrease of 33%). However, the daily use was reduced from 27.91 gallons/m^2 in 2008 to 20.49 gallons/m^2 in 2010. The retrofitting process led to the reduction of water use per person per day, from 23 to 18 gallons, a reduction of 22%.
- Water savings techniques were used to reduce the building consumption rate of 2008. These include flow reduction in water supply, irrigation schedule (was modified from three to one and a half times per day), and sensors were installed on urinals to control water flow.
- For waste management, the percentage of the EAHM building's waste generated and recycled of the following materials glass, oil, batteries, printers, paper, cans, and aluminum is 80%, 100%, 90%, 100%, 60%, and 80%, respectively (Figure 6.17).

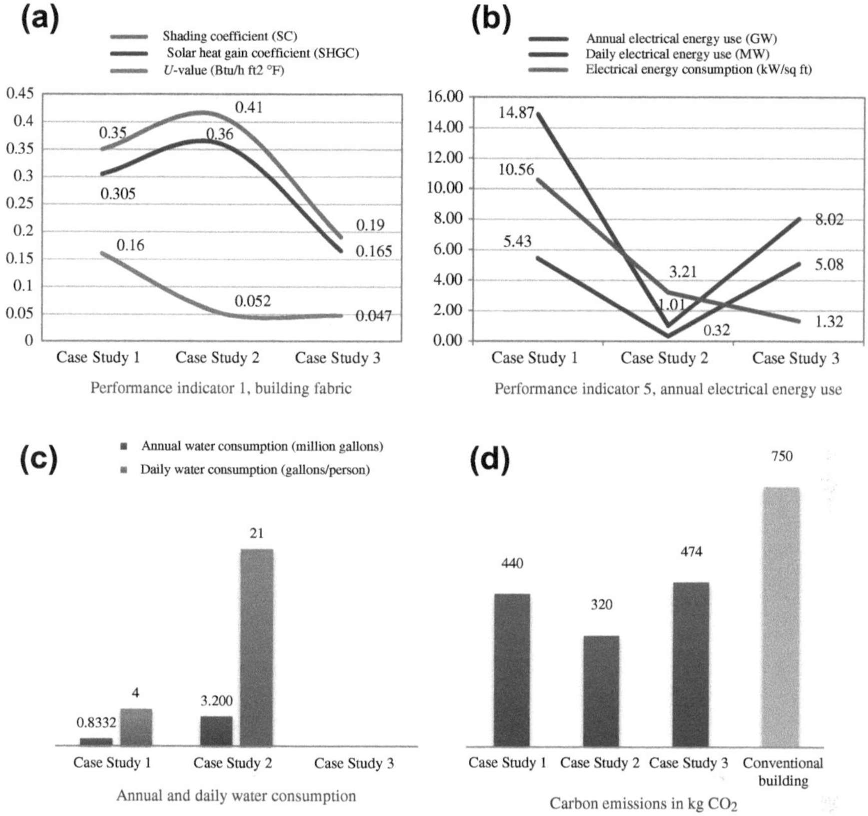

FIGURE 6.17 Comparison between the case studies (1, 2, and 3). *Curves & Charts Credit: Author.* (For color version of this figure, the reader is referred to the online version of this book.)

6.6.2.4. Special Features—Case Study 4 (Dubai Chamber)

Being a mixed-use building, the second phase of the renovation included two new parts: the customer services area on the ground floor and the banquet and event halls on the fifteenth floor (Figures 6.18 and 6.19). In the customer area, recycled floor tiles were used to cover the concrete base (Figure 6.18(c)) and movable daylight control louvers were fixed on the window to limit glare resulting from the normal sun beams impinging on the eastern façade at the early hours (from 7 to 10 a.m.), to lower excessive glare and reduce heat gain from radiation (Figure 6.18(a)). Also, light emitting diode lamps were fixed to save energy. Additionally, chairs were made from sustainable materials (recycled plastics), as shown in Figures 6.18 and 6.19.

In the external water feature, condensate water from AC was used to make up for the supplied water to the fountain (Figure 6.18). It is worth noting that in the customer services area and in the banquet hall, all lighting fixtures were arranged in a hexagonal pattern, which matched the furniture layout to maximize

External louvres to control glare | LED efficient lighting layout | Condensed water used in fountain

FIGURE 6.18 Green elements incorporated in the new customer service area, ground floor, and outside. *Photos Credit: Author.* (For color version of this figure, the reader is referred to the online version of this book.)

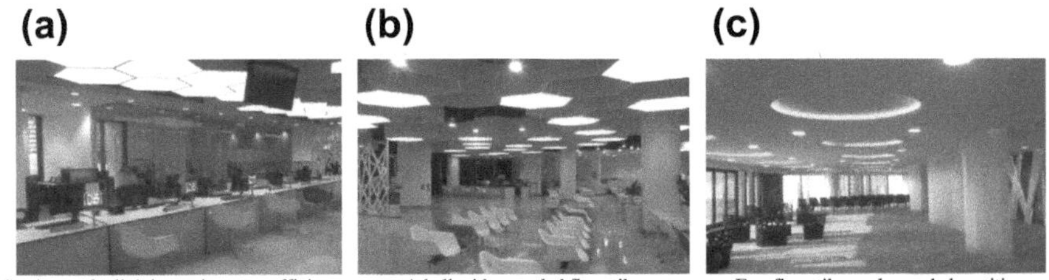

Seating under lighting to improve efficiency | A hall with recycled floor tiles | Eco-floor tiles and recycled partition

FIGURE 6.19 Coordinated lighting fixtures, seating for illumination effectiveness, and use of recycled partitions. *Photos Credit: Author.* (For color version of this figure, the reader is referred to the online version of this book.)

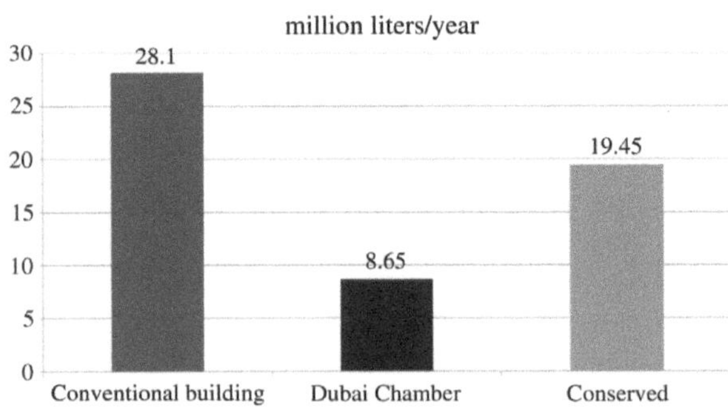

FIGURE 6.20 Overall water efficiency performance at Dubai chamber retrofitted green building. *Source: Survey Conducted by Global Tech Safety and Environmental Consultancy, 2007.* (For color version of this figure, the reader is referred to the online version of this book.)

illumination level effectiveness (Figures 6.18 and 6.19). Figure 6.20 illustrates the overall water-efficiency performance at Dubai Chamber retrofitted green building compared to a conventional building, a reduction of 19.5 million liters annually.

6.6.2.5. *Special Features—Case Study 5 (dnata Freight Gate-5 Warehouse)*

An assessment of the energy performance of the air cargo logistics warehouse (main hall) was conducted. It revealed that almost 50% of the warehouse carbon emissions come from electrical lighting, and the rest from forklift and operations (35% and 30%). All lighting fixtures were replaced with efficient fluorescent bulbs, and motion sensors were installed for some fixtures in low-traffic areas. Also, a sunlight duct was installed in areas that are dark and have no access to daylight. Hence, this eliminates the use of electrical lighting, unless it is totally dark outside. In addition, solar-powered lamps were installed in the parking lot to minimize the reliance upon electrical energy. Over 3 years, electricity charges were reduced sharply. For example, in the high summer months (from June to September), it dropped from US $32,679.41, $33,787.00, $35,967.30, and $33,242.51 in June, July, August, and September (2008–2009) to US $28,610.35, $26,435.20, $27,247.96, and $27,792.92 for the same months of 2010–2011, respectively. Figure 6.21 shows an average of 19% reduction in energy use.

6.6.2.6. *Special Features—Case Study 6 (EAHM Building)*

Being an educational institution, a Noise Risk Assessment was conducted by EAHM for the university in 18 main retrofitted space types to measure the level of noise in these spaces, according to international standards, and to compare the findings with the benchmark. This was pursued to ensure that building users perform their activities in a safer environment and that the measured values are in line with the recommended level in decibels (dB). Figure 6.22 shows measured and recommended noise level (dB) in such spaces during working hours. It illustrates that almost all spaces were in agreement with the safe noise level. Obliquely, Lighting Risk Assessment (LRA), using calibrated HIOKI 3423 Lux meters, was carried by EAHM in 22 main spaces. The purpose of the LRA was to measure the levels of illumination in various work areas (offices, classrooms, library, etc.) to ensure compliance with the standard recommended illumination level (Figure 6.23) and to ensure that the working spaces of students and faculty confirm with internationally recommended illumination levels to prevent eyesight damages. As shown in Figure 6.23, the assessment revealed that the illumination levels in two spaces (PI reception & PII reception—117 and 80 Lux) were below the recommended illumination level (250–500 Lux). In some spaces, the light source was moved to be closer to the workspace (finance room). As a result, additional lighting sources were installed to bring lighting to meet the benchmark.

6.7. CONCLUSIONS

The concept and manifestation of energy-efficient buildings is gaining momentum. Green building directives, standards, and regulations worldwide, including that of Dubai, UAE, have been illustrated and discussed.

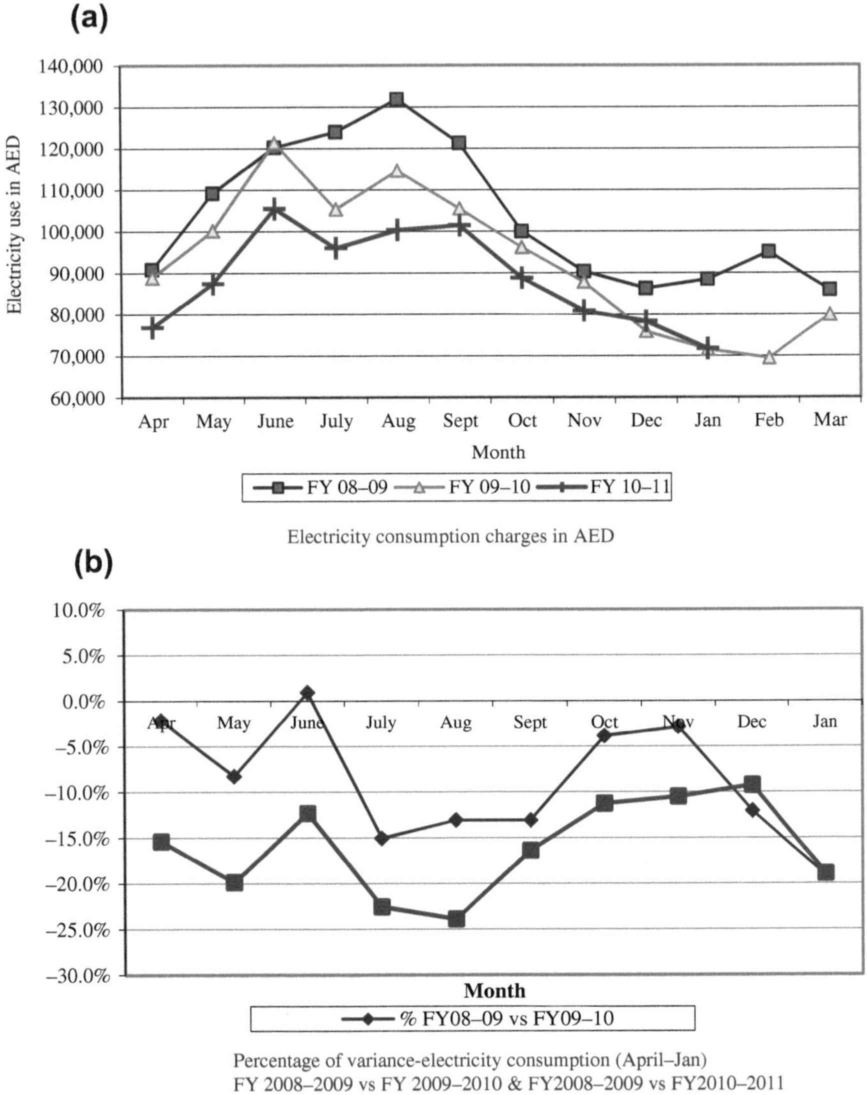

FIGURE 6.21 Energy savings and yearly performance of dnata Freight Gate-5 retrofitted building. *Source: dnata 2011.* (For color version of this figure, the reader is referred to the online version of this book.)

Sustainable buildings in Dubai (newly built and retrofitted existing buildings) were reviewed, examined, and presented. These have positioned Dubai as the world's ninth city to apply green building specifications and regulations, and the first city in the Middle East and North Africa region to adopt such policies and rules.

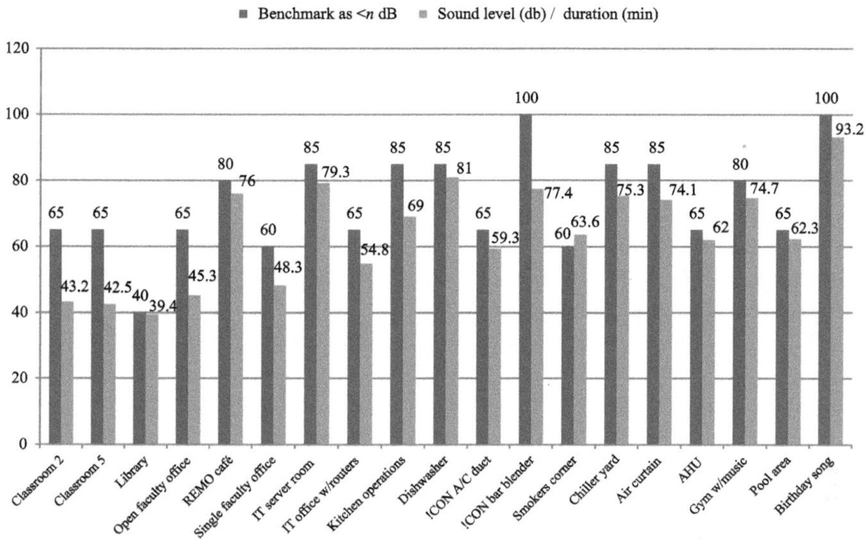

FIGURE 6.22 Measured and recommended noise level (dB) of EAHM spaces, normal working hours. *Source: EAHM 2011.* (For color version of this figure, the reader is referred to the online version of this book.)

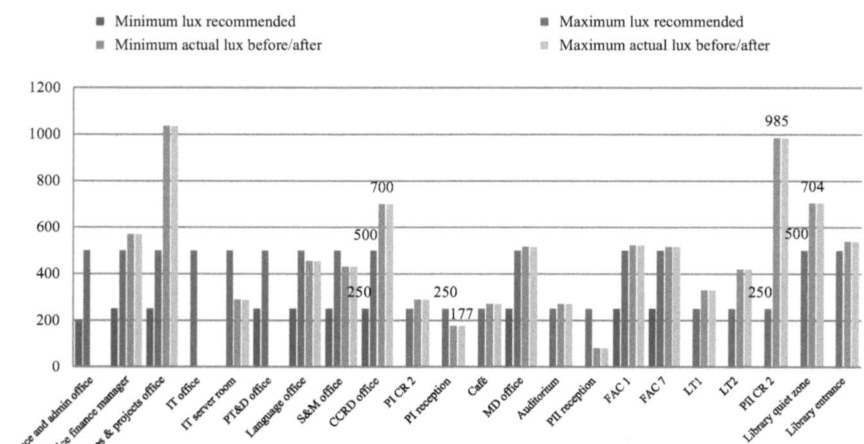

FIGURE 6.23 Measured and recommended illumination (Lux) of EAHM spaces, normal working hours. Note: Lighting in spaces (PI reception and PII reception) was improved by adding more lighting to bring these spaces to meet with the recommended level (250–500 Lux). *Source: EAHM 2011.* (For color version of this figure, the reader is referred to the online version of this book.)

This chapter portrayed six green building case studies in Dubai; three represented new buildings (Aviation College—Emirates, Pacific Controls headquarters, and METITO headquarters) and three demonstrated existing buildings that experienced retrofitting processes (Dubai Chamber, dnata Cargo's Freight

Gate-5, and the EAHM). It also gave a brief on Burj Khalifa, the tallest skyscraper in the world, depicting its environmental innovation in using the condensed water occurring on the outer leaves of the massive glazing areas (during the very hot-humid summer months) for the site's irrigation. These retrofitted processes and manifestations of many green features enabled these buildings to be classified or graded as follows: (1) Dubai Chamber, the First LEED Existing Commercial Building in the UAE and the Middle East; it was certified in 2009, before any retrofitting process was implemented; (2) dnata Cargo's Freight Gate-5, Dubai's first carbon-neutral warehouse; and (3) EAHM building as one of Dubai's sustainable educational facilities.

This study revealed valuable insights to the principles underlying the design of a sustainable building for a climatic zone categorized as hot-humid, with a clear rationalization and illustration of innovative design practices, for new as well as retrofitted buildings. Important sustainable issues and green building best practices in Dubai divulged the lessons to be learnt from these buildings (Tables 6.3 and 6.4).

The review and assessment of the first three case studies (1, 2, and 3) showed that most of the 12 selected green buildings' elements have been incorporated. However, in case studies 4, 5 and 6, not all the 12 elements were applied, i.e. glazing, thermal insulation, and roofs. These case studies focused mainly on energy (electrical energy efficiency and the lowering of carbon emissions) and water. Nonetheless, Case Study 5 (dnata Cargo's Freight Gate-5) focused on energy savings and carbon-neutral features as a main target alongside recycling, whereas case studies 4 and 6 (Dubai Chamber and EAHM) focused on energy efficiency, water conservation, and other targets, such as air quality, recycling, and site and landscape.

In the case studies 3, 1, and 2, the build cost per square feet is US $85, $93, and $164, respectively. This indicates that the construction of low-carbon office buildings in Dubai has been achieved at a reasonable cost. It was as low as US $85.00 and could be up to US $164.00 (an average of US $114.00 per square feet), whereas it was as low as US $81 and with an average of US $105 per square feet for the retrofitted buildings (case studies 5 and 6). It is imperative to mention that the cost of retrofitting per square feet in case studies 4, 5, and 6 was at an average of US $123 (US $173 (including FF&E), US $81, and US $115, respectively). Although it is an upfront investment as it is recovered in 2–4 years, it dramatically improved operational efficiency and reduced cost of electricity charges by 20%, and made the dnata Freight Gate-5 warehouse a carbon-neutral building.

For new buildings, the comparative analysis illustrates that Case Study 3 is the most efficient in terms of fabric efficiency, followed by Case Study 1. For electrical energy use, case studies 3 and 2 are less in consumption than Case Study 1. In terms of CO_2 emissions, Case Study 2 was less than case studies 1 and 3 (carbon emissions reduction average of 42% was achieved). As for retrofitted buildings, the study showed that green building principles have been applied to existing buildings (case studies 4, 5, and 6), and the retrofitting process illustrated that exiting buildings can be refurbished to be sustainable and energy efficient. However, in the case of Dubai Chamber, it was mainly the approach toward effective facilities management, which achieved the LEED certification in 2009 even before the retrofitting process commenced.

Finally, this chapter is intended to raise awareness among both public and private sectors. It is hoped that it would support the learning process, by helping both undergraduate and postgraduate students in the fields of Architecture, Sustainability, and Built-Environment. In addition, it gives insight on green and sustainable building to professional architects and engineers, facilities managers of commercial and institutional buildings as well as developers, researchers, and academics in this field. Furthermore, it adds value and supports policy makers in countries of similar climatic zones to formulate and put forward strategies, policies, and related indicators to reduce energy use, water consumption, and cut carbon emissions to minimize GHGs and, in return, counterbalance climate change and create better living environments.

REFERENCES

[1] OECD Work on Sustainable Development, OECD February 2011, Brochure was Prepared for OECD Meeting of the Council of Ministerial Level, 25–26 May 2011, Paris-France, pp. 3 and 8. http://www.oecd.org/greengrowth/48012345.pdf/, (accessed March 2012).

[2] OECD, towards Green Growth, a Summary for Policymakers. OECD Publishing, May 2011. www.oecd.org/greengrowth/ (accessed March 2012).

[3] OECD, Towards Green Growth, OECD Publishing, 2011. http://dx.doi.org/10.1787/9789264111318-en/, the OECD Meeting of the Council at Ministerial Level, 25–26 May 2011, Paris, (accessed September 2011).

[4] UNEP Green Economy Report, Towards a Green Economy: Pathways to Sustainable Development and Poverty Eradication, Introduction – Introduction: Setting the Stage for a Green Economy Transition, 2011, p. 16. http://www.unep.org/greeneconomy/Portals/88/documents/ger/ger_final_dec_2011/Green%20EconomyReport_Final_Dec2011.pdf/ (accessed March 2012).

[5] The United Nations Environment Programme. UNEP, 2010. http://www.unep.org/ (accessed March 2012).

[6] UNEP Green Economy Report, Towards a Green Economy: Pathways to Sustainable Development and Poverty Eradication, Part II – Investing in Energy and Resource Efficiency/Buildings – Andreas Bassi, MI, 2011, p. 362. http://www.unep.org/ (accessed December 2011).

[7] The Regional Environmental Centre for Central and Eastern Europe. REC Conference Centre Publication, 2011. http://www.conference@rec.org/ (accessed March 2012).

[8] N. Johnson, Green Horizon, 2010. http://www.greenhorizon-online.com/ (accessed March 2012).

[9] M.M. Aboulnaga, N.M. Al-Ali, Low Carbon and Sustainable Buildings in Dubai to Combat Global Warming, Counterbalance Climate Change, and for a Better Future World Renewable Energy Congress 2008 WREC-X, Inv89, The 10th World Renewable Energy Congress and Exhibition, 21–24 July 2008, Glasgow – Scotland, UK, 2008.

[10] L.R. Brown, Plan B.4.0, Mobilizing to Save Civilization, Earth Policy Institute Publication, W.W. Norton and Co., New York, 2009.

[11] M. Polase, Urban Population in 1900, Urbanization and Development, Development Express, No.4, 1997, U.N. Population Division, World Urbanization Prospectus: The 2007 Revision, UNPD, New York, 2009.

[12] U.N. Population Division, World Urbanization Prospectus: The 2007 Revision Population Database, Electronic Database. 2007. Available at: www.esa.un.org/unup/ (accessed March 2012).

[13] G. Metschies, Pain at the Pump, Foreign Policy, July/August 2007, World's Automotive Group, World Motor Vehicle Data, Southfield, Michigan, 2008.

[14] A.R. Migiro, The U.N. Report on Water Crisis, U.N. Symposium on Water Security, News Services, February 5, 2009. http://www.un.org/apps/news/story.asp?NewsID=29796&Cr=water&Cr1=agriculture/, (accessed March 2012).

[15] P. Anastasiadis, G. Metaxas, Environmental protection and the sustainable city, in: Proceedings of 1st World Conference on Technology and Engineering Education, Kraków, Poland, 14–17 Sept, 2010 WIETE, p. 91.

[16] M.M. Aboulnaga, S. Abdullah, 2011, Sustainable Cities: Strategy and Indicators for Healthy Living Environments, World Renewable Energy Congress, WREC 2011, Sweden, 8–11 May 2011 (full paper was accepted February 2011, published 9th May 2011).

[17] D. Anink, C. Boonstra, J. Mak, Handbook of Sustainable Building: An Environmental Preference Method for Selection of Materials for Use in Construction and Refurbishment, revised version, Earthscan, 1996. http://www.academia.edu/2432328/Ecology_and_Design_Ecological_Literacy_in_Architectural_Education (accessed March 2012).

[18] E.O. Cofaigh, J.A. Olley, J.O. LewisEnergy Research Group, University College Dublin-The Climate Dwellings: An Introduction to Climate-Response Residential Architecture, James & James (Science Publishers) Ltd, 1996 on behalf of the European Commission Directorate General XII for Science Research and Development, p. 8.

[19] Green Building, Why Build Green? Buildings and Their Impact on the Environment: A Statistical Summary Revised April 22, 2009, US Federal Environmental Agency (EPA) at Research Triangle Park, NC, the United States. http://www.epa.gov/greenbuilding/pubs/whybuild.htm/, (accessed March 2012).

[20] The Inside Story: A Guide to Indoor Air Quality. The U.S. EPA/Office of Air and Radiation, Office of Radiation and Indoor Air (6609J) Cosponsored with the Consumer Product Safety Commission, EPA 402-K-93-007. http://www.fusionsvc.com/The_Inside_Story.pdf/, (accessed March 2012).

[21] M. Sinclair, J. Anderson, D. Shiers, The Green Guide to Specification: An Environmental Profiling System for Building Materials and Components, third ed., Wiley-Blackwell, 2002. http://www.wiley.com/WileyCDA/WileyTitle/productCd-0632059613.html (accessed September 2011).

[22] Energy: 20-20-20 Targets; 20% Energy Savings by 2020, Sustainable, Secure and Affordable Energy for Europeans, Completed in February 2012 and Published in 2013, European Union (EU): European Commission. http://europa.eu/pol/ener/index_en.htm/, (accessed March 2012).

[23] B. Shui, M. Evans, S. Somasundaram, Country Report on Building Energy Codes in Australia Prepared by U.S. Department of Energy (DoE) – Contract DE-76RL01830, Asia-Pacific Partnership on Clean Development and Climate, Pacific Northwest National Laboratory, March 2009.

[24] Highlights: Dubai Strategic Plan 2015. The Executive Council, Government of Dubai, 2007. http://www.deg.gov.ae/SiteCollectionImages/Content/pubdocs/Dubai_Strategic_Plan_2015.pdf/ (accessed March 2012).

[25] HH Sheikh Mohammed Issues Directive to Adopt 'Green Building' Model in Dubai. Ameinfo-Dubai, October 25, 2007. http://www.ameinfo.com/135876.html/ (accessed March 2012).
[26] D. Carrington, 'Green deal' for energy-efficient homes begins parliamentary journey, Wednesday, 8th December 2010 issue, Environment – Energy efficiency. http://www.guardian.co.uk/, (accessed March 2012).
[27] P. King, Commercial buildings must disclose their energy ratings, The Guardian, Wednesday, 2 March 2011 issue. http://www.guardian.co.uk/, (accessed September 2011).
[28] NFEE, NFEE Stage One, 2008. www.nfee.gov.au/about_nfee.jsp?xcid=66/ (accessed December 2008 and March 2012).
[29] ABCB, Australian Building Codes Board – about the ABCD, 2008. www.abcb.gov.au/go/about-abcb/about/ (accessed January 2009).
[30] ABCB, Australia Building Codes Board: Energy Efficiency General Information, 2008. www.abcb.gov.au/go/whatweredoing/workprogram/projectsae/energy/eegeneral/ (accessed October 2008).
[31] Australian Government, Securing Australia's Energy Future, 2004. www.efa.com.au/Library/CthEnergyWhitePaper.pdf/ (accessed January 2009).
[32] International Energy Agency (IEA), Energy Policies of IEA Countries – Australia (2005 Review), International Energy Agency, 2005. http://www.iea.org/publications/freepublications/publication/australia2005.pdf/ (accessed May 2013).
[33] Geyser building: New Zealand's first 6 Green Star – Office Design building takes shape in Auckland's Parnell, 25 Jan 2012, ARCHITECTURENOW. http://architecturenow.co.nz/articles/geyser-building/#img=1 & http://www.geyserparnell.co.nz/, (accessed September 2011).
[34] Green Building Regulations 2011, Government of Dubai. http://www.dewa.gov.ae/images/greenbuilding_eng.pdf/ and greenbuilding@dm.gov.ae/, (accessed September 2011).
[35] Electricity and Water Tariff, Dubai Electricity and Water Authority (DEWA). Government of Dubai, January 2011. http://www.dewa.gov.ae/tariff/tariffdetails.aspx/ (accessed March 2012).
[36] Architect's Guide to Glass, USA. 2011. http://www.glassguides.com/index.php/archives/1115/ (accessed September 2011).
[37] Burj Khalifa: Towering Challenge for Builders. www.GulfNews.com/, 4 January 2010, retrieved 10 February 2010. http://gulfnews.com/business/property/burj-khalifa-towering-challenge-for-builders-1.561802, (accessed March 2012).
[38] Escaping the Dubai Downturn: Voltas's Latest Engineering Feat. Wharton, University of Pennsylvania, April 20, 2010 retrieved 21 May 2011.
[39] Air Conditioning in Burj Khalifa. Timeoutdubai.com, retrieved 17 April 2011.
[40] J.L. LaSalle, On Point, Dubai City Profile, September 2011. http://www.joneslanglasalle-mena.com/ (accessed March 2013).

SUGGESTED READING

[1] M. Kļaviņš, W. Leal Filho, J. Zaļoksnis (Eds.), Environment and Sustainable Development, Academic Press of University of Latvia, Riga, 2010.
[2] T. Jackson, Prosperity without Growth: Economics for a Finite Planet, with Forwards by Herman Daly, Bill Mickbben, Mary Robinson and Pavan Sukhdev, Earthscan Publishing for a Sustainable Future, London, 2009.
[3] L.R. Brown, Plan B.4.0, Mobilizing to Save Civilization, Sustainability Revised, Earth Policy Institute, W.W. Norton & Company Inc., New York, USA, 2009.

[4] L.R. Brown, Plan B.3.0, Mobilizing to Save Civilization, revised and expanded ed., Earth Policy Institute, W.W. Norton & Company Inc., New York, USA, 2008.
[5] M. Hegger (Ed.), Sonnige Zeiten-Sunny Times, Solar Decathlon House Team Deutschland, Verlag Müller+Busmann, Germany, 2007.
[6] C.J. Kibert, Sustainable Construction: Green Building Design and Delivery, Wiley, John Wiley & Sons, Inc., USA, 2005.
[7] D. Gottfried, Green to Green: The Transformation of an Industry and a Life, Forwarded by Paul Hawken, World Publishing, a Division of WorldBuild Technologies Inc., USA, 2004.

Chapter 7

The LED Lighting Revolution

Nada El-Zein
LED Light Energy LLC, Chicago, IL, USA

Chapter Outline

Sustainability, Energy and Architecture. ISBN: 9780123972699

Chapter Outline

7.1 INTRODUCTION

7.1.1 History of LED (Light Emitting Diode) Technology and a Brief Technical Background

Lighting has evolved through many incarnations over the past 100 or so years. Most of those incarnations have involved the heating of a filament of some kind (tungsten, ceramic) or burning a gas (fluorescent and metal halide lamps). However over the past few years, a new technology which uses completely different physics has emerged as a clear winner in the category of 'high quality and energy efficient' lighting. This new technology is 'Solid State Lighting' (SSL) and it relies on light emission from a semiconductor or a Light Emitting Diode (LED). The progression of lighting is shown in Figure 7.1.

The earliest (but not very practical) LED is attributed to H. J. Round (1907) and was made of silicon carbide [1]. This SiC LED produced some electroluminescence when forward biased. Dr. Nick Holonyak Jr. invented the first practical LED in 1962; this was made of Gallium Arsenide Phosphide (GaAsP) and emitted red light. These types of semiconductors are called 'compound semiconductors' and have since that time provided the foundation for the expansion of LEDs into many different applications. The first LEDs were used in very low-light applications such as indicators and calculators, but the materials from which the 'phosphide' family of LEDs kept improving through optimizations and several technological breakthroughs, and so the amount of light emitted by them kept increasing at an almost exponential pace. In fact, just as Moore's law [3] predicts a doubling of the number of transistors in computer chips (which are made out of silicon) Haitz's law predicts that the light output (or flux measured in lumens) of LEDs will double every 18–24 months. It has held true (or been exceeded) for the past four decades and this has allowed the use of LEDs in high brightness applications such as lighting (see Figure 7.2).

In the early 1990s another breakthrough in material science allowed 'white' light to be produced by LEDs. This breakthrough came from Japan, through the development by Dr. Nakamura, then at Nichia (now in UCSB), of the 'nitride' material system [4]. The compound materials mentioned earlier are assembled (or grown) into layered crystals in a rather complex chemical process called Metalorganic Chemical Vapor Deposition or MOCVD [5,6]. Each layer that is grown has a specific role in forming the diode or p-n junction that will emit light. The elements introduced into the crystal will determine the color emitted by that particular diode. As mentioned earlier, the first LEDs were made out of

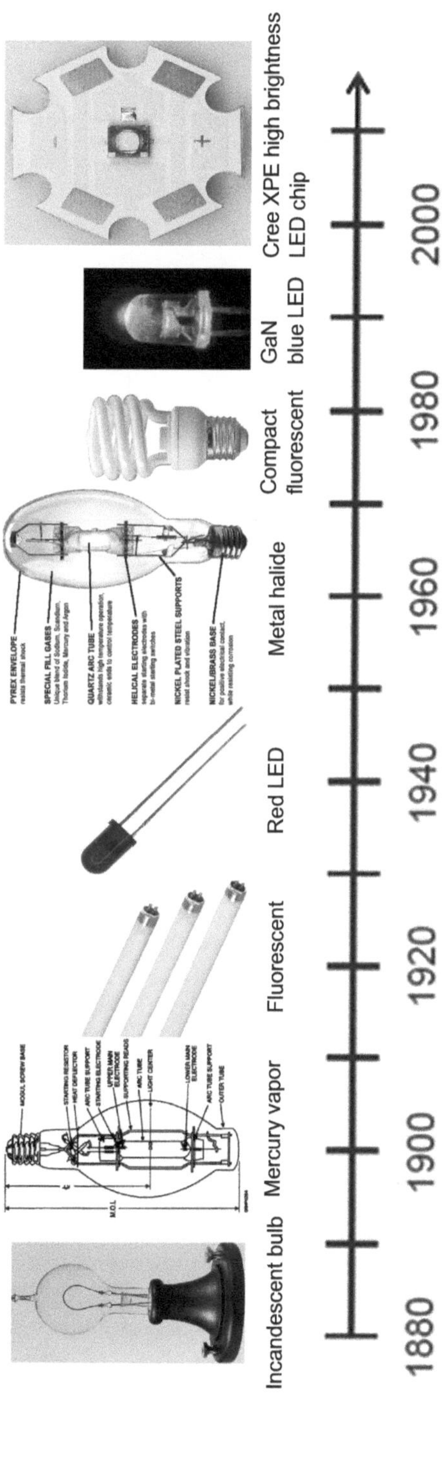

FIGURE 7.1 Lighting technology progression over the past 130+ years.

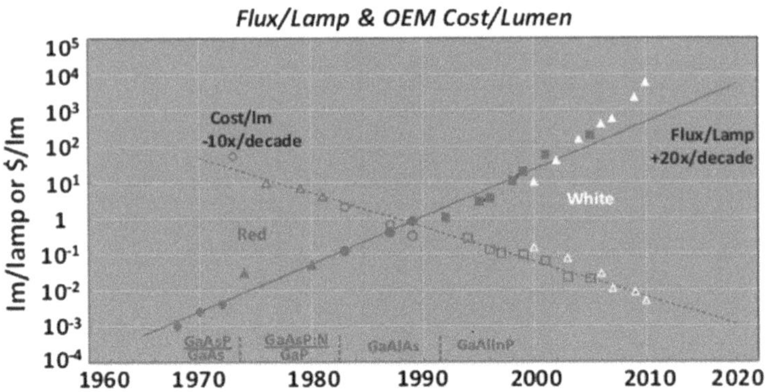

FIGURE 7.2 Haitz's law shows the increase of LED flux for the past four decades. Data was compiled by R. Haitz from HP (Hewlett Packard) (now Lumileds) historical records. This figure was taken from a Sandia Labs report referenced in [3].

materials containing phosphorus (phosphides) and those particular crystals emit red and amber light. These red or amber LEDs contain a combination of indium, gallium, aluminum and phosphorus. The compound semiconductors developed by Dr. Nakamura were based on nitrogen or 'nitrides', and these crystals emit blue and green light. The blue/green LEDs contained the following elements: indium, gallium aluminum and nitrogen. Examples of these layered 'stacks' of materials (called epitaxial layers) are shown in Figure 7.3, each stack results in a different wavelength or color of emission.

Now that some of the history of the creation of LEDs has been described, a discussion of the nature of the device itself is in order. What is an LED? An LED is simply a diode (an electrical device that allows current flow in one direction only – think of it as electrical valve) that emits light when it is biased. 'Normal' diodes are heavily used in the design of electrical circuits, most commonly in the conversion of an AC signal into a DC signal and in rectifiers. In the electrical engineering world, they are referred to as 'p-n junctions', because a diode is made out of the junction of materials which contain a 'positive' (p) charge on one side and a negative charge (n) on the other. LEDs are p-n junctions that emit light when current flows through them; this is shown by the schematic diagram in Figure 7.4. The specific color of the light emitted depends on the material they are made from, as explained earlier in this section. A solar cell is the exact inverse of an LED: it is a p-n junction that absorbs light and produces electricity in return. Because LEDs are semiconductor devices and these devices belong to a field of physics called solid state physics, you will often hear LED lighting referred to as 'Solid State' Lighting (SSL).

Of course in this chapter we are interested in general illumination. Illumination to us means white light, and a very specific white light at that, since our eyes are tuned to the sun's light. Essentially this means that we are used to 'whites' with a Correlated Color Temperature (CCT) between 2600–6500K depending on the time of day. CCTs in the 2600–3300K are usually called 'warm white'

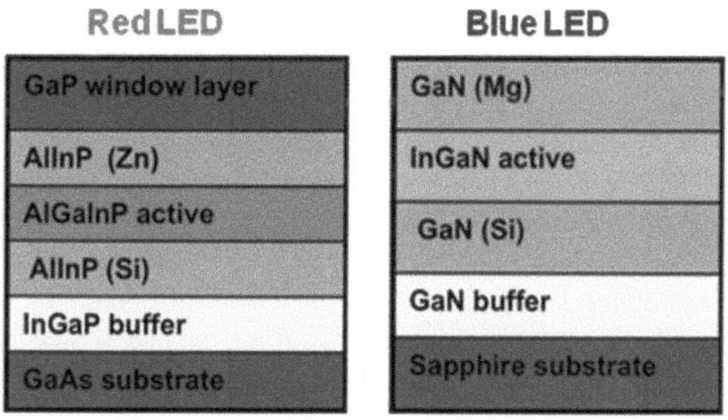

FIGURE 7.3 'Epitaxial' stacks showing the different materials that form LEDs that emit different colors. Red and Blue stacks are shown in this figure.

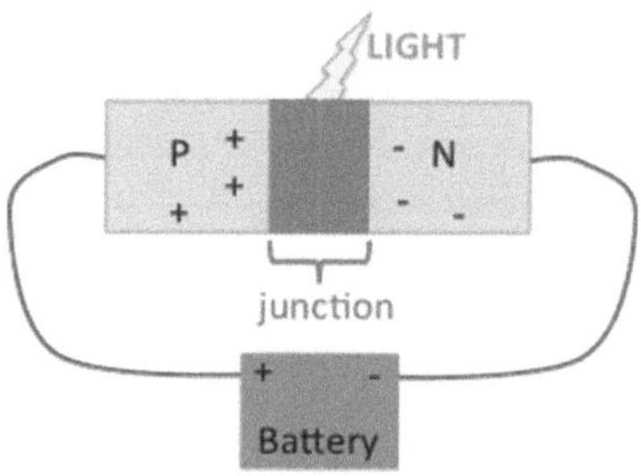

FIGURE 7.4 A schematic showing how an LED works.

and are preferred in residential and hospitality environments. CCTs in the range 4000–6500K are called 'cool white' and tend to be used in office and outdoor applications.

With the development of blue diodes, the creation of white light became possible, by two methods: 1) the use of phosphors, or 2) the combination of red, blue and green diodes. Phosphor-coated blue diodes are more efficient, reliable and, most importantly, 'broader' spectrum white LEDs. This leads to higher color rendering which is very important in retail and hospitality applications. Essentially the LED chip is coated by an appropriate phosphor compound, which converts the blue light emitted by the LED to white light. Depending on the composition of the phosphor, the light emitted can be 'cool white' which is close to daylight or fluorescent white (4000–5000K), or warm white which is close to the color of incandescent or halogen light (2700–3200K). A lot of work has been carried out on optimizing the phosphors to ensure that they do not degrade over the useful life of the fixture and to maintain the CCT as it was originally designed. Because the human eye is extremely sensitive to CCT changes (it can detect changes in the 50–100K range), phosphor coating uniformity on the LED chip and chip to chip coating control has improved tremendously to the point where large installations with multiple (thousands) of fixtures do not show detectable CCT variations.

An additional critical variable that has been optimized over the past decade is the ability of the LED light source to render true color. The ability of a source to show the 'true' colors of whatever it is shining on is determined by measuring the Color Rendering Index (CRI), Ra, scaled from 0 to 100, with 100 being the best [7]. Early LEDs had too much 'blue' color leaking through the phosphor coating, which resulted in poor CRIs. Today there are LED lamps that can deliver CRIs as high as 98. This was achieved by improving the phosphors used for the conversion of the blue light to white light. The National Institute of Standards and Technology is trying to encourage the use of a new color rendering measurement method called Color Quality Scale (CQS) because CRI sometimes does not adequately reflect how well colors show up under a light source. For more information on that, the reader might refer to the following Ref. [26].

The development of 'high brightness/power' LEDs finally happened through the optimization of the packaging of the diode itself, as well as the manipulation of the LED chip post MOCVD growth. The old 'lamp style' package, shown in Figure 7.5(A), was not appropriate for general lighting applications. LEDs used for illumination need to be driven at higher power densities (higher currents and voltages) and therefore need to be protected from overheating. These new packages, shown in Figure 7.5(B), were developed in the late 1990s in order to accommodate the requirements of the lighting market. The first high power LEDs were made from the red/amber material system since it was well understood [8]. Soon afterwards (1996), high power LEDs made from the blue/green material system entered mass production [9]. At the time these were 1W LED modules. Now 3W and even 5W modules are commercially available, from companies such as Lumileds (Philips), Cree, Osram and many others. The lumen (flux) output of LEDs has jumped from <10 lm/W in the

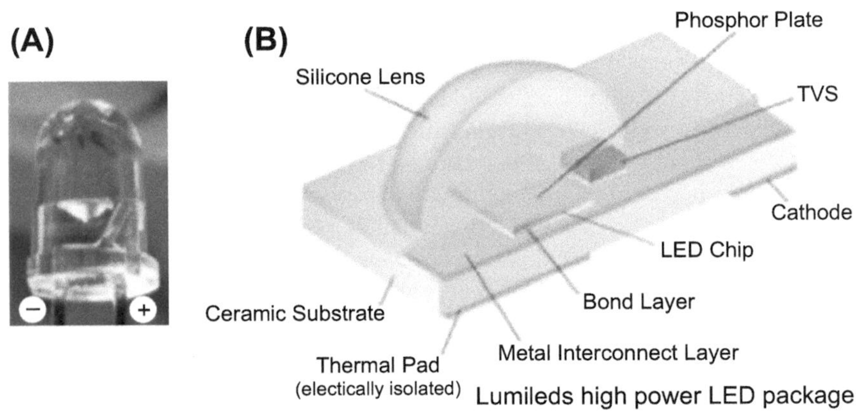

FIGURE 7.5 (A) The typical lamp style 5mm package. (B) All the elements that constitute the higher brightness/power LED packages of today's illumination LEDs. Part (B) is taken from www.philipslumileds.com/technology/thermal, and shows the packaging of a typical Luxeon® rebel LED package from Lumileds.

1990s to >200 lm/W today (announced by Cree early in 2011). This allows high brightness LEDs to compete against all other lighting technologies in terms of performance and energy savings. The high power LED modules are also used by lighting fixture manufacturers to produce fixtures/lamps such as: A-shaped (traditional Edison incandescent) lamps, PAR (Parabolic Reflector) lamps, AR111 lamps (Aluminized Reflector lamp (the 111 is reflective of the diameter of the lamp)), down-lights, floodlights, MR16s (Multifaceted Reflector lamp (the 16 is reflective of the diameter of the lamp)), streetlights etc...

Here are some of the applications that use LEDs today:

General indoor illumination
Outdoor illumination (street lighting and traffic lighting)
Architectural illumination (wall washing and color changing)
Automotive (headlights, brake lights, interior lights)
Mobile devices (cell phones, iPad, iPod etc.)
Flat screen backlighting (TVs, computer screens)
Theatrical lighting
Channel letter lighting (logos and names of stores)
Signage (advertisements and signaling on roads)
Boat and cruise boat lighting (makes it easier all DC - direct current)
Medical applications (operation room lighting, dental applications)

7.2 FROM LED CHIPS TO FIXTURES

The LED chip or module (or an array of them) is just one of the many components that goes into a full LED fixture. Of course, it is the most important one because it is the component that produces light. However, we now need to examine what makes a 'good' LED fixture.

7.2.1 Thermal Management

The second most important aspect of an LED fixture is thermal management. Like any semiconductor electronic device, an LED produces heat when it is turned on (when current flows through it). This heat is detrimental to the lifetime of the LED, and if it is not removed, this will be severely shortened. Unlike filament-based bulbs which radiate heat, LEDs do not radiate the heat they produce, so this needs to be 'conducted away'. If you examine your computer, which contains many semiconductor chips (processors, RAM, drivers etc.), you will notice that there is a fan that cools the electronics, in order to protect them from the heat they generate when they are in use. In addition to the fan you might also notice a heat sink, which is basically a thick piece of metal with fins, onto which the fan blows. Essentially the heat sink 'soaks up' the heat produced from the electronics, and the fan blows on it to circulate air and keep it cool. This describes the two processes used to cool off electronics: 1) heat conduction away from the electronics via the heat sink, and 2) air 'convection' which carries the heat away from the heat sink. The same methods are used with LEDs, however because having fans in fixtures is not always desirable, the main method is the use of heat sinks. This is called passive cooling. Essentially the high power LED package is attached to a heat sink through thermally conductive materials, with the heat sink being designed in accordance to the heat profile of the LED. The heat sink absorbs the heat that is produced, and so protects the LED from any degradation. The design of the heat sink will also depend on the lighting application: is it going in a down-light can? Is it a track light? Is it a floodlight? Special care has to be taken where convection (i.e. air movement) is limited, in which case the heat sink might need to be larger than in 'normal' conditions.

A typical LED, if heat sunk properly, will have a long lifetime; around 30,000–50,000 hours for most applications. The lifetime indicates the L70 point, which is the point at which the LED lumen output is 30% below the level where it started. However if the heat sink is not well designed, the LED will degrade faster because heat will build up around the p-n junction. A 'badly' designed heat sink does not have enough metal mass; i.e., it is too small to handle the wattage of the LED(s) used to create the fixture, or it may not allow any convection through it. The testing protocol that is used to test an LED chip is called the LM-80 test [10]. The Illuminating Engineers Society designed the test, which summarizes the lumen output of an LED at different ambient/junction temperatures under different bias conditions. The results of the LM-80 test are used to design the heat sink to ensure that the LED does not exceed a certain temperature and that the lumen output is indeed maintained at <30% degradation for 50,000 hours. Usually the test shows 6,000 hours of measurements and the performance is extrapolated from that data. However, keep in mind that LM-80 tests reflect the performance of the LED chip and not the LED lamp or fixture. The LED chip data helps in the design of the fixture. Another standard, TM-21 [11], is in development and it is supposed to recommend how best to extrapolate the LM-80 data to predict long-term performance. A typical set of LM-80/TM-21 data can be found on most LED manufacturer's websites [12].

7.2.2 Drivers (also Transformer and Power Supply)

LEDs are DC devices, and therefore they cannot be plugged into the 'normal' mains in our homes and facilities, which, depending on the region you live in, are 120V AC or 220–240V AC. For that reason LED fixtures will always have an AC to DC driver associated with them. This driver can be integrated within the fixture or can be separate and connected to the LED array via wiring. It can be constant current or constant voltage. In the retrofit market, where one is replacing existing lamps, fully integrated fixtures are obviously the preferred choice.

7.3 OPTICS

The light emerging from an LED chip has no specific beam shape. The index of refraction of compound semiconductors is >3 and that of air is 1, therefore when the light exits the semiconductor it 'refracts' (which means changes its direction) and what one ends up with is a point source, from which light travels all directions. Obviously in order to make the LED useful for lighting applications, it is lensed to the desired beam angle. Most LED chip manufacturers (like Cree, Philips, Osram, Epistar) will deliver LED chips with a 'standard' lens, which produces a beam of around 120°. The fixture/lamp manufacturer will then add more lenses to produce tighter beam angles if needed (10°–60°). Also depending on the desired look of the fixture and the application, sometimes diffusers are used to 'hide' the LEDs from view or make the light more diffuse in nature. So, generally the optics applied will consist of lenses and diffusers.

7.4 FIXTURE BODY

All these components will then be assembled in the fixture body, which will depend on the application. If it is a retrofit bulb, such as a PAR or MR16s lamp, then the heat sink will be shaped into a PAR or MR16 shape and the driver circuitry will be integrated inside the lamp. If it is a down-light, then the LED fixture will be made to fit in a can (or the heat sink itself will be part of the can) and the driver will either be integrated within the fixture or wired to it within the ceiling. Local electrical regulations will sometimes dictate how the driver will be attached to a particular fixture.

A schematic showing all the components in an LED fixture is given in Figure 7.6. An LED fixture is a system of interconnected components, the reliability of which depends on how well all of these components are designed and how well they are connected together. This is expressed as follows:

$$R_{system} = R_{thermal} * R_{LEDS} * R_{optical} * R_{electrical} * R_{connections} * R_{mechanical}$$

where 'R' is reliability.

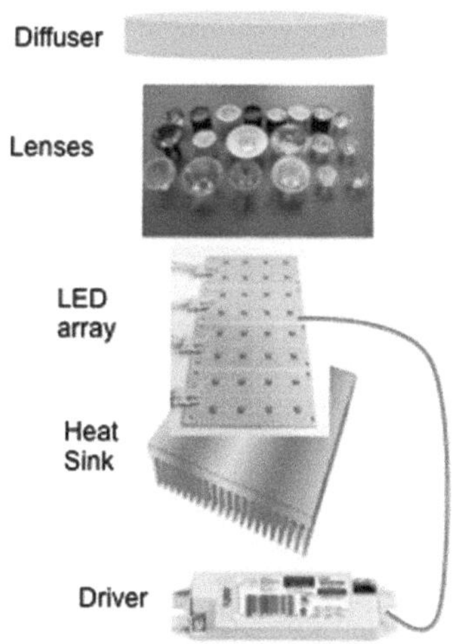

FIGURE 7.6 **A list of the main components that are needed for an LED fixture.**

As was mentioned earlier, heat sink design is critical ($R_{thermal}$). The heat sink needs to be large enough to conduct enough heat away, and the fin design should optimize air movement through it.

In addition to heat sink what additional design parameters need to be considered?

The LED chips (R_{LEDS}) must be of high quality, with LM-80 numbers that show long lifetimes if the junction temperature is maintained per the manufacturer's recommendations (<85 °C usually). Also, it must have the desired CCT and it should be able to maintain that CCT during its operation, at least within limits that are not detectable to the human eye. The CRI should preferably be >80 for average applications and >90 for retail or gallery applications.

The efficiency of LEDs is often mentioned as a huge advantage for LED lighting; however the other components in the fixture need to be well designed so as not to compromise this. The driver ($R_{electrical}$) should have a high efficiency (minimum losses), high power factor (>0.9) and low harmonic distortions. In some applications it must have a very compact design to fit within the fixture. The lenses ($R_{optical}$) used should also be of high quality and designed well so as not to distort the beam or introduce too many optical losses. Finally, the whole assembly and the connections between these components have to be put together flawlessly ($R_{connections} * R_{mechanical}$), so that the body of the fixture protects the components inside it, and there are no loose connections inside the fixture to cause problems later. Having studied several failure mechanisms in fixtures, it is not too surprising to note that the failures are often not related to the LED, but rather to one of the other components in the system.

The same thinking has to be applied to the efficacy of the fixture. Often, less than reputable manufacturers will quote the efficacy of the LED chip used in the fixture as the efficacy of the actual fixture. That is absolutely incorrect as several other efficiencies have to be taken into account: LED efficacy × Optical efficiency × Thermal efficiency × Electrical efficiency gives the total fixture efficiency. Typical optical efficiency can be higher than 90%, driver efficiency will vary between 80–92% and thermal efficiency is usually around 85%. So one could start with LEDs that have a 100 lm/W efficacy, use them in a fixture where all the above-mentioned efficiencies are >90% and so end up with a fixture efficacy of 761 m/W.

With all this in mind, how is a consumer to choose a 'good' fixture without becoming a very technical person? The governments of several regions are putting together testing and labeling requirements that will make it much easier for the consumer to choose a well-manufactured fixture. In the US for example, the Department of Energy (DOE) has encouraged manufacturers to use the 'Lighting Facts' label [13]. This label lists the results of a specific test called LM79-08. The test is performed by a third party lab, which is approved by the DOE. The LM-79 standard [14], devised by IES, measures the photometric performance of an LED fixture (total lumen), the wattage (power consumption), CRI and CCT. From this data one can calculate the lumen/watt efficacy of the fixture. The label is then affixed to the box containing the product. Also, certain rebate programs which encourage consumers to use LED lighting will require Energy Star rated fixtures [15,16], for which the rating is obtained only when the fixture achieves a minimum efficacy requirement as well as a minimum lifetime requirement. Each fixture is tested for 6,000 hours (about ten months of data) and that data is used to extrapolate a lifetime. Note that this is different from LM80, because in this case the fixture is being tested and not just the LED. The average consumer can now find hundreds of qualified fixtures on the Energy Star website.

7.5 ADVANTAGES AND FEATURES

LED lighting has many advantages over traditional lighting, the main one being substantial energy savings. Today's LED fixtures can save between 20% and 80% of the energy consumption of fluorescent tubes, compact fluorescents (CFL), incandescent bulbs, and halogens. In fact, a report prepared for the DOE by Navigant Consulting, which analyzed the energy savings potential of solid state lighting from 2010–2030, predicts that by 2030 the savings will be in the range of 190 TWh and that SSL will have replaced most existing lighting technologies [17].

In addition to energy savings, LED technology has many benefits, as described in the following sections.

7.5.1 Long Operating Life

Commercial and industrial consumers are interested in a lighting fixture that is highly reliable and lasts a long time. Frequent bulb changes can be costly (particularly in situations with high ceilings). In fact, we find that in certain applications maintenance is the main reason a consumer chooses LEDs. As explained earlier, a well designed fixture will have a lifetime of 30,000–50,000 hours (when the light is reduced by 30% – not dead!).

7.5.2 Environmentally Safe (no Mercury)

LEDs contain no toxic materials and emit no harmful UV (Ultra Violet) rays. All metal halide and fluorescent lamps, including CFLs, contain mercury (considered a toxic waste everywhere in the world).

7.5.3 Significantly Reduced Heat Radiation

Because LEDs are highly efficacious (the latest numbers reported by Cree for example exceed efficacies of Fluorescent and HID (High Intensity Discharge) sources), a higher percentage of electricity is converted to light and less into heat. This means that LEDs do not radiate heat in the living/office space. This can help lower AC energy consumption because the lights do not contribute to 'heating' the interior of rooms. A 100W incandescent bulb will radiate approximately 95 Watts (95%) of heat into the room it is lighting.

7.5.4 Flicker Free and Instant Turn on

LEDs are flicker free and turn on instantly. All fluorescent bulbs flicker during start up. In addition, metal halides and some CFLs need time to warm up to reach their full intensity (sometimes as long as 5 minutes).

7.5.5 Unaffected by Frequent on/off

LEDs are not affected by being switched on/off frequently (as will happen if occupancy sensors are in use). Frequent switching significantly reduces the lifetime of fluorescent and incandescent lamps.

7.5.6 Dimmability and Controllability

LED fixtures are easily dimmable; the driver will contain a dimming capability in most currently available fixtures. Several Energy Star rated fixtures are dimmable by default. The dimming protocols available are varied, from 0–10V, to phase dimming (leading or trailing edge) to PWM (Pulse Width Modulation). Care must be taken when choosing an LED fixture in a retrofit project to make sure that the existing dimming system is compatible with the LED dimming protocol. For new fixtures there is no such issue, because an appropriate dimming system can be chosen for the specific fixture. In retrofit projects, one of the most common issues with dimmers occurs because the LED load is so much lower than that originally planned, so it falls below the minimum required wattage for the dimmers. This will usually cause flickering. If the particular installation will not accept a new dimmer system, the easiest way to resolve this is by keeping a couple of the original bulbs, in order to keep the load above the minimum requirement.

In the near future the possibility of integrating functionality into LED fixtures will become more likely. Since LEDs are semiconductor chips, it will be easy to integrate electronic circuitry within the fixture to give products never thought of before. Examples include:

- Fixtures which have wireless functionality and can interact with a computer or smart phone to can turn on/off or dim the lights;
- Integrated occupancy sensors;
- Integrated daylight controls and light harvesting;
- Speakers with integrated lighting (a real product, see www.musiclites.com/t-allcategories.aspx).

7.5.7 Durability

As explained earlier in the chapter, LED technology uses a fundamentally different mechanism to produce light to that of incandescent lighting. It is essentially a layered crystal that produces light. This makes LEDs very resistant to vibrations and impact, making it ideal for certain applications, such as roadway and elevator lighting, where vibration is a major issue.

7.5.8 Minimal Light Loss

There is another efficiency factor that is often forgotten when discussing the advantages of LEDs. They are chips mounted on ceramic substrates, and therefore emit light in a much more directional fashion than traditional lighting sources. Most of these traditional sources require reflectors to direct the light to the desired area; this is particularly true for down-lights, spot lights, track lighting, roadway lighting etc. The use of reflectors introduces losses into the fixture and often allows light to escape behind it (light pollution). LEDs are naturally directional and therefore ideal for such applications and eliminate the use of reflectors and the associated losses.

There is one caveat to the directionality of the LEDs if one requires a 270+ degree light, as is emitted by A19, candelabra, or globe bulbs. Several design tricks have to be implemented to obtain omni-directional lighting as opposed to something that is more 'focused'. Also the incorporation of the heat sink into such small footprint bulbs can be a challenge. However, there have been successful designs that reach 75 lm/W and offer a very warm, high CRI, long living, Energy Star rated, LED A19 lamp [18].

7.6 COMPARISONS WITH TRADITIONAL LIGHTING

7.6.1 Comparison with Halogen and Incandescent Lighting

The energy savings achieved by using LEDs rather than traditional bulbs and fixtures has improved rapidly since 2006, because their efficacy has increased at an exponential pace (as shown by Haitz's law in Figure 7.2). Retrofits that reduce the power/energy consumption by >80% of a halogen or incandescent equivalent can now be easily found. This will be shown in the case studies discussed later in this chapter, in which several halogen retrofits will be explored. PAR retrofit lamps in the US have become particularly attractive, with LED PARs on the market having CCTs in the 2700K–3000K range, CRI > 85 and wattages that tend to be 80% lower that the halogen equivalent. Care must be taken in the choice of beam angles in order to get comparable illumination to the original lighting. The toughest retrofit remains the 50W halogen MR16, because it has a small footprint and incorporating a moderately high LED wattage (8–10W) with an appropriately sized heat sink in such small space is difficult. Philips has recently introduced an actively cooled 10W MR16, which has passed Energy Star testing and which is an excellent replacement for a 50W MR16. Several companies (CRS in Canada and MSI in the US) are also claiming that they will be releasing 50W equivalent LED MR16s, which are in the 8W range. For retail applications we have seen several consumers express their surprise and satisfaction with the appearance of color under LED lamps, particularly for men's clothing stores where grays, blues, blacks and browns dominate. Their main comment was that the halogen light, although it has a CRI of 100, made the merchandise look more 'yellow' and therefore customers found it hard to discern between navy blue and black for example. This is probably why NIST (National Institute of Standards and Technology) is trying to convince the lighting community to move to the new standard mentioned earlier (CQS).

For most halogens and incandescents, such as MR16s, PARs, AR111s, Rs, track lights and down-lights, retrofits are simple, easy to install, have the same form factor and generate significant savings. It is the main reason why several utilities in the US and elsewhere offer substantial incentives to change from existing inefficient bulbs to LED bulbs.

In cases where the fixture has halogen 2-pin >50W bulbs, clearly their small footprint makes a bulb retrofit almost impossible, and in that case, the whole fixture will need replacing instead of just the bulb. Generally, a full fixture replacement would always be more desirable, because every component in the fixture is optimally designed to maximize the life and output of the LED chips, unlike a bulb replacement which might have to use an existing low voltage transformer (for MR16 retrofit), or where the LED lamp might end up being installed in a can where air flow is not optimal. However, realistically, in the retrofit market the most attractive solution is a 'lamp' and not a fixture replacement because it makes the economic equation that much more attractive and the installation extremely simple.

For retrofits that replace halogens and incandescents, the savings are so large that even in places where the price of energy is relatively low, if the facility is on 10–12 hours/day the payback period is always < 2 years, and is often close to a year. In places with energy prices >$0.12/kWh, the payback is < 1 year for 12 hours/day operation. This will be made clear in the case study section of this chapter. In addition to the energy savings, one must take into account the elimination of 'bulb' change-outs for several years, since LEDs have much longer lifetimes, and the lowering of AC usage due to the fact that LEDs radiate much less heat. For example a typical LED PAR lamp has an average life of 40,000 hours, whereas the halogen equivalent has a life of 2,500 hours. That corresponds to a 15x improvement in lifetime.

7.6.2 Comparison with CFLs

Compact fluorescent bulbs have several advantages, but light quality was never one of them. In fact, in high-end retail they are almost never used for that reason. Also for directional applications they are not optimal replacements, as the fixtures they go into tend to lower the overall efficiency of the system. LED bulbs and fixtures are now much more efficient than CFL bulbs and fixtures. For example, some of the highest quality 6' CFL down-lights tested by the DOE (Department of Energy) in their yearly Caliper tests [19], at the time of this writing have efficacies of 40–50 lm/W, whereas the highest quality 6' LED downlight is approaching 90 lm/W. That is an energy saving of more than 40–50% compared to the highest quality CFL fixtures. In addition, even for bare bulbs (where the inefficiency of the fixture does not come into play) SSL replacement lamps are now exceeding CFL lamp efficacies. Generally LEDs will deliver a more efficient, better CRI, higher quality light, that is easily dimmable and has a longer life. Also when replacing a bulb, in cases where the CFL bulb has an integrated ballast, the bulb is larger than the original incandescent/halogen making it aesthetically unacceptable. Finally as mentioned above, the lifetime difference can still be significant as most CFL bulbs have lifetimes of ~6,000 hours compared to 30,000 hours or more for LED bulbs.

7.6.3 Comparison with Fluorescent Tubes

For the case of LED "tube" lights vs T8 (or T5) fluorescent tubes, the equation is tougher but improving. In early 2013 there were reports of >100lm/W tubes (Green Ray LED tubes for example, www.greenrayled.com), however a tube replacement is still not recommended, since the fixtures are designed with fluorescent tubes in mind and are not optimal for LED ones (which are directional). Although LED chips have reached >200 lm/W efficacy, these diodes are not yet produced commercially, and an LED tube will have all the components mentioned in earlier sections, and the 'inefficiencies' of those components will lower the overall efficacy of the fixture (in this case the fixture is the LED tube). LED tubes have been improving [19] and the expectation is that within the next two years or so, tube replacements will be possible. There are many facilities today that have opted to do an LED tube retrofit and are satisfied with the results. With today's LED tubes the savings are optimistically in the 20% range and with the price difference being quite large the payback is longer than acceptable (unless attractive local incentives are available). Also, the life of good quality fluorescent tubes can be as high as 30,000 hrs.

Although tube replacements will always be preferred by facilities that are cash strapped, the best way of replacing fluorescent troffers (rectangular recessed fluorescent fixture) with LEDs is to replace the full fixture with an LED fixture. This is mainly because the prismatic lensed and parabolic fixtures are optimally designed for fluorescent tubes and shape the pattern of the fixture light according to the light pattern from the tubes, which is omni-directional. LEDs are unidirectional (as explained in earlier sections) and therefore these fluorescent fixtures do not work well with LEDs. 2ft x 4ft (60cm x 120cm), or 2ft x 2ft (60cm x 60cm) LED fixture replacements that fit into ceiling tiles have excellent performance (100 lm/W from Cree for example [20]), are aesthetically pleasing, have CRIs of 92 (which is excellent for retail replacements), are easily controllable (dimmable and sensor equipped) and exceed the performance of a typical fluorescent troffer. Additional savings can often be obtained by use of the controls incorporated within LED fixtures, which are harder for fluorescents. The economical equation remains a bit challenging for pure retrofit projects if one wishes to change the fixture, but for new or renovation projects the payback can be < 3 years compared to an equivalent T8 fixture.

One of the main environmental reasons that some consumers might chose to move away from fluorescents (CFLs or tubes) is that these lamps contain mercury, and although recycling is encouraged it is unfortunately not as prevalent as one would like. Using LEDs instead eliminates that issue.

Another side comment about tubes: An application where T8 LED replacement has been extremely successful is in refrigerators (in grocery stores) and refrigerated warehouses, where, because of the low temperatures, the savings are substantial. The penetration of LED 'refrigerator sticks' as they are called is almost 100% in the US. If you walk into a Walmart, Target, Walgreens, Whole Foods and many other large chains you will only see LEDs in their fridges. In the UK, Tesco has retrofitted all its fridges with LEDs as well.

Although this is not the focus of this chapter, I would like to briefly address metal halide replacements, since they are becoming more prevalent. Street lights, flood lights and wall pack applications that use metal halide lamps are getting good competition from LED fixtures. In this case the replacement is rarely (if ever) an LED lamp, as the wattages required for the LED lamps are high (>30W for flood lights and >100W for street lights) and the heat sink has to be well designed and has to get enough air circulation, it is instead an LED fixture. There are some LED fixtures that can fit in an existing MH (Metal Halide) fixture (like a Cobra Head for example) but only a few are well designed. Generally the savings are in the 50% range. Several cities worldwide are running large LED street lighting tests in order to determine which types are optimal, including London, LA, San Diego, Raleigh, NYC and several large cities in China. The biggest challenges are encountered in places where temperatures get very high, such as the Middle East Gulf region, or Arizona and Nevada in the US. In these regions, night-time temperatures can remain quite hot and therefore the degradation of LED fixtures is likely to be faster, and so appropriate fixtures have to be chosen. Abu Dhabi in the UAE (United Arab Emirates) is planning to replace its traditional streetlights with LED fixtures and tunnel lights after running an 18-month test, which gave very satisfactory results.

The main point to be understood from this chapter is that there are many LED retrofits and fixtures that are excellent replacements for the incumbent halogen/incandescent lamps as well as other technologies, but as pointed out in reference [19] and earlier in this chapter, buyer beware! Make sure LM-79 data is available for the fixture, as well as lifetime data if possible, lighting facts label or Energy Star rating (if not, another good one is Design Lights Consortium).

7.7 ARCHITECTURAL/GENERAL ILLUMINATION APPLICATIONS

7.7.1 Color Changing/Outdoor Wall Washing

Color changing applications were the first lighting applications in which LEDs became dominant, because they naturally emit different colors (depending on the crystal) and mixing them can produce a very impressive color gamut. Some color mixing companies can boast millions of colors by using LED arrays (RGBA) and mixing them in different proportions. Most of these applications are for outdoor use, although a few are also indoor, especially in cove lighting where the color of the cove changes.

Outdoor wall washing or facade lighting often uses color changing schemes, but can also be done in white to highlight the facade of the building. Although both these applications are not necessarily encouraged from a light pollution or energy saving perspective, if one wishes to install such lights for marketing reasons, LEDs are fast becoming the preferred choice, certainly for color changing where the energy savings are large compared to the original plan (usually some form of incandescent or MH with filters) and offer a more architecturally pleasing solution. In Asia, color changing building facades are extremely popular and 100% of those installations are LED based.

7.7.2 Residential/Retail

For residential and retail applications, the CCT choice is mostly 2700–3000K (except grocery stores, who tend to prefer cooler ranges, of around 4000K). In retail, color rendering is extremely important because it highlights the merchandise and customers need to be able to tell the difference between burgundy and red, black and navy, black and dark browns. Also beam angles for retail tend to be tight (spots or narrow floods) in order to spotlight the merchandise. For residential applications, the desired beam angles are very wide because the aim is general illumination. For restaurants, dimming is key, as in these establishments, the lights are often dimmed to a low level to offer an intimate ambiance to the diners. For all these applications, LEDs are gaining significant penetration because they can satisfy all the requirements of the different applications. In retail they are often even replacing MH track lights, because of the stability of their CCT (MH can turn very green when they are close to failing) and the elimination of the ballast, which is often a pain to deal with. Even compared to self-ballasted MH lamps (which are good solutions), LEDs have a longer life and are competitively priced. In the US, the retail sector uses halogen lights heavily: MR16s and PAR lamps (like AR111s in Europe and the Mid-East) in particular. Today, very efficient LED replacements exist for these lamps, which deliver a high quality light (as explained in earlier sections). In jewelry displays, LEDs (especially if in the 4000K range) make the merchandise look very ‘sparkly’, hence they have become favored for use in jewelers’ display cases. LEDs are also gaining market share in gallery and display use, because they radiate little heat and no UV, and this protects paintings and objects of art from degradation.

7.7.3 Office Lighting

As mentioned earlier, tube replacements are still a challenge and often claims by manufacturers are somewhat inflated. The general recommendation is to move to a full LED fixture, especially for new or renovation projects. For retrofits, if the LED tube is UL/CE certified, has solid technical data (LM-79 for example) and comes from a reputable manufacturer (Philips for example now offers a 2ft and 4ft T8 LED equivalent), then a consumer can test it in a portion of their facility, and if this is satisfactory they can then expand to the rest of the facility. Often, facilities might chose LEDs because their controls are very good and do not affect the lifetime of the LEDs. For example, they can incorporate occupancy sensing and dimming to optimize the usage of the lamp when certain spaces are unoccupied. This is the area in which the largest expansion of LED use is expected in the near future, as fluorescent lighting is the main lighting used in offices.

7.8 CASE STUDIES

7.8.1 Argo Tea – Chicago and New York City, USA

Argo Tea is a chain of tea cafes in Chicago and NYC (www.argotea.com). The original lighting in all the cafes was track lighting using 50W halogen PAR20s in 75% of their locations (any new locations opening now are also using PAR20s) and 75W PAR30s in the other 25%. A month-long test was conducted in order to choose the desired PAR20 and PAR30 lamps, specifically to examine the 'look' of the lamp (are the diodes visible?) and decide on the right CCT (2700K vs. 3000K). After conducting surveys and listening to comments from customers, the 2700K lamp with the diodes behind a diffuser was chosen. The PAR20 wattage was lowered from 50W to 6.7W and the PAR30 wattage was lowered from 75W to 11W. They also had a few MR16s and PAR38s in a couple of their cafes, but by far the majority were PAR20s. They have retrofitted all their locations and this resulted in the following:

- 50.5 kW lower load across their stores (86% saving);
- 331505.112 kWh savings per year;
- CO_2 emission reduction of 99.5 tons/year approximately.

Calculations using an average of the energy price in Chicago and NYC show that this achieves a cost saving of $48,068.24/year. The cost of the lamps was $48,554.75 which means that the payback on energy savings alone is about 1 year. If the elimination of halogen bulb replacements and AC reduction is added, than the payback period drops to 9.5 months. (There were some subsidies involved and the payback was actually shorter.) Picture of the cafes are shown in Figure 7.7.

7.8.2 Shangri-la Hotel, Abu Dhabi, UAE

The Shangri-la in Adu Dhabi had several types of halogen spots in its corridors. Among them were 50W MR16s, whose only function was to produce a nice circular pattern along the floor of the hallway and provide wall washing effects along the hallway walls. There were approximately 1,000 of these in

	Total	Halogen (W)	LED (W)	kW savings
PAR20	678	50	6.7	29.3574
PAR30	268	75	11	17.152
PAR38	37	90	17	2.701
MR16	43	35	5.7	1.2599

LED PAR20

kW saved	50.4703
kWh saved/year	331589.871
Avg kWh price	$0.145
$ saved/year	$48,080.53
Project Cost	$48,554.75

FIGURE 7.7 **Before and after picture of the cafe (this particular one only has PAR20s), a summary of the savings, and the particular LED lamp used.**

the hallways, and although that seems like a small number, since it was not used for lighting (plenty of other lights for that purpose!) it was a costly way to achieve this effect on the hallway floors and walls. Essentially the total load of the MR16s was 50KW and they were switched on for 24hrs/day. Even taking into account the low energy cost in Abu Dhabi ($0.04/kWh) it still cost approximately $17,000 to have them on, without considering the cost of bulb change-outs and the effect on the AC. We suggested they try a 5W MR16, which would not supply the same amount of lumens, but would most certainly give the desired architectural and lighting effects. The management tried it in a whole corridor and decided to proceed because there were other lights to illuminate the corridors and the LED MR16s gave the exact effect they were looking for. This retrofit resulted in the following:

- 45 kW lower load (90% saving);
- 394200 kWh savings per year;
- Savings of $17,000 per year just on electricity ($0.04/kWh);
- CO_2 emission reduction of 98 tons/year, approximately.

The cost of the project was $21,800, which meant that the payback period for energy savings alone was 1.3 years. Including the reduced cost of change-outs and the effect on AC the payback is closer to 1 year.

Pictures and a summary table for this project are shown in Figure 7.8.

	Total	Halogen (W)	LED (W)	kW savings
MR16	1000	50	5	45

kW saved	45
kWh saved/year	394200
Avg kWh price	$0.040
$ saved/year	$15,768.00
Project Cost	$21,800.00

FIGURE 7.8 Pictures of the LED MR16 installed at the Shangri-la Hotel in Abu Dhabi; a summary table and a picture of the specific product used.

7.8.3 Sons of the Revolution Museum, NYC, USA

This is a small museum in New York City that is dedicated to educating the public about the American Revolution. In the display rooms, the lighting was mainly provided by MR16s, and the administrative offices mostly used fluorescent lighting. We will not address the fluorescent part, as this mainly involved swapping T12 tubes for T8 tubes. That in itself saved them 30% on their energy bills, while increasing the illumination levels in the rooms. In the display rooms, the MR16s were a mixture of 50 W, GU5.3 and GU10 bulbs, and these were replaced by 6W MR16 LEDs. The effect was quite satisfactory in all rooms, the temperature profile dropped significantly (visitors had complained it was hot) and the exhibits were highlighted as required. There had been concern that the LED GU10 MR16 may not fit properly in the track head because it is a bit longer than a halogen MR16 as the driver needed to be incorporated in the small fixture. However, they fitted fine, but the original glass cover sealing the fixture needed to be removed to accommodate them, and that would have been a recommendation anyway in order to allow the heat sink to 'breathe'. This retrofit resulted in the following:

- 6.6 kW lower load (88% savings);
- 36135 kWh savings per year;
- Savings of $6,865.65 per year only on electricity ($0.19/kWh);
- CO_2 emission reduction of 13 tons/year approximately.

The cost of the project was $6,900, which puts the payback on energy savings alone at 1 year. Including the change-outs and the effect on AC puts the payback period at even less than this.

	Total	Halogen (W)	LED (W)	kW savings
MR16	150	50	6	6.6

kW saved	6.6
kWh saved/year	36135
Avg kWh price	$0.190
$ saved/year	$6,865.65
Project Cost	$6,900.00

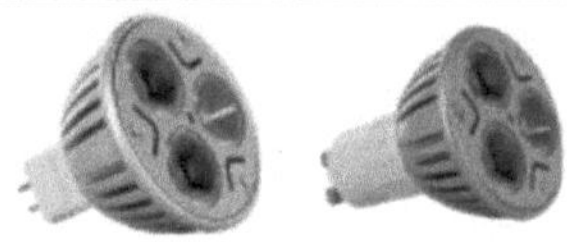

FIGURE 7.9 Pictures of the LED MR16 retrofit at the Sons of the Revolution museum in NYC; a summary table and a picture of the specific product used.

Pictures and a summary table of this project are shown in Figure 7.9.

7.8.4 Radisson Hotel, Dubai, UAE

The Radisson Hotel in Dubai had several outdoor 150W metal halide floods to illuminate the facade of the building. The hotel management was dissatisfied with the light from these floods because it was too cool, the bulbs failed too frequently and the energy consumption was too high. Hence the MH floods were replaced by 30W LED floodlights. This was a full fixture replacement (not a bulb replacement), in which the whole MH fixture was removed and replaced by an LED flood. The floods chosen had a very narrow beam angle (8°) in order to project the light upward through the floors of the hotel. The savings in this case are high, although usual savings involving MH replacement are around 50%, as mentioned earlier. However, the management were satisfied with the light provided, perhaps because the original MH floods were too bright. This retrofit resulted in the following:

- 6.24 kW lower load (80% saving);
- 27,331 kWh savings per year;
- Savings of $2,733 per year only on electricity ($0.1/kWh);
- CO_2 emission reduction of 15 tons/year approximately.

The cost of the project was $10,900, which puts the payback on energy savings alone at a little less than 4 years. Taking the cost of maintenance (scaffolding needed to deal with bulb change-outs) into account means that the payback period drops to 3 years.

Pictures and a summary table of this project are shown in Figure 7.10.

	Total	MH (W)	LED (W)	kW savings
LED flood	52	150	30	6.24

kW saved	6.24
kWh saved/year	27331.2
Avg kWh price	$0.100
$ saved/year	$2,733.12
Project Cost	$10,920.00

FIGURE 7.10 Pictures of the LED floods that were installed on the facade of the Radisson Blu in Dubai; a summary table and a picture of the specific product used.

There are several other comparisons and projects listed on the US DOE's gateway page, which showcase LED retrofit cases. They include a mix of outdoor and indoor lighting with lux measurements and very detailed financial calculations, which go beyond the scope of this chapter but might interest the reader. The webpage is given in reference [21].

7.9 FUTURE/NOVEL DESIGNS POSSIBLE WITH LEDS

The LED field is far from nearing its 'saturation' point – there is still a lot of work to do to improve their output, increase their lm/W further and decrease their cost. The penetration of LED lighting into the marketplace will gain momentum as pricing becomes more and more competitive. Over the past year or so, the market has moved from being demand constrained to supply driven, because several new companies have started LED manufacture and existing ones have expanded their capacity. The price per lumen ($/lm) has dropped as fast as the efficacy has increased, as is shown in Figure 7.2. Research into the LED field is mainly concentrating on the following directions:

- Substrate research: As shown in Figure 7.3, the substrate used for blue LEDs (used for white light) is sapphire. As you can imagine, this is an expensive material. Several alternatives are being pursued to make the substrate costs lower:
  - Increasing the size of the substrate from 2' to 8' (higher yield and more chips/production run) – see Rubicon Technology; a sapphire substrate manufacturer.
  - Growing the layers on a silicon substrate instead of sapphire. Silicon is abundant and Si substrates are well developed because of the computer industry – see Nitronex and Translucent; companies which produce GaN (Gallium Nitride) on Si, and Bridgelux, an LED chip manufacturer who announced good LED performance for devices grown on Si (bridgelux.com/media-center).
  - Using high quality (non-polar) GaN substrates to improve the performance of the diodes by eliminating certain defects that cause the light output to diminish.

- Light extraction efficiency: As explained earlier, the index of refraction difference between compound semiconductors and air causes light to become trapped inside the device. Several companies are using innovative schemes to extract the light from the semiconductor more effectively [22].
- Improving the efficiency of phosphors by using quantum dots to convert blue light to white [23].

Some of the hot topics in fixture R&D are:

- Optimization of driver design for long life, low harmonics, high efficiency and high reliability. Also for features such as dimming, sensing etc.
- Optimization of lenses for high optical efficiency.
- Fixtures with tunable color temperature (able to change cool white to warm white).
- Where suitable, the use of remote phosphors instead of phosphors integrated on the LED chip.
- Thermal management optimization through research on heat sink materials (carbon, resins and Al) and the incorporation of active cooling (forced air movement).

We would be remiss if we do not mention Organic LEDs (OLEDs) in this chapter. These are LEDs made from organic materials rather than semiconductors. Although they have made some inroads into lighting, their efficiencies are still too low to make a serious impact. However, they have been used in several small screen displays, such as cameras, phones and others. Research in this field has shown the potential to improve OLED efficiencies in the future. The attraction of OLEDs is that they can be deposited (or even printed) on very large plastic substrates, which produce fairly large panels of light. For some applications, this might be preferable to the point source nature of LEDs. For the interested reader, [24] is a reference discussing white light OLEDs.

7.10 CONCLUSIONS

LEDs are becoming a major player in the world of lighting, and their role in reducing the carbon footprint of lighting and its energy consumption is critical. They are an inherently more efficient process of producing light, and incorporate features such as tunable color temperatures and colors that traditional lighting technology does not have. They have already started making an impact in today's lighting, and their market share keeps increasing as their price becomes more competitive. This is because their supply is being augmented by the expansion of production capacity in Asia. Nations worldwide are conducting large scale LED municipal lighting studies, as they recognize the potential for energy savings and the budgetary savings that ensue. The use of LEDs in lighting is expected to grow at a compound annual growth rate of 39% through 2015, reaching $4.5 billion, from $888 million in 2010 [25]. As the title of this chapter says, let's join the LED lighting revolution!

REFERENCES

[1] H.J. Round, Electrical World 49 (1907) 309.

[2] N. Holonyak Jr., S.F. Bevaqua, Coherent (visible) light emission from GaAs1-xPx junctions, Appl. Phys. Lett. 1 (1962) 82–83. In 1995, Nick Holonyak received the Japan Prize for his seminal work on LED's and lasers.

[3] R. Haitz, J. Tsao, Solid State lighting: 'The Case' 10 years after and future prospects, Phys. Status Solidi A 208 (1) (2011) 17–29.

[4] S. Nakamura, GaN Growth using GaN Buffer Layer, Jpn. J. Appl. Phys. 2 (1991) 1705.

[5] H.M. Manasevit, Recollections and reflections of MO-CVD, J. Cryst. Growth 55 (1981) 1–9.

[6] G.B. Stringfellow, Organometallic Vapor-Phase Epitaxy, Academic, New York, 1989.

[7] G. Wyszeki, W.S. Stiles, Color Science: Concepts and Methods, Quantitative Data and Formulae, second ed., Wiley, New Yok, 1982.

[8] G.E. Hofler, D.A. Vanderwater, D.C. Fevere, F.A. Kish, M.D. Camras, I.-H. Tan, Wafer bonding of 50mm diameter GaP to AlInGaP light emitting diodes, Appl. Phys. Lett. 69 (1996) 803.

[9] Philips Lumileds, Lumileds lighting launches multi-format luxeon light sources, Compound Semiconductor Mag. 7 (6) (2001) 11.

[10] Illuminating Engineering Society Standard, IES LM80–08, Approved Method: Measuring Lumen Maintenance of LED Light Sources (2008).

[11] J. Jiao, TM21 seeks methods for lumen maintenance prediction, LEDs Magazin (no.40) (Feb. 2011) 37–39.

[12] Examples of LM80 & TM-21 data sets can be found on the Philips Lumileds website at: www.philipslumileds.com/support/documentation/lumen-maintenance.

[13] http://www.lightingfacts.com/

[14] Illuminating Engineering Society Standard, IES LM79–08, Approved Method: Electrical and Photometric measurements of Solid State Lighting Products.

[15] http://www.energystar.gov/

[16] http://www.energystar.gov/index.cfm?fuseaction=products_for_partners.showLightbulbs

[17] Energy Savings potential of Solid State Lighting in General Lighting Applications 2010 – 2030, Feb 2010. prepared for the DOE by Navigant Consulting, http://apps1.eere.energy.gov/buildings/publications/pdfs/ssl/ssl_energy-savings-report_10-30.pdf.

[18] A good example is the Philips EnduraLED 12W L-Prize winning A19 bulb on http://www.usa.lighting.philips.com/

[19] http://www1.eere.energy.gov/buildings/ssl/reports.html, at this time Round12 reports are the most up to date ones (2011).

[20] www.creeledlighting.com/Products/Architectural-Troffers.aspx

[21] http://www1.eere.energy.gov/buildings/ssl/gatewaydemos_results.html

[22] Y. Narukawa, M. Ichikawa, D. Sanga, M. Sano, T. Mukai, White light emitting diodes with super-high luminous efficacy, J. Phys. D: Appl. Phys. 43 (35) (2010) 1.

[23] Hsueh-Shih Chen, Cheng-Kuo Hsu, Hsin-Yen Hong, InGaN–CdSe–ZnSe Quantum Dots White LEDs, IEEE Phot. Tech. Lett.S 18 (1) (2006) 193–195.

[24] M.C. Gather, A. Köhnen, K. Meerholz, White Organic Light Emitting Diodes, Advanced Materials 23 (no.2) (2011) 233–248.

[25] Strategies Unlimited 2010 data published in the proceedings of the 'Strategies in Light' Conference in Feb. of 2011.

[26] Y. Ono, W. Davis, Rationale of color quality scale, NIST, June 2010.

Chapter 8

Minimum Energy Housing in Cuba

Dania González Couret
Faculty of Architecture, ISPJAE, Havana City, Cuba

Chapter Outline

8.1 INTRODUCTION

Reducing energy consumption and carbon dioxide emissions related to buildings and particularly in housing is essential for the planet's survival. These depend on the building materials and technologies used to build the house, their durability and recyclability. The project needs to be adapted to its context in order to make the best possible use of the available natural resources to minimize the negative impact of the energy needed by the family to live and the equipment they use. Of course, if the energy comes from renewable sources, the carbon dioxide emissions are reduced.

The amount of energy needed by humans to live is related to the many economic, social and cultural factors which determine their lifestyle. It varies between continents and from one country to the other, but also within the same country, between regions or towns and even from one family to another, related to their different social statuses.

Sustainability, Energy and Architecture. ISBN: 9780123972699

This chapter intends to discuss the main principles involved in reducing energy consumption and carbon dioxide emissions related to housing in Cuba. Cuba is a developing country located in a tropical island in the Caribbean Sea. The Cuban situation may be useful as a reference for other poor regions in warm and humid climates.

Traditional and modern Cuban architecture is evaluated, in order to find the key elements to use in present and future sustainable architecture.

This chapter should contribute vital knowledge to students everywhere in explaining the general aspects to be taken into account when designing minimum energy housing, particularly in tropical developing countries.

8.2 LIFE CYCLE AND SUSTAINABLE BUILDINGS

Energy consumption and carbon dioxide emissions in buildings should be considered on the basis of their life cycle. A building is born when the raw materials to produce its components are extracted and transported to the place where the components for construction will be manufactured. Afterwards, these components are taken to the site and located in their final position in the building, where they will stay for the duration of the useful life of the building.

During all the steps of the construction process, energy is needed and carbon dioxide is produced, in amounts that depend on what the extraction of the raw materials and the production of the building elements demand. The energy demand is also related to the transportation distances between the sources, the factories and the building site. All this energy is embodied in the components and will be released when the building dies, at the end of its useful life. Then, the building durability is directly related to the conservation of the energy embodied in its mass.

The period of time between the end of building construction when the family starts living in the house, and the moment in which it is no longer inhabited, or is demolished, is the useful life of the building. This is the longest part of the life cycle, and its duration depends on the design of the building and its materials, as well as any exploitation and conservation processes.

During its useful life, the building is inhabited and energy is consumed, in quantities which depend on its adaptation to the local context and the climate, as well as the availability and efficiency of the electronic domestic equipment in the house. Any conservation process required to ensure the building's durability also consumes energy, as does the systematic maintenance and rehabilitation of the building. These could happen several times, until their cost makes it inconvenient or too expensive to maintain the building. Depending on the energy sources involved, the building will be responsible for high or low emissions of carbon dioxide during its useful life.

When the building is so old that maintaining it is very difficult and expensive, it will die, completing its life cycle. Buildings in this condition are usually demolished, in which case all its components will become waste, which can

TABLE 8.1 Embodied Energy of Building Materials

	Density	Low value		High value	
Material	(Kgm^{-3})	Gj tone−1	Gj m3	Gj tone−1	Gj m3
Natural aggregates	1500	0.030	0.05	0.12	0.93
Cement	1500	4.3	6.5	7.8	11.7
Bricks	1700	1.0	1.7	9.4	16.0
Timber (prepared softwood)	500	0.52	0.26	7.1	3.6
Glass	2600	13.0	34.0	31.0	81.0
Steel (steel sections)	7800	24.0	190.0	59.0	460.0
Plaster	1200	1.1	1.3	6.7	8.0

Gj: giga joule, a unit of energy, 1Gj: 278 kWh
Range of published figures
Source: Building Research Establishment, cited by Roaf, Sue, Manuel Fuentes and Stephanie Thomas, *Ecohouse 2. A Designs Guide,* Architectural Press, Oxford 2003, p. 51.

affect the environment. Alternatively, sustainable approaches promote deconstructing the building at the end of its life, which means separating and classifying its components so that they can be reused or recycled. The availability of processes to facilitate deconstruction should be taken into account in the building design.

Durability is directly proportional to sustainability. When the useful life of a building is long, more advantage is taken from its embodied energy, and more time passes before the building components become waste and impact the environment, and new raw materials are extracted to build a new house.

The building's durability depends on the materials from which it is made and whether or not it was designed to make best use of them, and to adapt well to the local climate, among other factors. Sometimes, buildings last for a shorter time than their materials' potential, because of changes in people's preferences, way of life, fashion, scientific and technical advances, or because of economic reasons, such as land speculation.

As a house is for a family, its duration should ideally coincide with the family's lifetime (approximately 60 years)[1]. This would allow a new house to be built to fit the new context, conditions and requirements of the next generation,

1. The estimation considers around 20 years as the minimum age to create a new family and an expectance of life of about 80 years. See González Couret, D., *Economía y Calidad en la vivienda. Un enfoque cubano,* Editorial Científico Técnica, Havana, 1997.

but the question then arises of how to preserve the built heritage as a cultural value for future generations.

In any case, the useful life of a house is the longest part of its life cycle, in which energy consumption related to its use is directly related to its design and the local microclimate – so the role of the architect is decisive. That is why the discussion of the minimum energy house in this chapter will focus on the useful life of the house, and therefore on bioclimatic design.

8.3 DESIGN STRATEGIES IN WARM AND HUMID CLIMATES

The energy consumed by a house during its useful life depends on its energy efficiency and the appliances within it, but also to a great extent on whether the architectural design incorporated as much daylight as possible, and took account of the local climate to reduce energy use in the thermal conditioning of the indoor environment.

The possible use of renewable energies is other important contribution to the building's sustainability that could be provided at urban or architectural scale. In which case, this should preferably be harmoniously incorporated into the design of the building.

Architectural design is therefore directly related to local climatic conditions. Cuba has a warm, humid climate characterized by high day time temperatures and relative humidity throughout the year, as shown in Figure 8.1. There are no seasons (winter, spring, summer and autumn) as one might find further from the equator, but the year has two periods identified by rainfall (rainy and less rainy).

Because of the combination of high temperatures and relative humidity, it is difficult to achieve comfortable conditions, even outdoors. This is only possible at night in some months (from December to February) (Figure 8.2)[2]. The high nocturnal temperatures and their small amplitude of variation during the day invalidates the possible use of thermal inertia as a bioclimatic design strategy, because it is not desirable to store heat in the thermal mass of the building during the day as this would be released into the indoor environment at night, when air temperatures are still above the comfort zone.

This situation also makes it difficult to use night cooling ventilation or to take advantage of the thermal gradient of the ground as design strategies. On the other hand, the high relative humidity considerably reduces the potential for evaporative cooling. The main design strategies to be followed in warm, humid climates are to provide maximum protection against solar radiation, and to allow as much natural ventilation as possible.

In these climates, the building should not contribute to increase the air temperature or perceived temperature by thermal radiation, but on the contrary should absorb radiation in order to produce as much radiant cooling

2. Op. Cit.

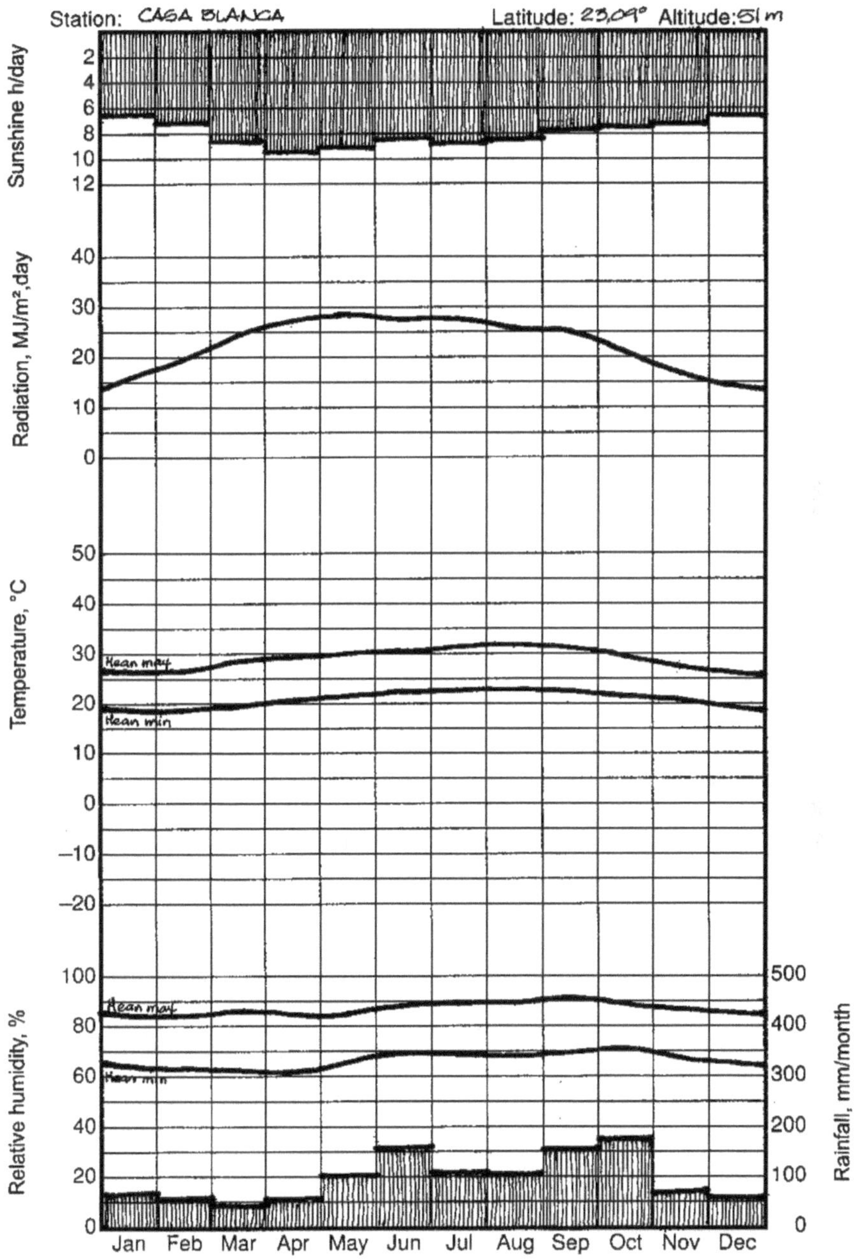

FIGURE 8.1 Climatic Data for Havana. *(Graphic from González Couret D., "Progressive Social Housing in the Historic Centre of Matanzas City", Architecture Energy and Environment, Housing Development and Management, Lund University, 1997)*

Bioclimatic Diagram (Givoni)

Location	HAVANA
Longitude	82.20°
Latitude	23.09°
Altitude	51.0 m

Climatic data

	Jan	Feb.	Mar	Apr.	May	Jun.	Jul.	Aug.	Sept.	Oct.	Nov.	Dec.
Monthly mean max. Temp	26.0	26.2	28.0	28.9	30.0	30.5	31.3	31.6	31.0	29.2	27.4	26.5
Monthly mean min RH	65	63	63	61	67	70	69	68	70	71	68	66
Monthly mean min Temp	18.5	18.4	19.5	20.8	22.1	23.4	23.7	23.9	23.7	22.8	21.0	19.3
Monthly mean max. RH	85	84	86	84	86	89	89	89	91	89	87	86

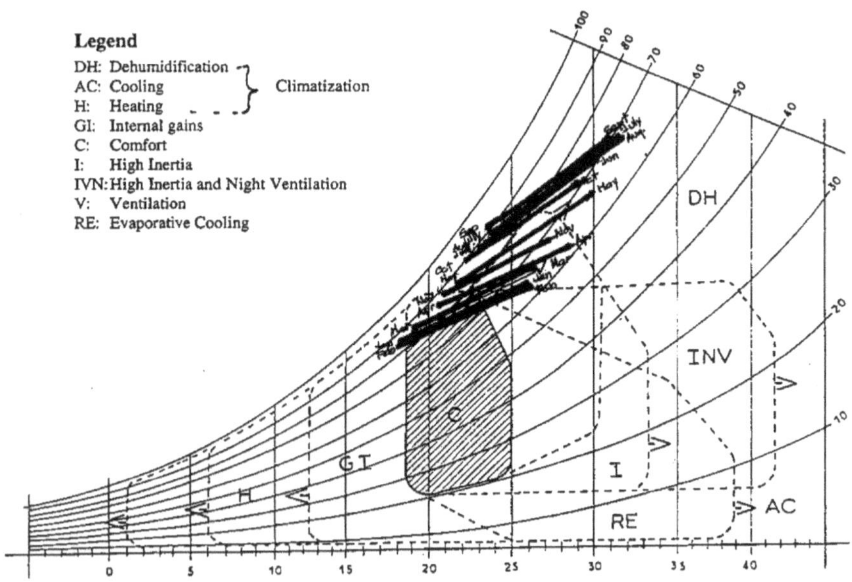

Recommendations

FIGURE 8.2 Maximum and minimum values of temperatures and relative humidity in Havana are mostly out of the comfort zone; see the Givoni Bioclimatic Chart.

as possible. This means that the building envelope should be protected from direct solar radiation, preferably by live green shade. Reflecting solar radiation from external surfaces could be undesirable in the urban context, but if it is not possible to achieve total solar protection, low absorptivity surfaces and low transmittance building elements should be used on the roofs and external walls.

Movement of air in contact with a person's skin is the main mechanism of heat exchange between the human body and the environment. Evaporation by transpiration accelerates with increasing air velocity, which is possible by cross ventilation. This is why traditional design strategies for warm humid climates recommend the use of light buildings well protected against sun and rainfall, but unable to store heat in their envelope. On the contrary, they can be cooled quickly, and are as open to the wind as possible in order to achieve high indoor air velocities from cross ventilation, and so improve the inhabitant's thermal comfort. However, some authors recently insist on the importance of thermal mass, even in warm, humid climates.[3]

In order to take advantage of daylight to reduce energy consumption in buildings, direct solar radiation should be avoided, not only because of the heating effect, but also to reduce glare, especially from glass windows.

However, people living in tropical regions and in warm, humid climates are generally poorer, so their houses are generally not artificially air-conditioned. In such situations, people have to adapt in order to live in an environment uncomfortable enough to affect their productivity, but not their life. They often have air conditioning equipment (AC) for the main bedroom, and fans whose energy consumption and carbon dioxide emissions are not as high as the AC. In these cases, any bioclimatic design strategy will have a great impact on people's quality of life, but less on the energy consumption of the house.

The architectural typology described above is mostly only possible in open or rural areas, where buildings can be far one from one another and do not affect each other's privacy or security. In contrast, urban architecture should contribute to the city's sustainability, not only by reducing energy consumption and carbon dioxide emissions, but also by making the best possible use of the urban land. Also, the urban island heat effect and the wind characteristics in the city require some variations to be made to these traditional strategies.

8.4 THE URBAN MICROCLIMATE

The urban morphology and thermal environment affect the microclimate which surrounds a building in the city. The urban island effect increases night-time temperatures, especially as the city becomes more compact (which usually occurs in the central areas). The buildings and pavements absorb and store more heat in their mass, depending on their thermal inertia, which is released after sunset. However, green areas and trees provide shade, which helps to reduce this effect.

In the urban context, streets act as canyons to funnel the wind, depending on their cross-section and orientation[4]. In addition, the permeability of the built mass, the distance between buildings and the combination of their heights in the blocks affects not only the amount of solar radiation that reaches their external

3. See, for example, Szokolay, S., "The role of thermal mass in warm humid climate housing", *XX PLEA*, Santiago de Chile, 2003.

4. Johansson, Erik, Urban Design and Outdoor Thermal Comfort in Warm Climates. Studies in Fez and Colombo, Housing Development and Management, Lund, 2006.

surfaces, but also the flow of wind inside the urban area, and hence the levels of indoor ventilation[5].

The morphology, compactness and density of the urban area not only influences the thermal environment, but also the levels of daylight that are available. As the height of the buildings increase, and the distances between them diminish, the possibilities for indoor lighting by passive means diminish[6].

This complex phenomenon is still being studied today, and despite much effort, no definite conclusions have been reached regarding the way these climatic effects should influence architectural design and construction of indoor environments in warm, humid, urban contexts. A detailed site analysis is necessary for any specific project, in order to make best use of local potential to solving the problems which arise from local factors.

8.5 VERNACULAR ARCHITECTURE IN CUBA

Vernacular architecture is of course, low energy, because it is generally made from natural, local materials, using traditional manual technologies, and follows design models based on living in the natural environment that have been improved on over centuries.

The original vernacular Cuban architecture was developed by its inhabitants before the colonizers arrived. Different rural models existed, all of which were based on the use of natural available fibers, were very permeable to the wind and were protected from sun and rainfall (Figure 8.3).

These models have been followed in rural traditional housing until the present, with building materials and spatial solutions varying according to available resources and local culture. A common feature of all of them, however, is the use of the trunk and leaves of the royal palm, a natural resource, and the architectural permeability of their layout (Figure 8.4). Rural houses belonging to wealthier families generally also include a surrounding gallery, together with openings in the upper part of the roof to enhance convective ventilation (Figure 8.5).

These features can also be seen in housing located on the seashore, where porches and arcades are common, as well as the elevation over the sea by pilots. Fixed blinds allow the wind and light to get in, while avoiding direct solar radiation (Figures 8.6 and 8.7).

Traditional wooden housing in small towns is either constructed as isolated buildings or row houses. The isolated buildings follow the principles formerly described, such as tilt roofs with openings on top, porches to protect the walls, and fixed blinds to allow permeability (Figure 8.8).

---

5. Coca, O., D. González Couret, R. Sotolongo y M. Pulido, "Comportamiento del viento en zonas urbanas: Habitabilidad Vs. Vulnerabilidad, Medio Ambiente Construido, ISPJAE, Havana, 2011.
6. González Couret, D., Aprovechamiento del suelo y ambiente interior como variables contrapuestas para la sustentabilidad de la vivienda urbana, Editorial CUJAE, Havana, 2008.

FIGURE 8.3 **Población Indoantillana, Roberto Mateizán.** *(Taken from Revista Arquitectura y Urbanismo, No 2, 1985)*

FIGURE 8.4 **Traditional rural housing (Bohío) in Viñales.**

Row houses have been influenced by the compact model imposed by outside and by land speculation. They are generally very economic houses built on long, narrow plots, in which indoor spaces are located one behind the other, and connected to a lateral yard in an 'L' or 'C' shaped plan, depending on whether or not there is a back yard. They always have a high front porch with a roof tilted to the front (Figure 8.9).

Despite the tradition of isolated houses built from natural fibers, the urban model introduced by the Spanish colonizers corresponded to the compact

FIGURE 8.5 **House surrounded by a corridor and enhancing convective ventilation in 'Oriente'.**

FIGURE 8.6 **Housing in Granma Key.**

Mediterranean city, which was structured in irregular blocks incorporating the streets, containing houses with attached courtyards (Figure 8.10).

This model characterizes the urban centers of the main Cuban cities, where the indoor spaces are connected to the outdoors mostly by internal yards. This arrangement has high thermal inertia and limited cross ventilation. More recently, colonial architecture has adapted better to the local climatic conditions; windows have become bigger, movable blinds as well as colored glass have been added, and porches and balconies have appeared (Figure 8.11).

At the beginning of the twentieth century, the wooden attached house was substituted by a solid one, built of masonry and according to the eclectic style, but keeping the lateral yard in the long and narrow plot (Figures 8.12 and 8.13).

FIGURE 8.7 **House by the sea in Punta Gorda, Cienfuegos.**

FIGURE 8.8 **Traditional wooden house in Calabazar.**

Some investigations carried out in the 1980s showed that this compact urban model was very different to that traditionally recommended for the Cuban climate, and indoor thermal conditions could be better in Old Havana than in the new urbanizations developed according to the modern urban model in the 1970s[7]. The indoor climate could be much improved by optimizing the dimension, orientation and amount of vegetation used in the yards, squares and streets alongside the houses.

7. See Colectivo de autores, "Estudio de los factores físicos para la valoración higiénica del medio residencial, *Selección de Artículos,* No 10, Centro de Información de la Construcción, Havana 1989, and Alfonso, A., G. Díaz y A.M. De la Peña, "Por el rescate de la tradición", *Revista Arquitectura y Urbanismo* No 2, ISPJAE, Havana, 1989, pp. 2 – 7, among others.

FIGURE 8.9 Wooden row house in Cojímar.

FIGURE 8.10 Courtyard of a colonial house in Old Havana.

Since that time, several studies have been carried out in Cuba and abroad to determine in more detail how urban morphology influences architectural design and the indoor environment in warm, humid conditions[8]. No final conclusion has been reached, but it seems that intermediate solutions, neither completely compact nor totally open, could be better. This takes the form of a semi-compact urban model, in which buildings are separated by shaded corridors through which wind can flow. Of course, this is not a good solution for daylight.

8. See Johansson, E., *Urban Design and Outdoor Thermal Comfort in Warm Climates. Studies in Fez and Colombo,* Housing Development and Management, Lund, 2006; González Couret, D,. et. al. *Vivienda apropiada para la ciudad de La Habana,* Editorial ISPJAE, Havana, 2005; González Couret, D., et. al. *El edificio de apartamentos en Cuba,* Editorial ISPJAE, Havana, 2010.

FIGURE 8.11 **Colonial house with added porches in the Cathedral Square.**

FIGURE 8.12 **Eclectic row housing in Pinar del Río.**

8.6 MODERN ARCHITECTURE IN CUBA

The extension of the original compact cities started to following new urban patterns related to the hygienist model in the twenty first century. At that point, the urban grid was opened; the buildings became detached one from the other, and were separated by corridors, back yards and gardens. The internal courtyard disappeared because it was not necessary anymore, and the relationship between indoors and outdoors was set by the exposed building envelope. At the same time, amounts of vegetation in the city increased according to the garden city paradigm; so gardens, green parks and trees were used plentifully along the streets (Figure 8.14).

The architecture was also transformed at the beginning of the twentieth century; first formally, following styles such as Eclecticism, Art Nouveau and

FIGURE 8.13 Lateral yard in traditional row housing, Pinar del Río.

FIGURE 8.14 Aerial view of 'El Vedado'.

Art Deco, but keeping the same spatial design. It then started to assimilate the Modern codes initially in the 1940s and though into the 1950s. This meant not only a formal transformation, but an important volumetric and spatial change in architecture.The isolated houses of the wealthier people were characterized by their almost total relationship between the indoors and outdoors, by whole planes of fixed or movable blinds and glass, protected by overhangs or vegetation.

FIGURE 8.15 **Farfante sisters' apartment house, Arch Frank Martínez, Nuevo Vedado, 1955.**

The indoors were predominantly spatially continuous and naturally well illuminated and ventilated (Figure 8.15).

Buildings occupied by other social classes copied these architectural solutions, but using poorer quality designs based on the same spatial spirit – which was very appropriate to Cuban climatic conditions. The house plan was no longer a succession of linearly arranged spaces, but these were mostly centered and zoned. The more private areas were placed out of the visitor's sight, and bedrooms were connected to the bathrooms and toilets. All indoor spaces were directly connected to the outdoors in order to use natural daylight and ventilation, preferable in a cross way.

Despite the fact that this urban and spatial solution was more favorable in its use of natural daylight and ventilation, the reduction of the roof height and the substitution of the tilted wooden roof by a horizontal concrete flat roof was a backward step in architectural evolution, since this increased the radiant heat produced indoors.

The use of concrete facilitated the construction of multistorey apartment buildings containing one or more dwelling per floor, depending on its social status. The architectural designs of the better buildings followed the same principles that were used for individual houses, using volumetric and spatial solutions which promoted solar and rainfall protection while retaining maximum permeability to daylight and ventilation by an arrangement of continuous spaces, well related to the outdoors (Figure 8.16). However, economically speculative building promoted high density use of urban land, which affected the indoor environment. Some interior spaces were not directly connected to the outdoors, or did so only by very small yards in a conduit shape.

Apartment buildings were initially built on empty plots existing within consolidated urban areas, but in 1950, the first open plan urbanization was developed, incorporating typical housing blocks for workers (Figure 8.17). This model of repetitive social housing has been followed up until today, following

FIGURE 8.16 **Interior of an apartment in Miramar, Arch. Mario Romañach, 1956.**

FIGURE 8.17 **Open urbanization based on typical housing blocks for workers, Arch. Martínez Inclán, Romañach, Quintana and Mantilla, Guanabacoa, 1948.** *(From Rodríguez, E. L., The Havana Guide. Modern Architecture 1925 – 1965, Princenton Architectural Press, New York, 1999)*

an industrialization approach that impedes any use of architectural design to minimize energy consumption.

However, research and experimental work in this field has continued with the participation of university students, in competitions and experimental projects, some of which have been selected for discussion here, to show the possibilities for developing minimum energy housing in current Cuban conditions, even though these have not been put into practise.

8.7 PRESENT AND FUTURE

Research related to architecture and climate started in the Faculty of Architecture in Havana in the 1960s, and the subject 'Climatic Conditioning' was included in the curriculum in the 1970s. However, investigations of energy issues in housing,

FIGURE 8.18 **Prototype 1 of a bioclimatic and solar row house. South façade of the project.**

including renewable energy applications, did not begin until the early 1980s, and it was not until the end of this decade that the first important research related to urban microclimate was carried out[9]. Approaches to teaching architectural climatic conditioning varied from quantitative approaches that provoked student's rejection, to more qualitative ones which used the results of recent research. Today, computer simulation facilitates the design process, and research focuses on the influence of the urban context and urban architecture on the indoor environment in order to design more appropriate future buildings and cities[10].

The architectural examples illustrated here are classified according to their context, from isolated houses for rural and suburban locations to multistorey apartment buildings for urban contexts, including compact central urban areas.

8.7.1 Isolated, Rural and Suburban Housing

The 'bioclimatic solar house' was the first experimental project developed as the conclusion of research carried out from 1981 to 1984 and awarded in 1985. The aim was to show that bioclimatic design was not as expensive as some people thought, and that it was possible to use renewable energy even in less expensive houses. However, the relationship between bioclimatic architecture and the specific design for a site was not clear enough at that time. This is why the research ended up offering two prototype projects, to be located in a context specified

9. It is referred to in the research "The microclimate of the compact city", carried out by Alfonso, A, G. Díaz, and A.M. De la Peña, A.M., finished in 1989 and awarded by the Cuban Academy of Science in 1990.

10. See recent papers presented by González Couret in Passive Low Energy Architecture (Santiago de Chile 2003 and Geneva 2006), in the World Renewable Energy Congress (Cologne, 2002; Denver, 2004; Florence, 2006, and Linköping 2011), and CISBAT, Lausanne 2005. Also see Díaz G. and A. M. De la Peña, PLEA 2006, and Tablada, PLEA 2003 – 2010.

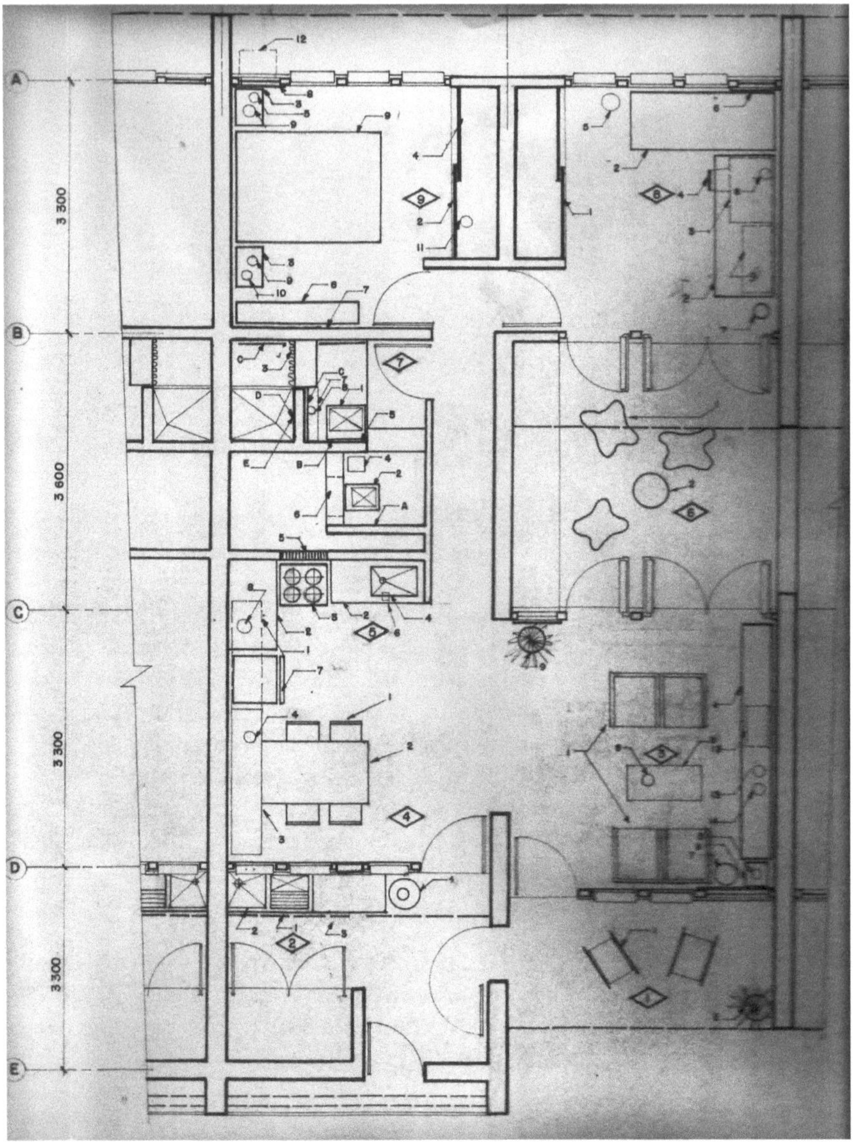

FIGURE 8.19 Prototype 1 of a bioclimatic and solar row house. Project plan showing the internal yard as well as porch and service yard with solar dryer to the south.

afterwards, taking into account only their recommended orientation (Figures 8.18, 8.19 and 8.20)[11].

11. Investigación y proyectos desarrollados por Alfonso A., D. González Couret y M. Matamoros con la colaboración de E. López, A. Alemany y A.M. de la Peña.

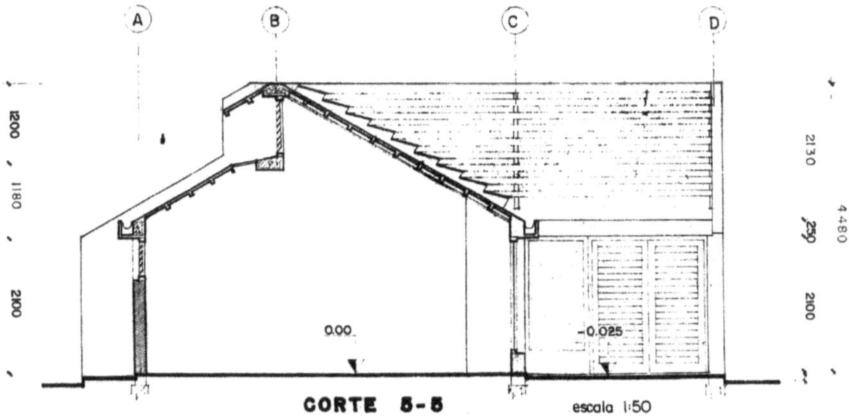

FIGURE 8.20 **Prototype 2 of a bioclimatic and solar row house. Section showing double auto-shading roof and protected upper windows for convective ventilation.**

Both projects were for attached row houses in order to reduce land use and solar gain by the external walls. Because of its proportions (depth), one of the projects required an internal yard that connected almost all indoor spaces to the outdoors.

Each project was conceived using different building materials, in order to show that specific or special building materials do not need to be used in bioclimatic architecture, but, on the contrary, the design should be able to adapt to use the available resources. In particular, the roof was solid and heavy in one case and light in the other, but both were tilted, and protected against the sun by a double skin with fixed shading devices. These prevented solar radiation reaching the external roof surface, and allowed convective ventilation for roof cooling and to protect the impermeable finishing.

The building should be oriented in such a way to allow most of the tilted roof surface to face north, where the average incident solar radiation is less throughout the year. Only a small proportion of the surface would be oriented to the south (30 degrees), for the installation of the solar water heater and the photovoltaic panels to provide energy in periods of higher daily consumption. The service yard located to the front was closed by a solar cloth dryer also oriented to the south.

Tilted roofs allow convective ventilation to remove hot indoor air through upper windows, which appropriately protected from the sun, but allow diffuse daylight to enter. This is reflected from the internal surface of the tilted ceiling to enhance indoor natural illumination. The tilted roofs also allowed rainfall collection to be used in the toilet. This consumes 50% of the domestic water.

Those designs were never built, but some of these principles were applied afterwards in many others, such as the awarded architectural solutions proposed

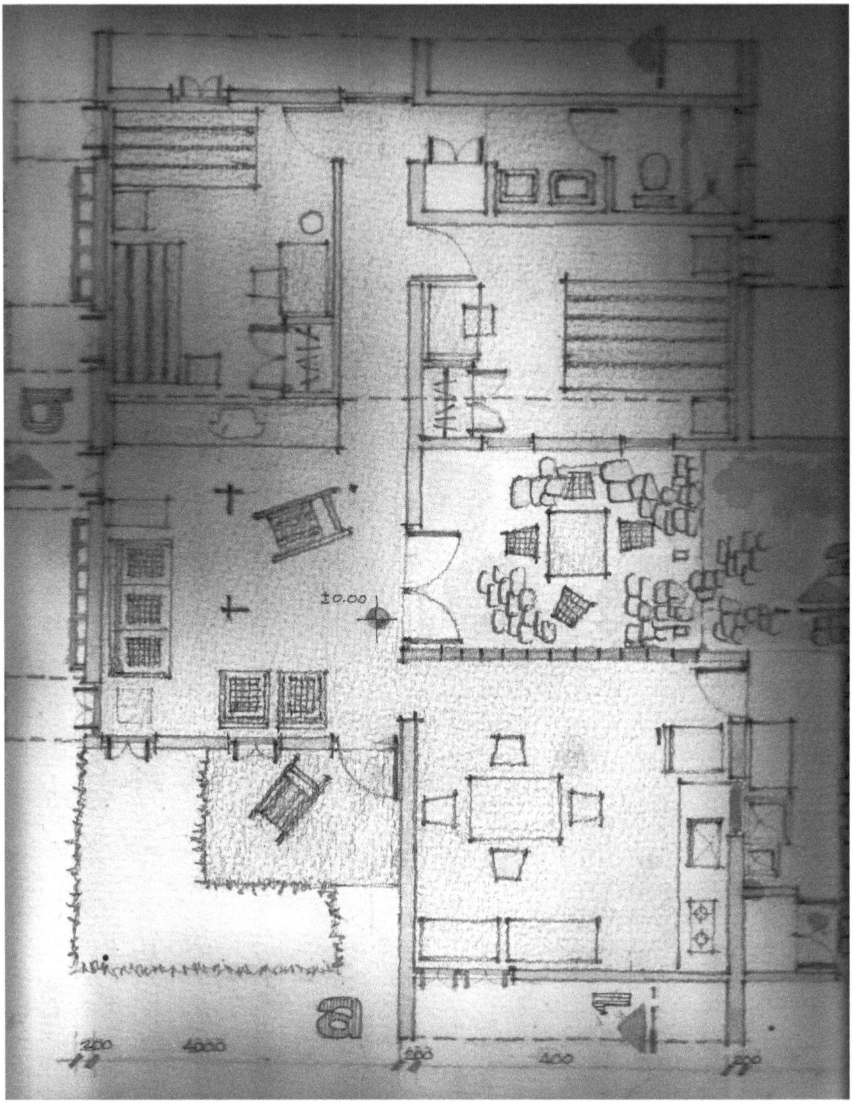

FIGURE 8.21 House in the mountains. Spatial solution.

for housing in the mountains (1988, see Figures 8.21, 8.22 and 8.23)[12]. It was a system that could be adapted to local topography and orientation, keeping most

12. Project developed by González Couret, D., M. A. Estivil, R.M. Castillo y M.E. Lopez, and awarded in the National Competition "Three Designs to Improve Living Conditions in the Mountains", Havana, 1989.

FIGURE 8.22 **House in the mountains. Lateral view, solar protection and water heater.**

FIGURE 8.23 **House in the mountains. Fitness to topography, double shading roof and convective ventilation.**

of the roof surfaces to the north and the solar water heaters to the south, and allowing convective ventilation and daylight entry by openings in the upper part of the roofs.

Because they were not attached houses but isolated ones, their lateral walls were exposed, and so were protected from solar radiation by external shading devices or vegetation. Total solar protection of the external walls was not necessary because of the lower overall temperatures and greater variation in these mountainous regions. The projects were designed to be built from local materials, such as stone, brick, stabilized soil or small format precast elements

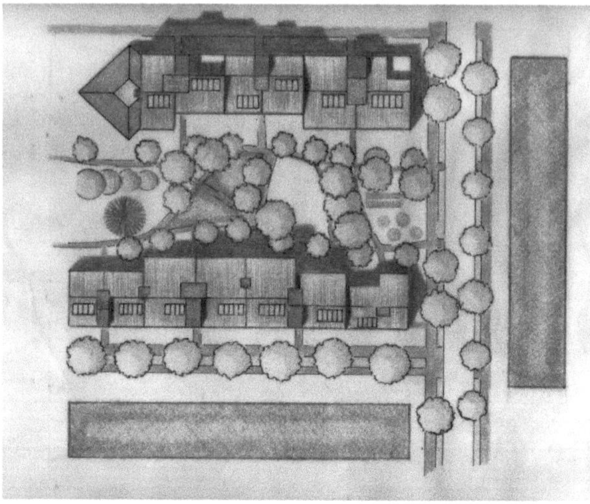

FIGURE 8.24 Location of the 'Umbrella House'.

FIGURE 8.25 Model of the 'Umbrella House'.

locally produced. Wood was only proposed for windows and was not considered as a building material because of its scarcity, although it is a renewable resource.

Similar principles were applied in the design of the 'Umbrella House' (Figures 8.24 and 8.25) presented to the international student's competition 'Eco-house 2003' in Oxford[13]. Located at the top of a multistorey building in a peripheral

13. Students: Y. Jiménez, A. Zorrilla, H. Wells, H. Gómez, A. de la Cruz and S. Morales. Professors: D. González Couret and A. Portero.

FIGURE 8.26 A new housing building in the historic center of Matanzas.

urban area, this two storey house is open to the outdoors in three directions. Air is sucked in from these three sides by the solar chimney located on the upper part of the umbrella roof. The roof is protected from solar radiation by a double skin of movable shading elements, which could be photovoltaic panels.

The open connected space, like in the better Cuban architectural traditions, is well ventilated, illuminated and protected from the sun. The umbrella roof also allows the collection of rain water for various purposes. The solar cloth dryer helps in delimiting the private open space of the house, which is complemented by a humid soil system for grey water treatment.

8.7.2 Multifamily Urban Housing

Apartment buildings are generally located in urban areas where water, sewage and power supply is guaranteed by city infrastructure. In these areas, proposals from experimental projects have focused on volumetric, spatial and constructive design for minimum energy consumption and maximum quality of life in the urban context, following the better experiences from traditional and modern Cuban architecture.

The project 'bioclimatic and progressive social housing'[14] (Figures 8.26, 8.27, 8.28, 8.29, 8.30 and 8.31) awarded in the National Competition for Habitat Design in 1999 consisted of an apartment building located in the historic center of Matanzas city, taking advantage of the remaining wall of the preexisting building as a double skin to protect the new one, preserving its embodied energy and avoiding demolition costs[15].

14. Project developed by D. González Couret, F. del Valle, G. García and L. Capote, 1999.
15. See González Couret, D., "Design Criteria for Warm Humid Climates. A Case Study", World Renewable Energy Congress, Brignton 2000.

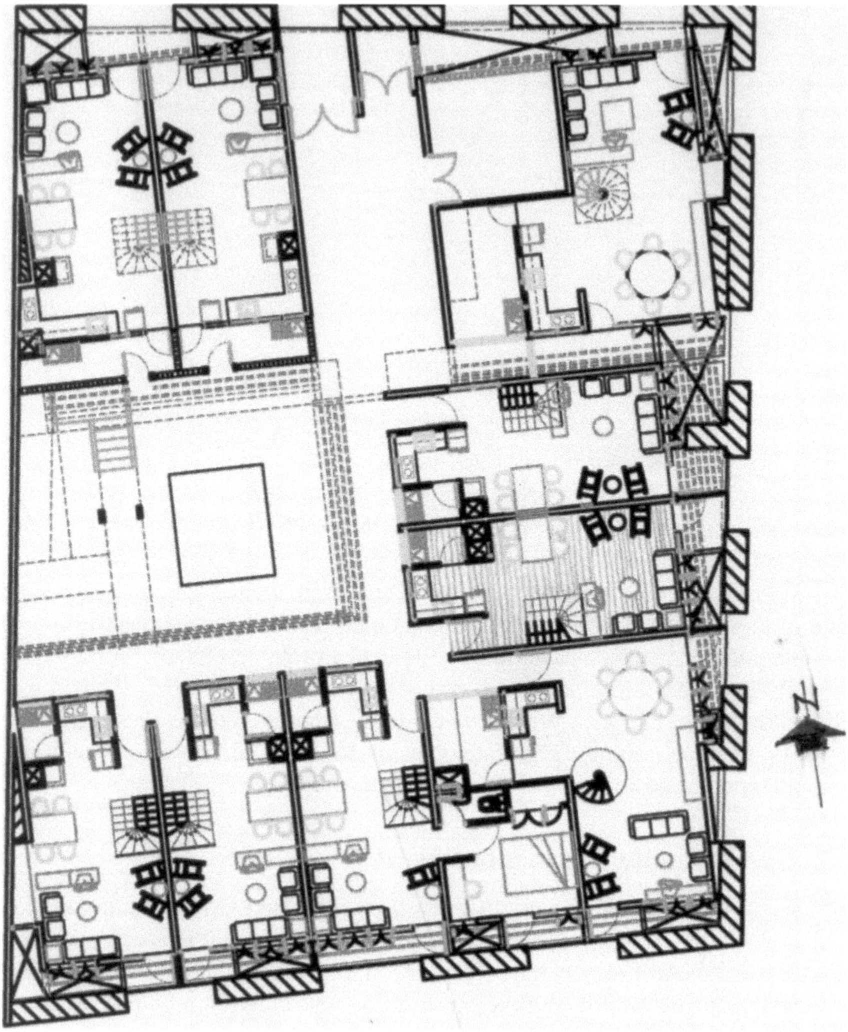

FIGURE 8.27 General plan of the new housing building.

The new building followed the idea of the attached houses, having an internal courtyard. The duplex dwelling was capable of being developed progressively according to the necessities and possibilities of the family, without affecting its urban image, based on the 'shell' typology of progressive housing.

The stairs connect to horizontal corridors located on two storeys, which enable access to the apartments without affecting the privacy of the occupants, as the bedrooms are located on the second floor. The living and dining room on this level are directly related to the street. On the second floor of each apartment the bedrooms are directly related to the outdoors by the street or the courtyard, with

FIGURE 8.28 External wall of the preexisting building.

FIGURE 8.29 Model of the building, incorporating the preexisting wall as a double skin.

FIGURE 8.30 Model of the building showing the courtyard as well as the ventilation and daylight conduits.

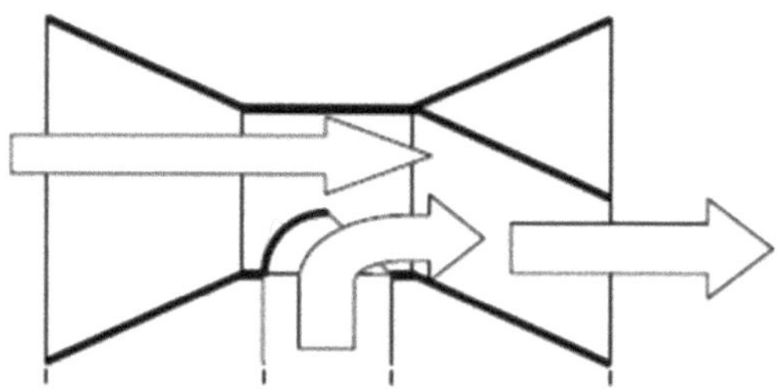

FIGURE 8.31 Schematic of the Venturi extraction device on top of the natural hygienic ventilation conduit.

the bathroom located between them, connected to outdoors by a vertical conduit for hygienic ventilation and a daylight conduit[16].

The ventilation conduit is connected at its upper part to the roof by a Venturi device which promotes the air extraction for the hygienic ventilation of the bathroom. The daylight conduit catches diffuse light from outside and carries it indoors via reflective internal surfaces. In this way it acts as a complementary light source for the second level bathroom and the kitchen and the dining room on the first floor of the duplex apartment.

The use of these specialized conduits for hygienic ventilation and complementary daylight allow high land use without affecting the indoor environment. Their advantage over the traditional small yards or 'wind boxes' is that, despite occupying less space, they provide higher privacy because their subdivision and

16. See González Couret, D., "Daylight and Ventilation Conduits for Housing in Compact Urban Areas", *World Renewable Energy Congress,* Cologne, 2002.

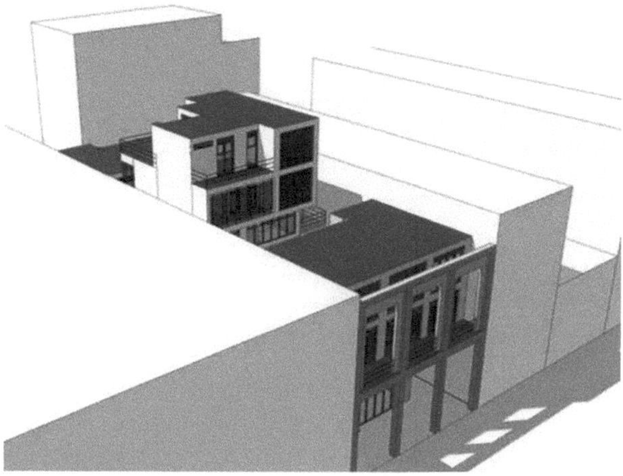

FIGURE 8.32 **Volumetric solution for the new housing building in Havana Center.**

FIGURE 8.33 **Relationship indoors – outdoors.**

specialization (daylight and ventilation separated) do not allow the noise and smell transmission and also are not directly visible from the outside.

The project presented to the Student's International Design Competition Eco-house 2007 'Eco-house in Havana Center'[17] (Figures 8.32, 8.33 and 8.34) was intended to show that sustainable housing should also contribute to the rehabilitation and preservation of the historic city and its patrimonial values, while at the same time making good use of urban land and guaranteeing an appropriate indoor environment for its inhabitants.

The new housing building is inserted in three connected empty plots on an existing block in Havana Center, a central, compact, run down, urban area. Applying the results of former research[18], the built volumes are two spans in

17. Students: D. Gelabert, C. García and H. Milián. Professor: D. González Couret.
18. González Couret, D., et. al., "Vivienda apropiada para ciudad de La Habana", Editorial ISPJAE, Havana, 2005.

FIGURE 8.34 Spatial solution and materials.

depth, in order to guarantee a direct relationship between all indoor and outdoor spaces, and be separated from each other by yards that are big enough to allow enough daylight to enter the center of the interior spaces located on the ground floor. The yards' depth depends on to the buildings' height, which is variable, in order to stimulate turbulence, and so enhance ventilation.

The principles of spatial continuity, permeability and solar and rainfall protection are again applied here. Movable blind windows, overhangs, open ground floors, terraces and balconies are other architectural elements used in bioclimatic design for the warm and humid climate. The use of ceramic hollow blocks for building walls, flats and roofs reduces the thermal transference by the envelope, and several recycled elements, sourced from the demolition of nearby buildings, are utilized. The solution is complemented by rain water collection and urban agriculture based on the use of permaculture techniques.

Among other projects developed for semi-compact urban zones in Havana is the Apartment building in Vibora Park (2007–2010)[19] (Figures 8.35, 8.36, 8.37, 8.38 and 8.39), located in a large plot in the corner of the block. In order to take as much advantage as possible of the land, while keeping the direct relationship between the indoor and outdoor spaces, the project comprised four volumes (two spans in depth) articulated in their corners and surrounding an atrium. The open ground floor in the buildings is facing the streets and the aperture on the articulate corners stimulates air movement and therefore, ventilation.

Lower volumes are located to the interior of the plot, to make a transition to the rest of the buildings in the context and their roofs are tilted to facilitate water collection. The taller volumes placed to the streets have flat roofs to be used for social and service functions. Rain water is collected in tanks located on the internal square, to be used for car washing in the semi-underground parking area, which is illuminated by skylights which also form benches in the square.

This building also has maximum window area, closed by movable blinds, and protected from rainfall and solar radiation by overhangs and balconies.

19. Project team: D. González Couret, A. Zorrilla, D. Gelabert, P. Rodríguez, C. García, C. Fernándes de Aballí, G. García Pol, L. González.

FIGURE 8.35 **Top view of the new housing building with solar hot water systems on each humid nucleus.**

FIGURE 8.36 **View of the building from the corner.**

FIGURE 8.37 **View from Kessel Street to the Eastern building façade.**

Other shading devices include a double skin which complements the solar protection according to the orientation, and solar clothes dryers are placed to close the service yard, oriented to the south and the west.

Installation for water supply and evacuation are concentrated in vertical and registerable conduits to facilitate solar hot water supply, using the compact solar collectors currently produced in Cuba, without pumping energy. The decentralized

FIGURE 8.38 **View from Porvenir Street to the Northern building façade.**

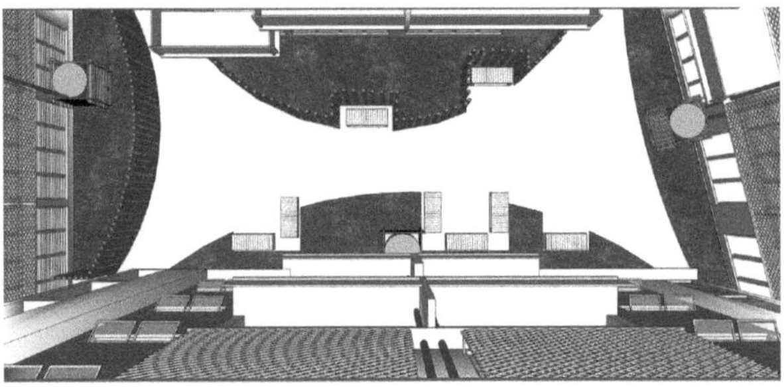

FIGURE 8.39 **Top view of the internal square with the rainwater tanks and the benches/skylight.**

use of these collectors in a thermosyphonic system needs the general water tank to be located in an elevated position, which generates technical volumes on each sanitary nucleus.

A final example of apartment building in Old Havana[20] comprises three volumes connected by staircases, surrounding a courtyard in order to guarantee natural ventilation and daylight (as shown in Figure 6.40). Solar protection is provided by a double skin (Figure 6.41) and some ecological systems are incorporated (Figure 6.42) such as solar water heating, rainwater collection and water reuse.

8.8 FINAL REMARKS

Energy consumption in housing and carbon dioxide emissions are not very high in poor tropical countries such as Cuba, where it is possible to live in the natural environment, and so the houses are not usually artificially air-conditioned.

However, as human thermal comfort is difficult to achieve by passive means in warm humid climates, attention in building design should be focused on the

20. Student: Arnaldo Hernández. Professor: D. González Couret, 2009.

FIGURE 8.40 Articulates volume to the courtyard.

FIGURE 8.41 Double skin for solar protection.

maximum reduction of solar gain, and in provision of as much daylight and ventilation as possible.

Minimum energy housing is therefore relatively easily to achieve during the useful life of the building, simply by taking into account some bioclimatic strategies and developing specific projects which aim to make best possible use of the local conditions. It is possible to develop sustainable buildings using local

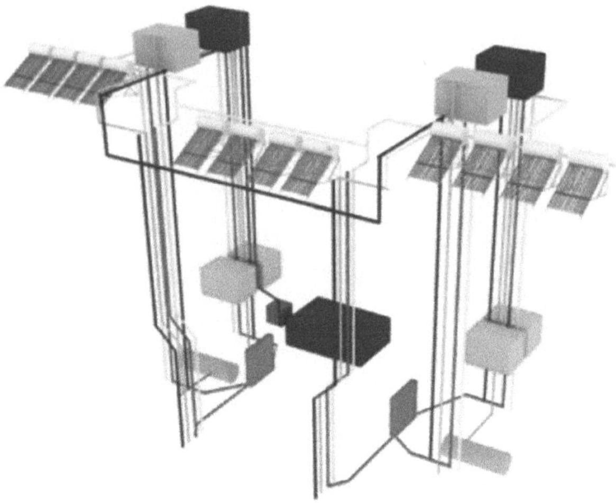

FIGURE 8.42 Water systems in the building.

building materials and appropriate technologies. However, continued research is needed to clarify the influence of urban morphology on architectural performance, and the combined effect of both on human comfort in those climatic regions.

Traditional strategies of bioclimatic design for warm and humid climates are not very abundant, and other contemporary cooling technologies are expensive. Because the countries located in these regions are commonly developing ones, with constrained budgets, the application of these technologies, use of renewable energies and other eco-techniques will require financial support. This could be sourced from cross subsidies, international collaboration, and/or taxation stimulation policies.

Chapter 9

Daylighting

Helmut F.O. Müeller
Environmental Architecture, Department of Building, University of Dortmund, Germany

Chapter Outline

9.1 INTRODUCTION

Daylight is necessary for the comfort of human beings, who often occupy buildings for most of their lifetime. We have adapted through evolution to natural light and therefore we need it as a basic condition in our built environment during the daytime. Its dynamic change is the stimulus for our daily biorhythms [1], and

Sustainability, Energy and Architecture. ISBN: 9780123972699

it can influence our mood and health. Daylight defines the range of brightness and color composition for our vision, it gives us orientation in time and space, and it is precondition for our perception and evaluation of the built environment.

In addition to comfort and visual performance, daylighting contributes to sustainability. It reduces electricity consumption by artificial lighting and the heat gains by lighting during daytime. Thus an overall reduction in energy consumption will be achieved if daylighting is appropriately integrated into the design process.

The means of daylighting cannot be isolated to a single building component – i.e., windows – but all aspects of light transmission between the indoor and outdoor environment need to be taken into account. Therefore daylighting is a matter of integrated design of the building area, building, rooms, components, building services and controls.

Urban design defines the external lighting conditions. Building design concerns the size and location of rooms and their light openings. Details of components, including room surfaces, glazing and shading modulate the characteristics of a room's illumination. Artificial light has to assist daylight when it is not sufficient, and replaces it at night, opening a wide field of lighting design. Control devices are necessary for natural and artificial light, for heat gains of light sources, as well as to regulate the energy consumption of lighting, heating and cooling.

There are no standard solutions for good daylighting, as the conditions such as geographic and climatic situation (clear/covered sky conditions), type of building (residential, office, school, etc.), performance and cost requirements, as well as regional technologies and standards have to be considered. For each task there exists a great variety of solutions, and technological developments are changing continuously.

The aim of this chapter is to discuss the basic principles of daylighting and guidelines for an efficient design approach. It starts by defining boundary conditions and performance requirements. The following steps of the work flow are closely connected to the building design, moving from general and comprehensive concepts (building area, building) to increasingly practical and detailed layouts (rooms, components). In order to find the best possible solution, a large variety of possibilities have to be generated, which are then reduced by assessment. Principal solutions, advanced developments and built examples are presented, together with tools for daylighting evaluation. References and hints for more detailed information are presented at the end.

9.2 CHARACTERISTICS AND AVAILABILITY

The nature of light is described as a wave-particle duality (Albert Einstein), as it has both a particle nature and a wave nature. Various experiments with subatomic photons bring out one or the other. This uncertainty, which is not consistent with our usual pattern of thought, is deepened by findings of other physicists, including Werner Heisenberg´s uncertainty principle in quantum mechanics and Albert Einstein's theory of relativity, which states that the speed

FIGURE 9.1 Illumination of a Chinese temple by daylight. Incense from joss sticks contributes to the light distribution.

of light in vacuum is a constant, as the maximum speed correlated with the variable quantities of time and space.

Our practical experience with light is mainly influenced by our visual perception, but the theoretical findings contribute to a deeper understanding of the observed phenomena. We can see light when it is emitted or reflected by the surface of a body. The transport of light through space, e.g. through air, is not visible to us. Figure 9.1 demonstrates how complex the distribution of light in our environment can be, and the way that the illumination of architecture creates an overwhelming abundance of impressions and feelings.

9.3 PHOTOMETRIC UNITS

The human eye is sensitive to light in the wavelength ranges between approximately 380 nm (violet) and 780 nm (red), which happens to be the band of solar radiation with the highest intensity. Visible light accounts for about 65% of total solar radiation reaching the ground (global radiation); the remainder being ultraviolet (UV) and infrared (IR) (Figure 9.2). The maximum light sensitivity of the eye decreases from a maximum at about 560 nm (green) to the boundary values given above. Because of this variable response of the eye to different wavelengths, the following SI photometry units are used to measure light (Table 9.1).

Luminance is the characteristic for measuring the image seen of an illuminated surface or room. It is used to assess illumination quality by brightness, darkness and contrast of areas seen. Equipment for point measurements may be helpful, but luminance cameras (High Dynamic Range Photographs) give a more comprehensive survey [2] (see Figure 9.3).

Illuminance describes the light incident on a surface, and often is used for standards (e.g. minimum Ev on horizontal areas (desk level) or on vertical areas (walls in museums)). It cannot characterize an image of illumination, which is influenced not only by the amount of incident light on a surface but by reflected light as well.

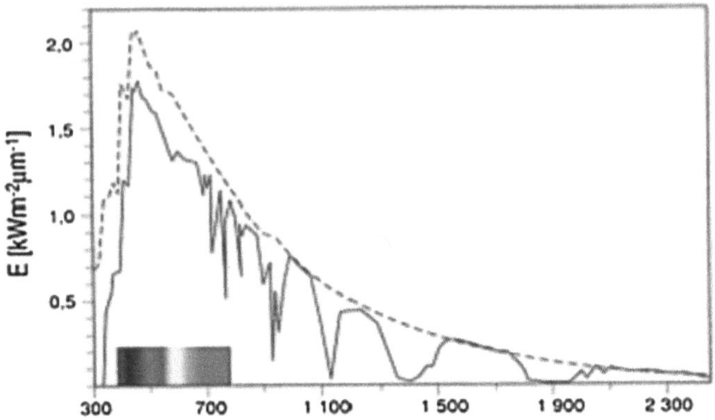

FIGURE 9.2 Solar radiation E [kWh/m²] over wavelength [nm]. Dotted line: Extraterrestrial. Straight line: Terrestrial (on the ground). Spectral colors of light.

TABLE 9.1 SI Photometry Units

Quantity	Symbol	SI unit	Abbreviation	Notes
Luminous flux	F	lumen(= cd.sr)	lm	also called luminous power
Luminous intensity	I_v	candela(= lm/sr)	cd	a SI base unit
Luminance	I_v	candela per square meter	cd/m²	used for light seen on a surface
Illuminance	E_v	lux(= lm/m²)	lx	used for light incident on a surface
Luminous efficiency	------	lumen per Watt	lm/W	used for light sources

The luminous efficacy is used to evaluate the energy demand and sustainability of light sources. Incandescent lamps, for example, deliver only 10–15 lm/W compared to 75–90 lm/W for luminescent lamps and light emitting diodes (LEDs); which is one reason why incandescent lamps are no longer on the market in many countries nowadays. Daylight, which has 90 (sun) to 120 lm/W (clear sky) is much more sustainable because of its low heat gains/cooling loads and lack of any CO_2 emissions.

9.4 COLORS

The perception of color by the human eye is performed by sensors on the retina, called cones, with sensitivity peaks at short (S, 420–440 nm), middle

FIGURE 9.3 Comparison of a conventional photograph and a High Dynamic Range Photograph, showing the luminance distribution in false colors (museum room).

(M, 530–540 nm), and long (L, 560–580 nm) wavelengths. If the light is not bright enough and the stimulus for the cones too small, another type of sensor, called rods, takes over, which results in vision without color – i.e., in black and white. That's why we can see colors only in the range of 10^{-2} to 10^5 cd/m², while at lower luminances down to 10^{-6} cd/m² we see only black and white images.

Because color perception differs from person to person, an objective definition of color is necessary for lighting design and for color reproduction. The CIE 1931 color space chromaticity diagram is used for this purpose, based on the red, green and blue sensitivity peaks of the cones [3]. An alternative way of defining colors, by their Black Body Temperature in degrees Kelvin [K] is also shown in the diagram of Figure 9.4.

9.5 DAYLIGHT AVAILABILITY

The source of daylight is, of course, the sun. Its incident light at the earth's surface is influenced by daily and annual cycles of movement, and the degree to which the light is diffused by atmospheric conditions. Continuous changes in intensity and color composition, influenced by solar position as well as clear or cloudy sky conditions, are typical of the daylight of our natural environment. Maximum horizontal illuminances exceeding 100 klx under clear sky conditions (Figure 9.5), and mean values are found to be about 10 klx under overcast sky conditions (Figure 9.6). Depending on the climate zone, either clear skies (e.g. hot arid climate) or overcast skies (e.g. moderate climate) predominate, and this is decisive for daylighting.

The daylight illuminating a given point in a room may reach it in the following ways (Figure 9.7):

- Direct sunlight travelling in a straight line from the sun through a window to the given point.
- Diffused light from overcast sky coming through a window.

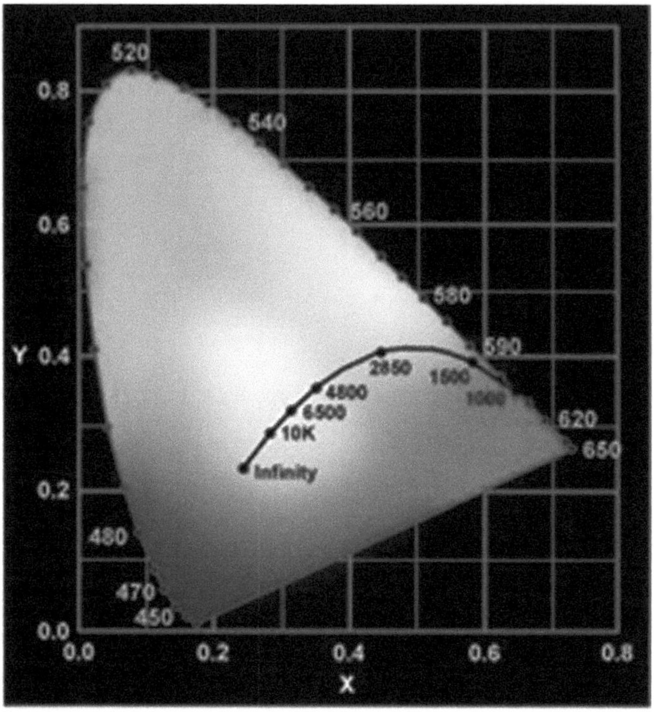

FIGURE 9.4 CIE Color Space [4]. The outer curved boundary is the spectral (or monochromatic) locus, with wavelengths shown in nanometers. The Black Body Temperatures of colors in degrees Kelvin are given on the black curve.

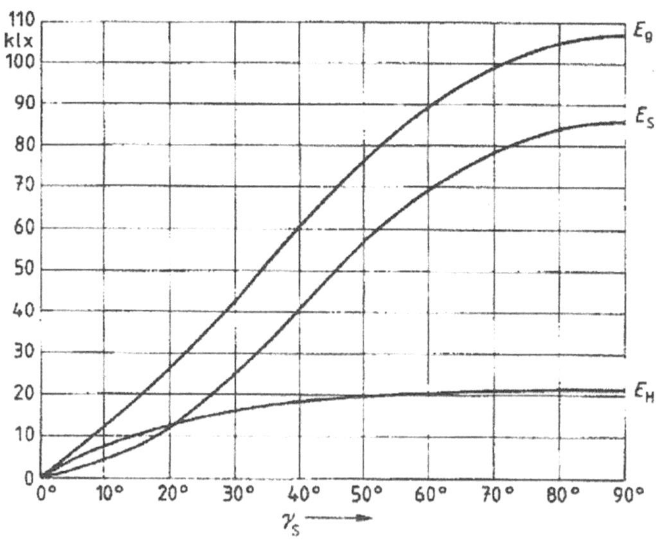

FIGURE 9.5 Horizontal illuminance for clear sky conditions vs. solar altitude. E_S by the sun, E_H by the sky, E_G total illuminance. Total turbidity factor (Linke) TL = 4,9.

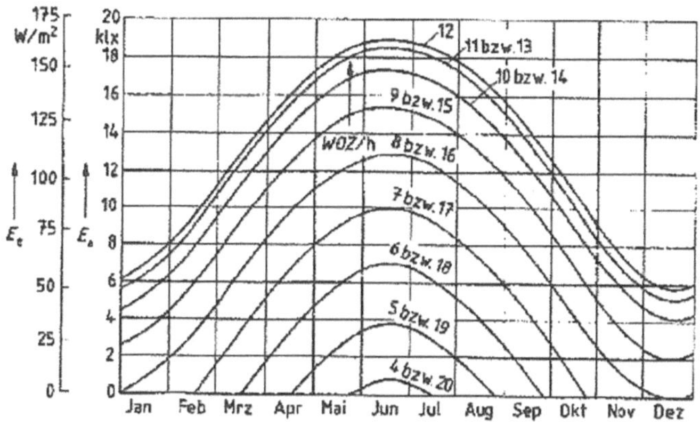

FIGURE 9.6 Horizontal illuminance [klx] and solar radiation [W/m²] for overcast sky conditions at 51° northern latitude vs. time (months, hours), [5].

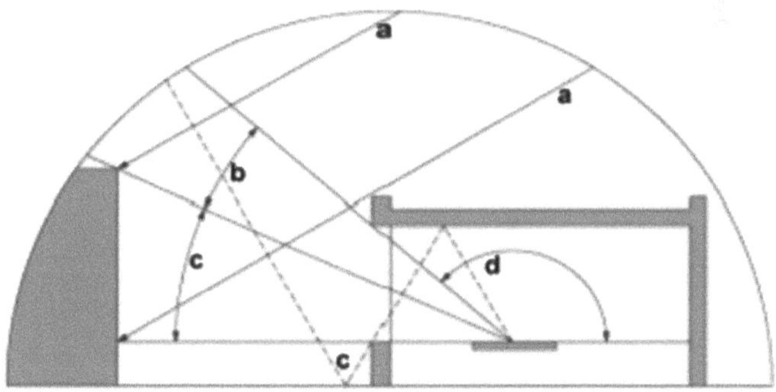

FIGURE 9.7 Daylight entering a building. a: direct sunlight. b: diffused or skylight. c: externally reflected light. d: internally reflected light.

- Externally reflected light (from the ground or other buildings) coming through a window.
- Internally reflected light, travelling from the floor, walls, ceiling or other internal surfaces.

These components of daylight are influenced by climate and environment. The sun has a luminance of 1,650,000,000 cd/m², which is unbearable to the human eye. The luminance of clear sky can vary between 2,000 and 12,000 cd/m², depending on atmospheric scattering effects. Covered (cloudy) sky has values ranging from 1,000 to 6,000 cd/m² depending on the altitude of the sun and the density of the cloud cover. The CIE has defined standard skies for daylight calculations; e.g. the CIE Standard Overcast Sky with a homogeneous luminance distribution from horizon to zenith has a factor of three. Several sources of information on daylight availability exist (e.g. www.meteonorm.com). Levels of externally and internally reflected light are influenced by the reflectivity of the surface.

9.6 PERFORMANCE OF DAYLIGHTING

The performance of daylighting involves many aspects, including comfort and health, visual function, energy use and economy, all of which are of high priority. None can be ignored and therefore daylight cannot be substituted by artificial light completely during the day.

9.7 COMFORT AND HEALTH

Solar radiation and light is necessary for our physiological and psychological wellbeing, as we evolved in an environment controlled by the light of the sun. As we now spend most of our lifetime in buildings, they have to offer, according to their use, sufficient daylight quality, which of course cannot equal outdoor conditions [6]. Buildings also have to protect our bodies appropriately from the harmful effects of solar radiation.

9.7.1 Circadian Effects

The 24-hour cycle of darkness and light is a fundamental need, used by the body to regulate sleep, hunger, body temperature and alertness in conjunction with hormone production. This circadian rhythm is influenced partly by light, which stimulates non-visual photo sensors on the retina of the eye. These sensors differ from cones and rods in having their peak sensitivity at wavelengths in the green-blue part of the spectrum [1]. Illuminances of more than 1000 lux on the retina for more than one hour are necessary in order to trigger the circadian response appropriately.

9.7.2 Seasonal Affective Disorder

The lack of sufficient daylight for long periods, e.g. during winter, can cause Seasonal Affective Disorder (SAD). It occurs predominantly at latitudes above 50°. The symptoms can be relieved if the sufferer is exposed to a bright (artificial) light source, and they tend to disappear in the brighter seasons [7].

9.7.3 Vitamin D

Exposure to sunlight is the natural means by which the human body produces vitamin D. A deficiency in this vitamin causes rickets and poor bone growth.

9.7.4 View Out

Visual contact with the outside world is vital for building users. A view of the outside is necessary for information and orientation, as well as to supply the minimum daylight requirement described above. The location and size of windows should allow for views which include ground, horizon and sky.

9.7.5 Glare

Too much light, which causes glare, should be avoided in buildings. Direct sunlight is intolerable to our eyes. Glare from reflected sunlight and from high contrasts in the field of view is classified as Disability and Discomfort Glare according to Hopkinson [8]. However, there are no satisfactory models to predict and evaluate glare from daylight. Some attempts include the Daylight Glare Index, DGI, Unified Glare Rating, UGR, [9], and maximum contrast ratio recommended by IES [10].

A new glare rating formula, the Daylight Glare Probability, DGP, by Wienold and Christoffersen, [11], which is based on the vertical eye illuminance as well as the glare source luminance, its solid angle and its position index, shows a very strong correlation with the user's response regarding glare. For tasks involving computer screens, Gall, Vandahl et al. have produced recommendations for maximum background luminances to avoid glare from reflections off the screen [12]. Depending on the type and class of screen, they recommend 1,000 to 2,000 cd/m^2 for positive polarity (black letters on a white background). Practical procedures on how to identify glare and faults in workplace lighting are described by Tregrenza, [2].

9.7.6 Color

Color rendering of daylight is important for many visual tasks and for a realistic contact with the outside world. Therefore the spectral composition of daylight should not be reduced or changed significantly by glazing and shading systems.

9.8 VISUAL PERFORMANCE

Visual function depends on the specific task and the parameters of luminance, contrast and color rendering. Orientation, e.g., while walking through a circulation zone, requires lower visual performance than accurate work on objects which have small scale detail for long periods of time. The variation in individual visual capability, e.g., its decrease with age, also needs to be taken into consideration.

Basically, visual performance improves with illuminance, and how well the task in hand is lit. Illuminances of 1,000 to 6,000 lux are recommended for high visual performance (Lange, 2002). This only applies to situations where the contrast is good, and produces a predominant direction of light and a resulting shadow. If the light is diffuse and without shadows, our three dimensional perception of objects is very poor.

Compared to these recommended values, minimum task illuminances for lighting in national and international directives, e.g., DIN EN 12464 [13], are much lower (100 to 1000 lux, depending on the type of task). They define the lowest acceptable standard, and take into account technical and financial inputs into artificial lighting. These values should not be taken as optimal or recommended levels.

Daylighting design and technology can realize much higher illuminances to promote optimal visual performance and comfort for many hours of the day, in economical and sustainable ways. Artificial light has to be used at night, and during the day when daylight levels fall below the minimum requirements. Such lighting should of course supply better than standard minimum illuminances for reasons of improved comfort, health and visual function.

9.9 DAYLIGHT FACTOR

It is not useful to define standard illuminances for daylighting, as the available daylight is continuously changing. But in some countries, which have predominantly overcast skies, minimum daylight factors are needed [14]. The values vary with building types and window position (roof, wall).

The daylight factor is defined as the ratio of horizontal indoor to outdoor illumination by daylight under continuously overcast sky conditions, expressed as a percentage. As an example, for a mean outdoor illumination of 10,000 lux, a daylight factor of 5% results in an illuminance of 200 lux at a task height of 80 cm in the middle of the room. The advantage of daylight factors in building regulations is that they provide minimum daylighting standards.

9.10 THERMAL COMFORT AND ENERGY USE

Windows transmit daylight as well as solar heat gains. During heating periods these thermal gains are welcome, but during cooling periods they increase the cooling load and hence the temperature of the room. Therefore windows need light (glare) control as well as solar (thermal) control, especially for sunny, clear sky conditions. These are defined in many directives by reduction factors (%) of light transmission and solar heat gain. Control of glare and heat gains should not reduce daylight levels. This is an issue of comfort and energy use and a challenge for integrated building design.

The energy used by artificial lighting will be decreased by sufficient daylighting, in combination with efficient lighting controls. It can be described in terms of kWh/(m^2a) for lighting electricity and/or by daylight autonomy (sufficient daylighting of annual working hours as a percentage).

Additionally the energy needed for heating and cooling is influenced by solar and lighting heat gains. Optimized daylighting and solar control systems will contribute to a significant decrease in energy use, i.e. running costs and CO_2 emissions.

We can conclude from this discussion that improved daylighting increases the quality, economy and sustainability of buildings.

9.11 DAYLIGHTING DESIGN

9.11.1 Urban Design

The daylighting of rooms is not only influenced by building and window design, but by urban design as well. The distance and height of neighboring buildings influences the amount of direct and diffuse daylight reaching the windows. The

TABLE 9.2 Means of Urban Design (b: Diffused or Skylight, h: Height, w: Width)

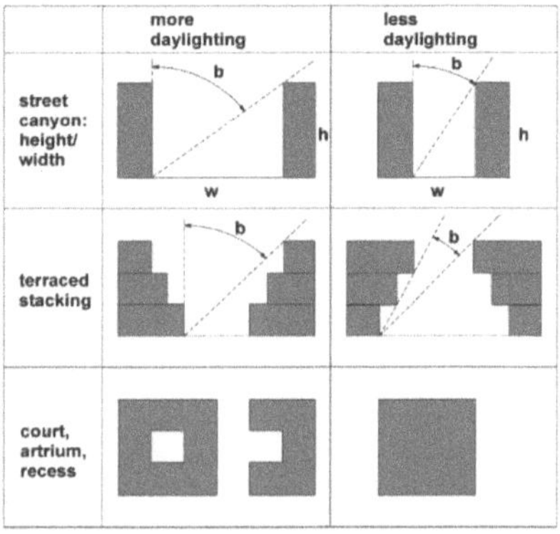

smaller the sky section that can be seen from a room through the window, the smaller the incidence of daylight will be (Table 9.2).

The orientation of the main window walls is also a significant design parameter. Polar-side orientation means that windows will only receive direct sunlight between the spring and autumn equinoxes, and from relatively low altitudes. Equatorial-side windows will recieve direct sunlight throughout all seasons, except in polar latitudes. East or west facing windows receive less direct light than equatorial facing ones, but in low latitudes they receive more light than equatorial-side windows. Thus east-west oriented streets with equatorial and polar-side window walls are preferable in many situations.

The ratio of building height to the width of the street canyon is another important design parameter for daylighting (and thermal heat gain). Climate zone, latitude and orientation have to be taken into account, as well as the reflectivity of walls and ground surfaces. Additionally the shadows of trees can influence the availability of daylight.

The definition of minimum sunlight hours per month for a dwelling or task zone may be helpful for urban design. In Germany, for example, there is a recommendation that dwellings [15] should receive sunlight on one window of a living zone for at least one hour in January (winter). But many urban structures exist which do not fulfill this requirement for daylight comfort and health (see Figure 9.8).

9.11.2 Building and Room Design

Buildings and their subdivisions (storeys, zones and rooms) receive daylight through openings in the building envelope. Single storey buildings and roof floors can be illuminated via windows (vertical openings in the walls/façades)

FIGURE 9.8 Narrow street canyons reduce daylighting (left: Shenzen, China, right: New York, USA).

and skylights in the roof, which offer a high degree of design freedom. In the following design recommendations, critical daylight availability under an overcast sky has always been considered.

Stacked storeys can only be illuminated via vertical windows. This reduces the potential design solutions for a given room depth, depending on the window (and storey) height. There is a rule of thumb that recommends the following room depth for adequate daylighting (daylight factor 1–2%) under overcast sky conditions:

- Room depth = 2.5 x window height above desktop for one window wall.
- Room depth = 5 x window height above desktop for windows in two opposite walls.

An office room, for example, 3 m high, with single windows 2.1 m high, will provide adequate daylight for a depth of 4.5 m. The deeper the room is, the smaller will be the section of the sky seen from its middle (Table 9.3). If the incident light comes from a lower part of the sky, it will have a lower luminance (ratio from zenith to horizon approximately 3/1 for continuously overcast conditions). Courtyards, atria or recesses of the façade can be incorporated in the design of the building to increase the depth to which daylight penetrates it (Table 9.2). Lighter colored room surfaces will result in more reflected light and higher daylight levels. Black surfaces are the most unfavorable.

9.11.3 Window and Skylight Design

The illuminance of rooms can be influenced by their size, location, orientation and shape of the openings. Skylights in a horizontal orientation are the most

TABLE 9.3 Means of Room and Window Design (h: Window Height above Desktop)

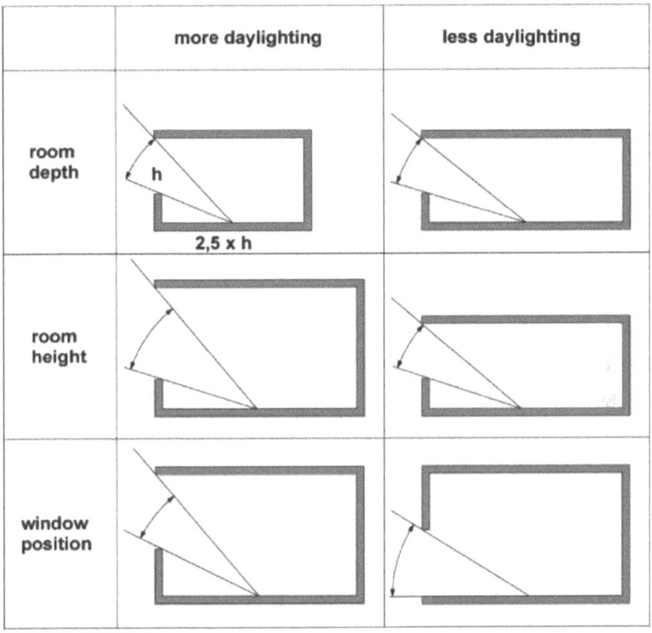

efficient for daylighting, as Figure 9.9 shows. This is because they face the full sky's hemisphere of 180°, receiving relatively high luminance from its zenith. For the standard room in Figure 9.9, the window area of a skylight need only be 20% of the floor area for a daylight factor of 5%, compared to 50–60% for vertical openings in roofs or walls. Because of their high daylighting efficiency, horizontal skylights should be used predominantly whenever possible. However, as they cannot provide an adequate view out, windows at eye level also need to be located in the walls.

For vertical windows in walls, the right mix of orientation to the sun is important. For thick walls, the sides of the window openings should be slanted, with the inner opening being larger than the outer one. Surfaces of the reveal and sill of the window should be reflective (light colors). The higher windows are located in the wall, the further into its center will the room be illuminated. Window openings below desk level do not contribute to room illumination, but may be important for the view from high rise.

The window area should be sufficient for daylighting, however, fully glazed façades are not the optimum. Glazed façade areas of more than 40 to 60% do

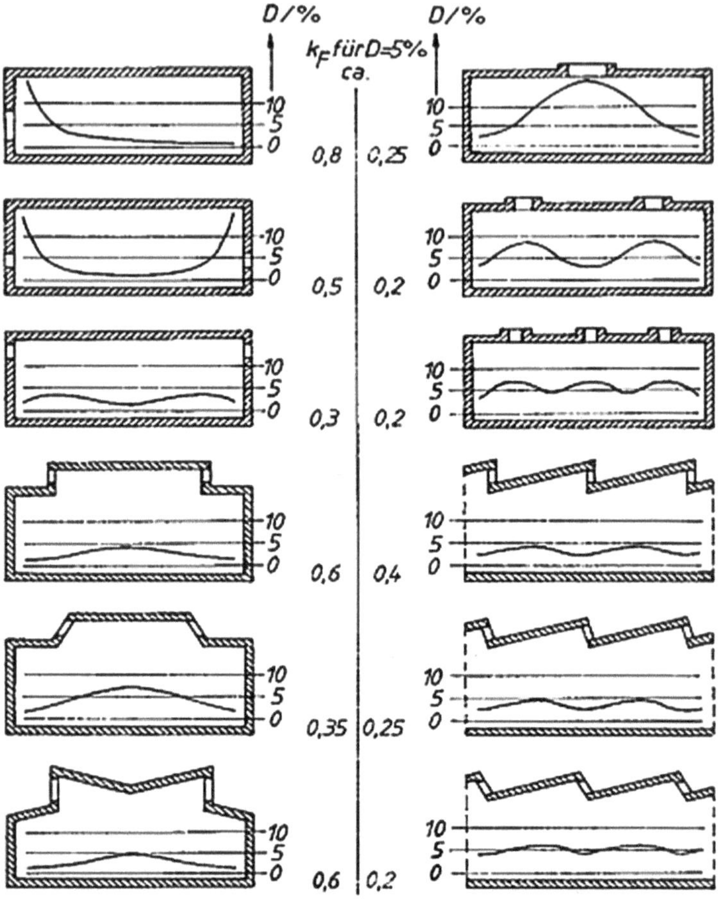

FIGURE 9.9 Influence of windows and skylights on the daylight factor D. Identical room-dimensions. Ratio of window area to floor area (KF) for mean daylight factor of 5% varies from 0.2 to 0.8. [16].

not improve daylighting and tend to create glare and solar heat gain problems (compare Figure 9.24).

Clerestory windows are placed in the upper part of walls above eye level in order to improve the illumination of the room's center. They can be tailor-made for daylighting performance, as a view of the outside is offered by windows in lower positions, which can be completely separate. Often, light shelves or other light directing devices are integrated into clerestory windows (Figure 9.10).

The light transmittance of skylights and windows is influenced by the type of glazing and the area of casement frames. Although additional requirements like thermal insulation and solar control have also to be fulfilled by the glazing,

FIGURE 9.10 Separate clerestory windows above eye level with special glass for redirection of sunlight deep into the room. Duesseldorfer Hypothekenbank, Germany, Koch Architects.

a high light transmittance and true color rendering are indispensable for daylighting.

9.12 DAYLIGHTING SYSTEMS AND SOLAR CONTROL

The requirements of daylighting and solar control seem to compete with each other, as natural room lighting can result in unwanted solar heat gains and efficient shading can result in darkening and turning on artificial light (see Figure 9.11). This is the case for windows or skylights with solar control systems which cannot be adjusted to deal with varying radiation conditions of clear and overcast sky, such as roof overhangs, fixed lamella and grid systems and conventional louver systems.

FIGURE 9.11 Fixed shading devices darken the room and artificial light is switched on: Vitra, fire station, Weil a.R., Germany, architect: Zaha M. Hadid.

In order to avoid this mismatch, daylighting and solar control should be combined in a compatible way, as is offered by many systems. A systematic survey of systems by IEA (2000) distinguishes the following principles:

- Diffuse skylight transmission
- Direct sunlight redirection
- Light scattering or diffusing
- Light transport.

Solar control glass can reduce daylighting as well. Therefore, selecting glazing with specific characteristics is an important step in daylighting design.

9.12.1 Glazing

The type of glazing used in a window or skylight influences the light transmittance considerably. A single sheet of glass transmits about 90% of the incident light, a double glazing with a low-e coating about 65%, and solar control double glazing from 65% to 30%. In climatic zones with significant heating demand, double or triple glazing with low-e coatings are used, to achieve low heat transmission coefficients (u-values).

In regions with significant cooling demand, double glazed units with low-e coating and external shading devices are very efficient, especially if they are adjustable for good daylighting. In buildings with large window areas or totally glazed façades, solar protective glass is often used in combination with internal glare protection. Body-tinted glass offers rather poor solar protection, as the radiation is mainly absorbed, hence increasing the temperature of the internal glass surfaces. Solar reflective coatings are more efficient. Nowadays, selectively reflective coatings are often applied, which are more effective for daylighting and solar control. These advanced materials predominantly reflect the infrared part of solar radiation, while transmitting the visible spectrum. Typical

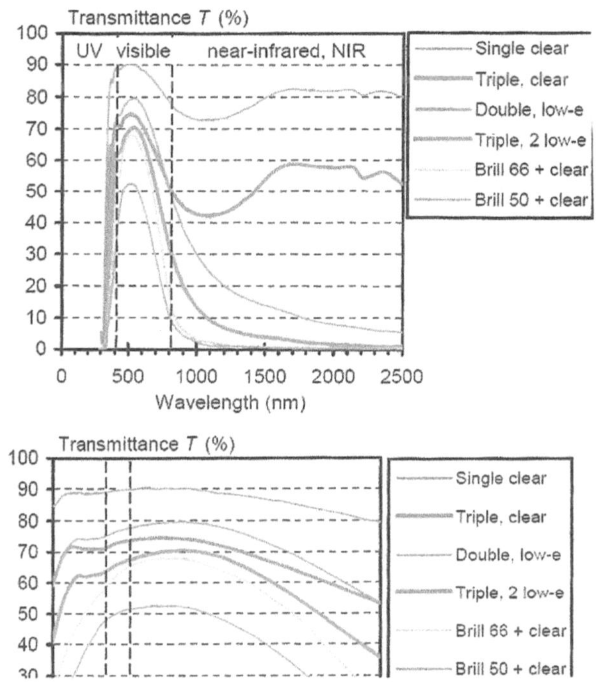

FIGURE 9.12 Spectral transmittance for some common glazing systems for the solar spectrum (above) and the visual spectrum (below), [17]. Ratio of light transmittance to solar heat gain coefficient: Single and triple clear glazing: 89/85 and 72/70. Low-e glazing of two types: Clear + low-e: 76/61. Low-e + clear + low-e: 66/48. Selective solar contol glazing of two types (Pilkington): HP Brilliant 66 + clear: 66/33. HP Brilliant 50 + clear: 50/25. HP Brilliant 50 + clear: 50/25.

ratios of light transmittance to solar heat gain coefficient are e.g. 66% / 33% or 50% / 25%. The reflecting characteristics of these glasses also avoid the typical mirror effect of conventional solar reflective coatings.

All the coatings described do not only change the light transmittance, but also alter the spectral composition of the transmitted light. Helena Buelow-Huebe [17] has shown in detail that coatings particularly filter out the red and the blue regions of the visible spectrum (Figure 9.12). The quality of color rendering is reduced in comparison to clear, uncoated glass and, last not least, the circadian effect by wavelengths from 446 to 477 nm, too.

9.12.2 Diffuse Skylight Transmission

These systems primarily use diffuse skylight for lighting, and reflect direct sunlight to avoid solar heat gain and glare. To gain as much diffuse skylight as possible, only the small area of sky around the sun should be blocked. Transparent lamelles have been developed to do this, by tracking according to the solar altitude, blocking light from just a small section of sky, and

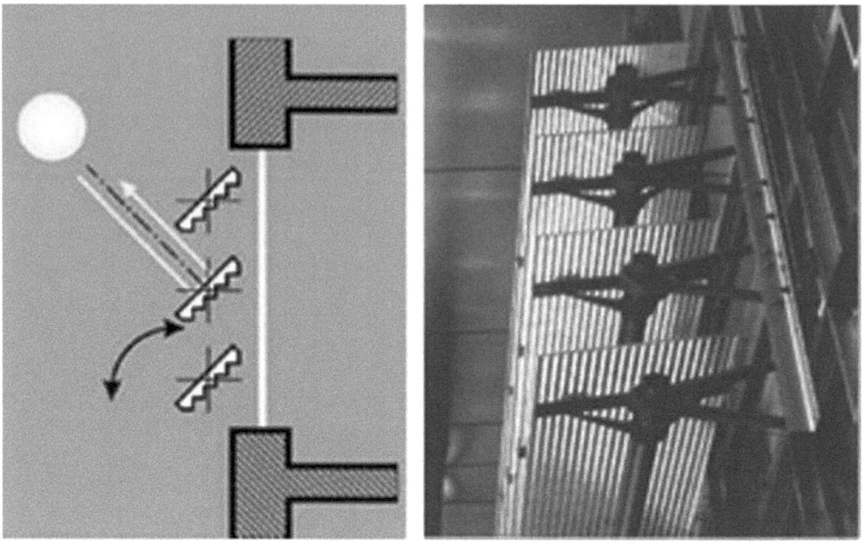

FIGURE 9.13 Prismatic louver system reflecting sunlight and transmitting diffuse skylight. Solar tracking for angle selective reflection [18].

FIGURE 9.14 Angle selective mirror elements reflecting sunlight (red) and transmitting diffuse skylight (white), [18]. Fixed Integration in gap of double glass unit. Application House of German History, Bonn, Germany. Architects: Ingeborg and Hartmut Ruediger.

transmitting light from all the rest. These are high-tech, expensive devices containing prismatic (Figure 9.13) or holographic panels, but simpler solutions are available, which are fixed and consequently mask larger areas of the sky. They are less efficient, and use prismatic panels or mirror elements (Figure 9.14) to reflect the sunlight. With the exception of holographic panels, all such systems cannot be looked through, and hence affect the view out of the building, which is why they are predominantly used in skylights and glass roofs.

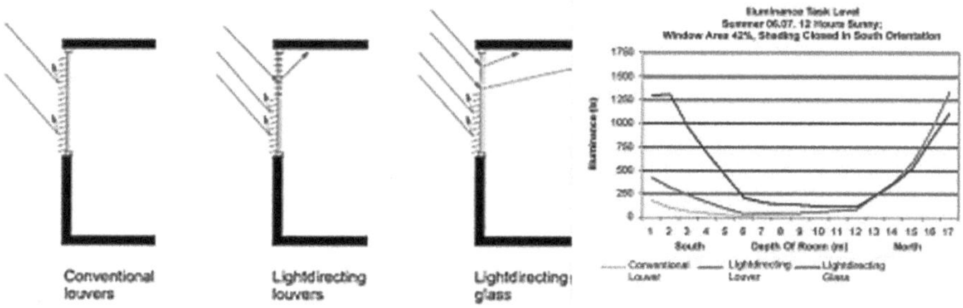

FIGURE 9.15 Comparison of three daylighting systems for office room in Dortmund, Germany (northern lat. 51°). Left: Three daylight and shading solutions: Conventional louvers, light directing louvers, light directing glass in upper and louvers in lower window area. Right: Illuminance (lx) and distance from window (m), south / north orientation, shading device closed in south, opened in north, [19].

9.12.3 Direct Sunlight Redirection

A small part of the transparent area of the building envelope transmits and redirects direct sunlight for lighting purposes, while the remainder is protected by conventional solar control systems. The potential daylight performance is relatively high because of the high luminance of the sun in comparison to the sky. On the other hand, there is a risk of glare from direct sunlight, which has to be avoided under all circumstances. A great variety of systems are available, which can be divided into movable devices (e.g. louvers, solar tracked reflective lamellas) and fixed ones, which redirect the light from all solar positions in the wanted direction (e.g. light shelves, light directing glass).

Figure 9.15 shows that the poor lighting performance of a conventional louver system can be improved by the relatively simple measure of adding light directing lamellas in the upper area of the louvers. A significantly higher increase of daylighting is possible by the use of light directing glass in clerestories.

Light directing louver systems are offered in many variations, with fixed or sun tracking lamellas for daylight transport. In order to avoid glare, only the upper part of the window above eye level should be used for sunlighting. Some louver systems have highly reflective lamellas, especially shaped for retro-reflection and daylighting (Figure 9.16). Often they are located inside, in combination with solar protective glass (spectral selectivity), or in the gap of a low-e double glass unit, to protect their surfaces against external factors, such as dust.

Redirecting systems which have not to be moved are more robust. Light shelves (Figure 9.17) have proved to be effective in climatic zones where clear sky conditions are predominant, but under overcast skies, the shading of the

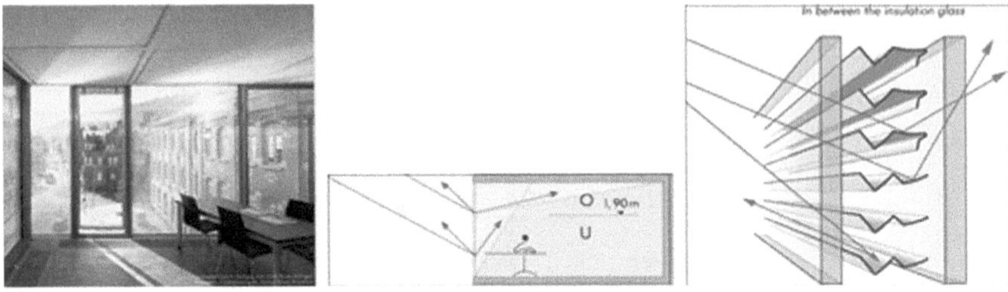

FIGURE 9.16 Louver system with retro-reflection and daylight transport integrated in double glass unit [20].

FIGURE 9.17 Light shelves for daylighting and shading. BP headquarters, Capetown, South Africa, Architects.

lower window area reduces daylighting. Table 9.4 gives an overview of the principal fixed systems for light redirection.

Solutions 2 to 4 were developed for the following boundary conditions: Redirection of direct sunlight in windows with east, west, and equatorial orientation, at solar altitudes up to 65°. Because of reduced radiation intensity and obstructions close to the horizon, a minimum solar altitude of 20° is assumed.

TABLE 9.4 Solution Principals for Redirection of Direct Sunlight in Fixed Components

Physical effect	Principle solution	No.	Drawing	Aspects of selection
Reflection	Diffuse reflector (Lightshelf)	1	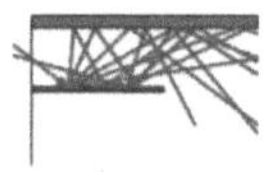	- limited reflection - limited transmission for covered sky conditions - glare for low solar altitudes
	Specular reflector (DLS-Fisch)	2	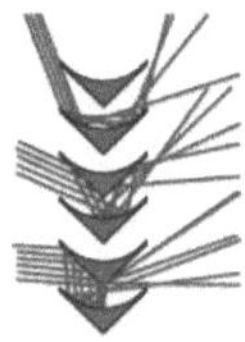	- limited reflection - limited transmission for covered sky conditions - dirt avoidance by enclosure in gap of double glazing - DLS-Fisch, Eckelt
Refraction (internal total reflection)	Light-conductor (Lumitop)	3		- reflection without loss - high transmission - dirt avoidance by enclosure in gap of double glazing - Lumitop, 11 mm wide light conductors in gap
	Pane with two structured surfaces (Lumilight)	4		- reflection without loss - high transmission - dirt avoidance by enclosure in gap of double glazing - Lumilight, 4 mm PMMA pane with micro structures

Most of the light has to be redirected above a horizontal level. The remainder of light, with an emergent angle less than 0°, must be less than 10% to avoid glare. Solutions 2 and 3 have been applied successfully in windows and skylights for more than 20 years (Figure 9.18), but solution 4, the advanced microstructure, has only been tested as a prototype so far (Figure 9.19).

Table 9.5 gives a survey of positioning light directing systems in the building envelope. Side windows and skylights can be used and the redirection can be vertical or horizontal. Even polar oriented windows can get sunlight, if external elements are applied.

FIGURE 9.18 Light directing glass in clerestory of Spherion office building, Duesseldorf, Germany, Deilmann Koch Architects.

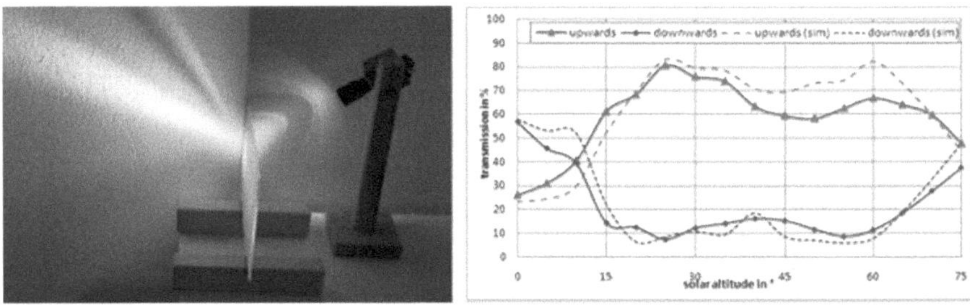

FIGURE 9.19 Panel with two micro structured surfaces for light redirection. Left: Sample, Right: Experimental results (solid lines) and numerical simulation (dashed lines). *Klammt et al. (2011).*

9.12.4 Light Scattering or Diffusing

These systems are used in skylights, glass roofs and windows to produce even daylight distribution by direct sunlight or diffuse skylight. The image of the aperture is white and bright; it cannot be looked through, and serious glare can result in task areas, depending on the light transmission. The scattering effect can be created by the glass (e.g. sandblasted, frosted, fritted, laminated with diffusing film) or by internal blinds or curtains, which can be moved (Figure 9.20).

9.12.5 Light Transport

Sunlight can be collected and transported over long distances into the core of buildings or into underground spaces. This can be achieved via hollow light tubes, lined with a reflective surface, or by fiber optics or acrylic rods. The entrance point has the function of collecting and reflecting the light, and the

TABLE 9.5 Locations for Redirection of Sunlight in Buildings
Window orientations for northern hemisphere, i.e. S = equatorial, N = polar.

Installation location	Redirection	Window orientation	No.	Drawing
Side window	Vertical	E/S/W	1	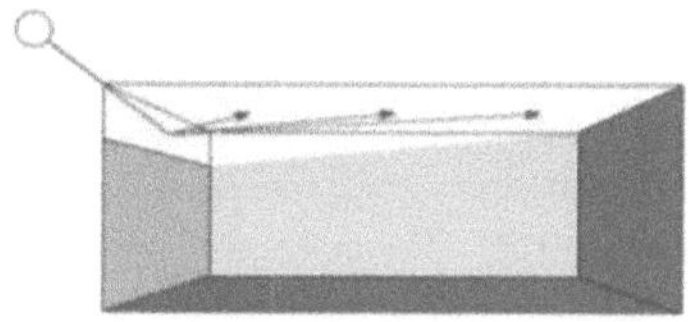
	Horizontal	E/W	2	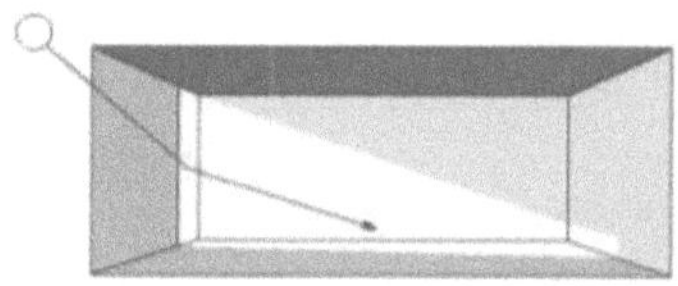
Skylight	Vertical	E/S/W 20° tilt	3	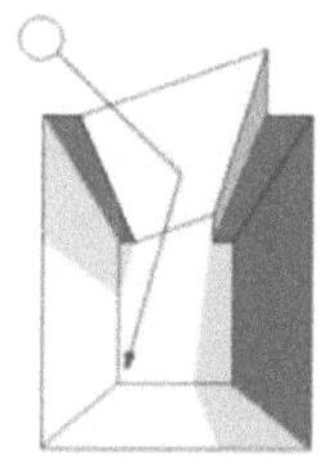
External components	Roof	N	4	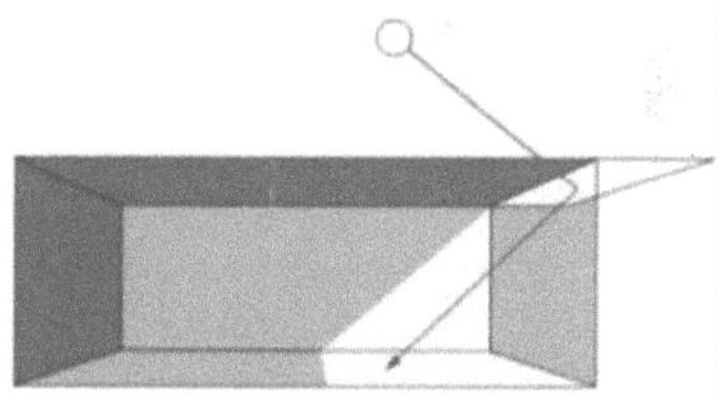
	Wall	N	5	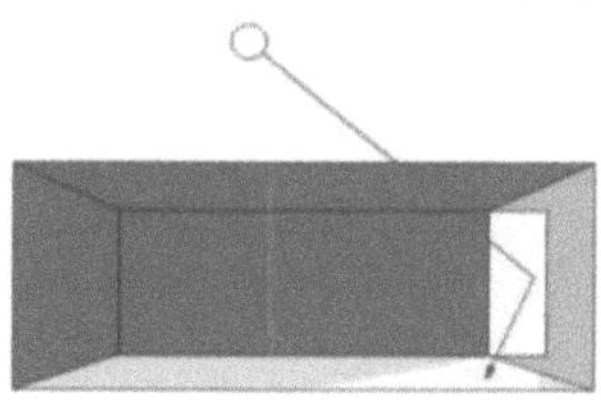

endpoint, that of distributing and diffusing it. Transparent light tubes can also be used for light distribution (Figure 9.21). Heliostats and other solar tracking systems can focus sunlight at the entrance of the light conductor. Systems are also available for adaptively mixing solar and artificial light.

FIGURE 9.20 Examples for light scattering windows. Left: Light diffusing curtains. Folkwang Art School, Essen, Germany, Sanaa Architects. Right: Fritted glass for illumination as well as solar thermal and glare control, Sedac production hall, Gersthofen, Germany. Architect: Thomas Herzog and partners.

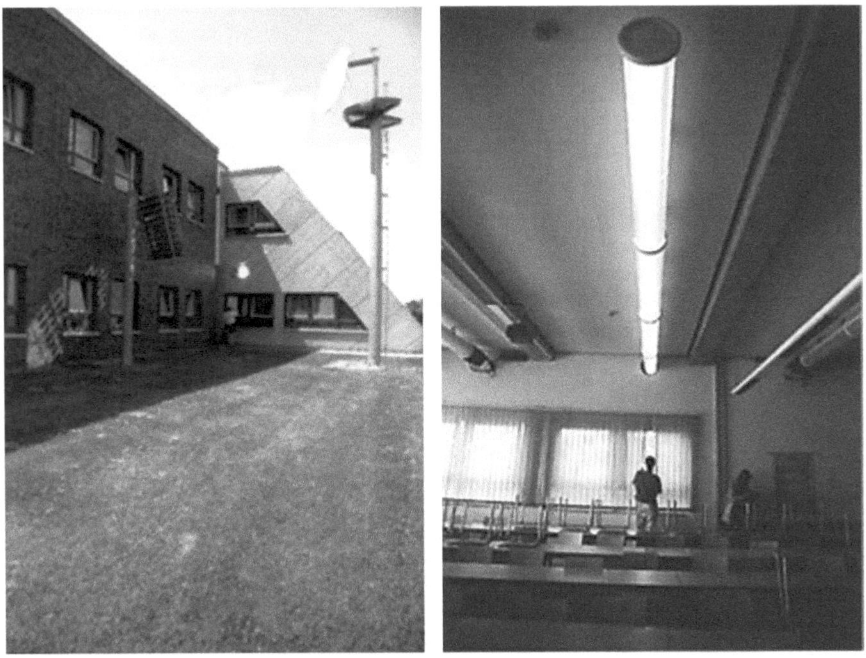

FIGURE 9.21 Heliostat and light tube at Solar Campus Juelich, Fachhochschule Aachen, Germany.

9.13 ENERGY SAVING AND DAYLIGHT RESPONSIVE CONTROLS

Electrical lighting, which has to be used at night and when daylight is not sufficient, contributes significantly to the overall energy consumption and CO_2 footprint of a building. In offices, for example, it can be 30% of the total. Higher levels of lighting for comfort and health increase the influence of daylighting on

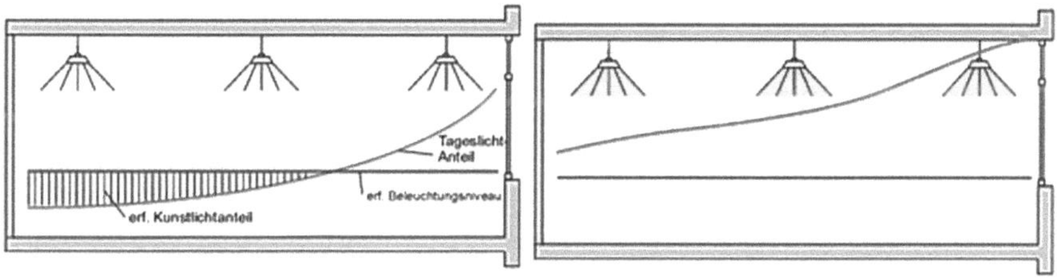

FIGURE 9.22 Switching of light. Left: Light switched off, required illumination (blue), available daylight illumination (red), required composite of artificial light (hatching). Right: Light switched on, required illumination (blue), available illumination (red).

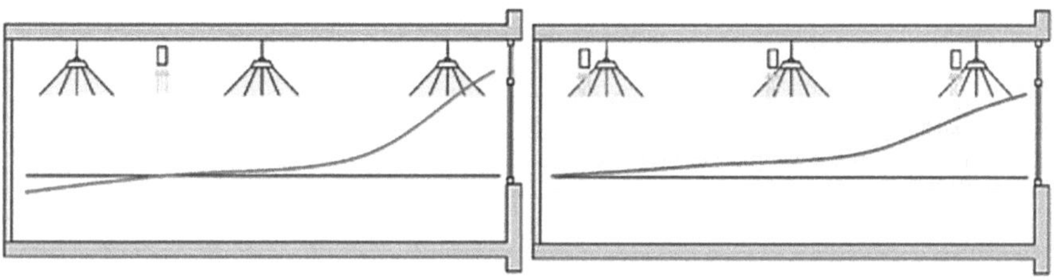

FIGURE 9.23 Dimming of light. Left: One sensor per zone. Right: One sensor per luminaire or row of luminaires.

energy savings. Additionally, the heat gains of daylighting and artificial lighting influence the amount of energy consumed by heating and cooling.

Thus the total energy consumption of buildings is influenced by lighting.

The key to sustainable energy use in buildings is the control of electric lighting and transmission of solar radiation through the openings. In residential buildings, manual control by the user is adequate in most cases, but in offices, schools and commercial buildings, automatic control is recommended, especially as it is economical and state of the art. Individual override should be possible, however.

9.13.1 Lighting Control

Lights can be switched on and off depending on the daylight illuminance and the occupancy of rooms or zones. Figure 9.22 shows how switching can result in an undercut or overshoot of the required illuminance. By the use of dimming, the artificial light levels can be adjusted more accurately to adjust to the available daylight and the required illumination (Figure 9.23).

9.13.2 Window Control

As described above, windows and skylights need to control solar heat gain and glare in ways that can be adapted to climatic and radiation conditions. Direct

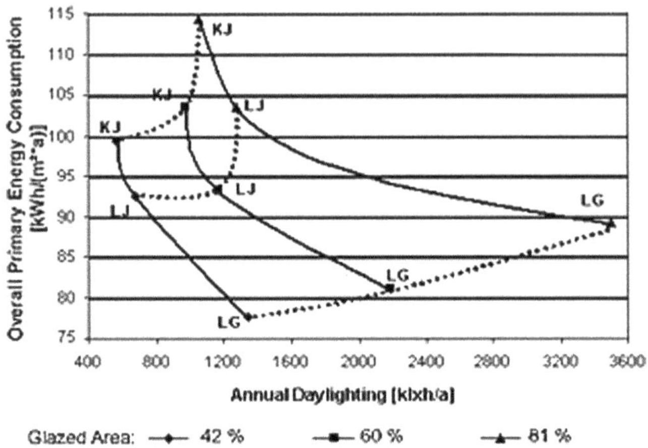

FIGURE 9.24 Overall primary energy consumption for lighting, heating and cooling (kWh/(m²a)) for a south facing office room in Germany over annual daylight illuminance (klxh/a) for three shading / daylighting systems (KJ standard louvers, LJ light directing louvers, LG light directing glass) and varying window areas [19].

sunlight causes glare, even if its transmission is reduced to less than 10% by special glass coatings or blinds. Therefore, means of reflection, redirection or absorption have to be applied, which must not be identical with thermal shading devices necessarily. While solar thermal protections are most effective, if they are located outside or in the gap of a low-e-coated double or triple glass unit, glare can be avoided by internal systems as well. In non-residential buildings the solar thermal control should be automatic. Individual override and adjustment for lighting atmosphere, glare control and ability to view out must also be possible.

9.13.3 Energy

Energy consumption by lighting depends on many boundary conditions (climate, type of building) and design parameters (percentage of glazed envelope, orientation, daylighting system, kind of electrical lighting and control). In offices, standard lighting systems have a wattage of 8–10 W/m², while in museums it can be up to 100 W/m².

A study of office rooms in middle Europe shows that, depending on daylighting design parameters, the annual lighting energy consumption can vary between 6 and 13 kWh/(m²a). The influence on overall energy consumption for lighting, heating and cooling and on comfort of daylighting is shown in Figure 9.24. Daylighting systems such as light directing glass can reduce energy consumption by up to 23%, and increase the comfort of annual daylighting simultaneously.

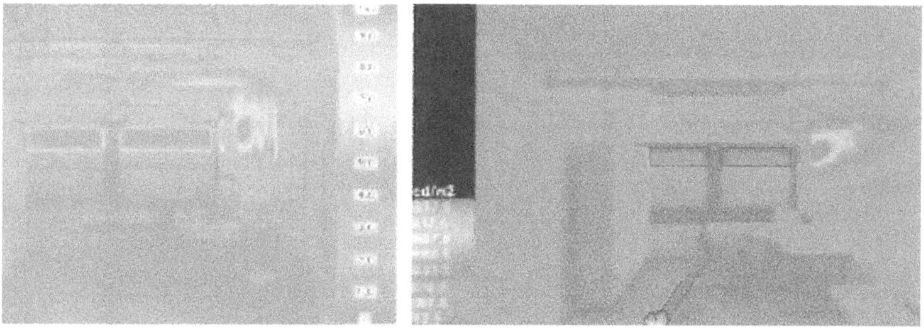

FIGURE 9.25 Comparison of measured and calculated luminance distribution in false colors. Left: Luminance measurement by High Dynamic Range Photograph. Right: Luminance simulation by Radiance.

9.14 DESIGN TOOLS

Several daylight design tools exist for each of the different phases of building design. For the early sketch design, simple graphical tools and rules of thumb can be used (see worksheets by [2]), or computer models for calculating daylight factors for simple room geometries under overcast sky conditions can be applied. Scale models give direct and realistic impressions of daylighting, either under real daylight conditions or an artificial sky. Facilities for creating these exist in many architectural schools and research institutions.

For the design and detailing stage, more sophisticated computer programs are available, which model luminances and illuminances for complicated room geometries, and can include many different surface characteristics in overcast sky and clear sky conditions. Photorealistic light and color renderings can be produced (Figure 9.25).

Two different techniques are used for calculation: ray tracing or radiosity. Ray tracing follows the light rays 'backwards' from the eye or camera, rather than into it (as actual light does in reality). This is an efficient approach, as the majority of light rays from a given light source do not make it directly into the viewer's eye. Radiance is a program based on ray tracing, which is applied in many different models, like Relux or Daysim.

Typical radiosity methods account for light paths which start at a light source and make multiple diffuse bounces before reaching the eye. They link the luminance of a patch of the sky to the illuminance it gives on a surface, and are used in many daylighting calculations together with the daylight coefficient method. Although there are several approaches to integrating other illumination effects such as specular and glossy reflections, radiosity-based methods are generally not used to solve the complete rendering equation.

If annual average daylight levels are needed, rather than just snap-shots of typical situations, a program called Daysim can be used. This calculates the daylight factor, daylight autonomy and the need for electric lighting [21]. It links the detailed simulation of light distribution performed in Radiance with a yearly calculation of interior daylight levels. Finally, it can calculate the annual requirement for lighting electricity under various lighting control strategies.

It would be nice to have a comprehensive program for dynamic light and heat modeling in buildings, which used the same input data for climate and building characteristics as well as occupation and operation profiles.

REFERENCES

[1] G.C. Brainard, et al., Action spectrum for melatonin regulation in human: evidence for a novel circadian photoreceptor, J. Neurosci 21 (2001) 6405–6412.
[2] P. Tregenza, M. Wilson, Daylighting: architecture and lighting design, Routledge, New York, 2011.
[3] CIE, Commission internationale de l'Eclairage proceedings, Cambridge University Press, Cambridge, 1931.
[4] DIN 50332, CIE standard colorimetric observers. International compliance: ISO/CIE 10527, CIE 527 (1972) 1991.
[5] DIN 5034-2, Daylight in Interiors, Part 2: Principles, Beuth Verlag, Berlin, 1985.
[6] M.S. Rea, Sustainable Lighting: Let's talk about Value. Proceedings of CISBAT 2011, EPFL, Lausanne, CH, 2011.
[7] P. Boyce, Human Factors in Lighting, second ed., Taylor & Francis, London, 2003, 584.
[8] R.G. Hopkinson, Glare from daylighting in buildings. Applied Ergonomics (3): 206–215. IEA International Energy Agency, SHC, Task 21, 2000: Daylight in Buildings, A Source Book on Daylighting Systems and Components, http://www.iea-shc.org/task21.
[9] CIE-117, Discomfort glare in interior lighting, Austria: Bureau Central de la Commission Internationale de L'Éclairage, Vienna, 1995.
[10] IES, Lighting Handbook: Reference and Application, eighth ed., Illuminating Engineering Society of North America, New York, New York, 1993a.
IES, American National Standard Practice for Office Lighting. ANSI/IESNA RP-1–1993, Illuminating Engineering Society of North America, New York, New York, 1993b.
[11] J. Wienold, J. Christoffersen, Evaluation methods and development of a new glare prediction model for daylight environments with the use of CCD cameras, Energy and Buildings, 2006. (in press, corrected proof, available online 24 April 2006).
[12] D. Gall, C. Vandahl, W. Jordanowa, S. Jordanowa, Light protection equipments in office rooms, The Shutters and Blind Expert. Magazine for Building Details, Kleffmannverlag Bochum, 2000.
[13] DIN EN 12464–1, Light and lighting – Lighting of work places – Part 1: Indoor workplaces, Beuth Verlag, Berlin, 2011.
[14] DIN 5034-3, Daylight in Interiors, Part 3: Calculation, Beuth Verlag, Berlin, 2007.
[15] DIN 5034-1, Daylight in Interiors, Part 1: General requirements, Beuth Verlag, Berlin, 2011.
[16] U. Ficher, Tageslichttechnik, Verlagsgesellschaft Rudolf Mueller, Cologne, 1985.
[17] H. Buelow-Huebe, Daylight in glazed office buildings, 2008, http://www.ebd.lth.se/fileadmin/energi_byggnadsdesign/images/Publikationer/Report_EBD_R_08_17_dagsljus.pdf.
[18] Siteco, Tageslichtsyseme, 2011, http://www.siteco.de/de/produkte/tageslichtsysteme.html.
[19] H.F.O. Mueller, et al., Sonnenschutz und Tageslicht in Bueroraeumen, in: Bauphysik-Kalender 2005, Ernst & Sohn, Berlin, 2005.

[20] H. Koester, Dynamic Daylighting Architecture, Birkhaeuser Architektur, 2004.

[21] C.F. Reinhart, Tutorial on the Use of Daysim Simulations for Sustainable Design, National Research Council Canada, Ottawa, 2005.

[22] IESNA, Reprint 1995. Lighting Handbook, in: M.S. Rea (Ed.), eighth ed., Illumination Engineering Society of North America, New York, 1993.

[23] H. Lange (Ed.), Handbuch für Beleuchtung, Ecomed Verlagsgesellschaft, Landsberg am Lech, Germany, 2002.

[24] C.F. Reinhart, Lightswitch 2002: A model for manual control of electric lighting and blinds, Solar Energy 77 (1) (2004) 15–28.

[25] C. Vandahl, D. Gall, M. Scholz, V. Dockhorn, Physio-psychologische Aspekte der Operationsfeldbeleuchtung, Zeitschrift LICHT 57 (4), (2005), S. 312–315.

Chapter 10

Vernacular Tower Architecture of Sana'a: Theory and Method for Deriving Sustainable Design Guidelines

Khaled A. Al-Sallal
Department of Architectural Engineering, UAE University, UAE

Chapter Outline

Sustainability, Energy and Architecture. ISBN: 9780123972699

10.1 INTRODUCTION

The way of life in the old settlements of Yemen has not changed drastically, and there one can find many vernacular houses and other buildings which are still in use. These vernacular houses are proven to perform better than most modern houses in Yemen with respect to climatic factors and thermal comfort, and they function efficiently to satisfy cultural and domestic needs. The new settlements, on the other hand, have been built according to foreign design patterns, rarely using the native construction methods or materials which have been developed in response to local environmental factors and society's needs.

The vernacular architecture of the historic city of Sana'a follows certain conventions which are well known and practiced by its inhabitants. These conventions are the result of history-long experience and have been developed throughout the life of the city to satisfy human socio-cultural needs and to respond to the requirements of climate and environment. The people's adherence to these conventions has resulted in a consistent architectural language that has produced a very unique identity for Sana'a architecture, and has preserved it for long time (Figure 10.1).

The vernacular tower architecture in Yemen is known for its strong character, deep cultural meaning, and ecological harmony, but today one can observe many cases of mediocre transformations in contemporary architecture. Regrettably, the contemporary architecture has no clear identity and lacks cultural meaning; it is a cheap mix of Yemeni and other styles. It lacks order and unity at all levels: from form/facade design to space design; from selection and use of construction materials to application of construction methods. It is alienated in its society and environment and gives little consideration to human functional and comfort (psychological, physiological) issues. It provides minimum consideration of socio-cultural needs such as privacy and sociability. It is not environmentally sustainable and its reliance on, or integration with, natural resources of energy, vegetation, or water, is minimal.

The vernacular architecture of Yemen is very distinctive; and its most unique characteristics are the slender vertical forms of the tower houses and the attractive designs of its windows. Al-Oulfi [1] has provided a historical background in which the Yemeni architecture is compared and related to other

FIGURE 10.1 **Panorama view of Sana'a.** *(Photograph by Daniel Pini, Courtesy of General Organization for the Preservation of Historic Cities of Yemen, GOPHCY.)*

architectures in the Arab region, and elsewhere in the world. His text analyzed the Yemeni architecture in different locations by giving a comparison of architectural forms and elements. It described the modern movements in Yemeni architecture and discussed the issue of how such architecture can be created in the modern era.

There is a need to understand how the traditional vertical architecture succeeded in keeping a balance between environmental influences and socio-cultural needs, and why modern houses have failed to do so. Learning from the past by exploring design-oriented guidelines from traditional tower architecture is necessary to reveal proven design methods that could be the basis of creating sustainable communities.

10.2 BACKGROUND

10.2.1 Geography and Climate of Yemen

Yemen, in the very south-west corner of the Arabian Peninsula, is situated between latitude 12° 40′ and 17° 26′ north of the equator and longitude 42° 30′ and 46° 31′ east of Greenwich. At the first glance it seems to be the continuation of the desert that exists in most of the Arabian Peninsula. Yet Yemen has a very different climate, mainly because of its altitude. In Yemen, one finds the highest mountains of Arabia, at a height of 3,720 m. Large basins above 2,000 m altitude and surrounded by the high mountains give the possibility of large settlements. Sana'a, one of these settlements, is situated in the geographic center of Yemen, in the middle of a vast plateau 2,500 m above sea level. The climate in Sana'a from May to September is warm and arid during the day and cool and moderately humid at night. From October to April, it is moderate and arid during the day, and cold and moderately humid at night. It experiences large diurnal variations in temperature in both winter and summer. The precipitation occurs mainly in May and August. The wind direction is mainly north-west, with an average speed of 2–4.5 m/s. The annual mean solar radiation on a horizontal surface is 145–225 $W.m^{-2}.day^{-1}$. The mean hours of sunshine are from 8 to 11 hours per day during the rainy months. Heating during the cold season and penetration of solar radiation to the interior are both desirable.

10.2.2 Architecture

Lewcock [2] described the tower houses of old Sana'a as impressing the visitor with their heights. They are usually more than five levels high, and the largest commonly have seven, eight, or even nine levels. A view of the city from a distance, with many hundreds of these houses soaring above the city walls, makes an unforgettable impression (Figure 10.1 and Figure 10.2) The streets of the old city of Sana'a are generally narrow and flanked by towering houses with almost no sight of vegetation or water to relieve the eye (Figure 10.2, right). Yet, this is just a momentary impression when visiting the old city for the first time. As one explores behind these towers, one will find large gardens extending right up to the towers (Figure 10.3).

FIGURE 10.2 Typical views of vernacular tower houses in Sana'a. *(© Khaled Al-Sallal)*

FIGURE 10.3 The traditional gardens in the old city of Sana'a occupy 20% of the city area, left image; 42% of a residential quarter, right image. *(© Khaled Al-Sallal)*

The gardens of the old city of Sana'a occupy 20% of the city's area, and 42% of each residential quarter; thus almost every house has a view through its windows into extensive gardens. These gardens have been an important resource of food for people throughout the long history of the city, as the inhabitants used them to grow their daily supply of vegetables and fruits, and to raise farm animals. The irrigation of these gardens depended on rain water collected and distributed through clever systems of irrigation cisterns and channels that provided equal shares to all inhabitants. In the past, Sana'a's inhabitants always relied on underground water by digging wells down to the water levels; water used to be abundant and available at higher levels than today because of the limited use and smaller population. Unfortunately, the present huge consumption of water has resulted in the water table lowering significantly, and in many springs drying out.

The urban pattern of the city comprises many housing clusters that are similar in the orientation and spatial organization of their elements, but different in shape and size. It follows a hierarchical system of organization that seemed to be very efficient in providing appropriate functional spaces for both public and private zones and the required separation between them. The roads divide the housing zones into quarters, and connect them together by public zones, such as the market and the large mosque. They are usually narrow, and open onto communal spaces used for gatherings and social events. Each housing quarter includes several mosques and is divided into a number of housing clusters, which are grouped around urban spaces. Urban spaces are of two distinctive types [3]. The first, called *Al-Sarhah* or the social square, is for public use and is usually connected to a mosque and used by people to gather together and for social interaction, especially on social or religious occasions. The second, called *Al-Bustan* or the urban garden, is for semi-private use and provides a space for a cluster of neighboring houses to use as a place for gardening, social interaction, and for children to play. The three main elements that shaped the general urban pattern of the old city of Sana'a are the city roads, the social square, and the urban garden. As seen in Figure 10.4, a typical housing cluster is centered on the urban garden (E) and contains several tower houses (C) that surround it and are attached to the social square (D) and the road. The road creates another row of houses, of the same typology, and connects the houses to the social square that includes the mosque (A) and other houses.

The tower house walls are built of 50 cm square, black, volcanic ashlar stone on the lower levels (i.e. up to approximately 6–10 m above street level) and 40 cm baked exposed brickwork above that. The roof consists of a frame of wooden beams set 50–60 cm apart, covered by branches and twigs, on top of which lie layers of finely sifted earth, wet and compact, up to a thickness of 30 cm. The lower floor is frequently double-storey in height and used for keeping animals and for storage. The higher floors are residential quarters, and include rooms that are functionally polyvalent and non-specific; rooms that can be used interchangeably for eating, sleeping, recreation and domestic tasks. This flexible use of living space is reflected by the absence of cumbersome furniture.

The high altitude of Sana'a's location makes it comfortable all year round, except for few months in the winter when the houses require some passive heating. To respond to this pleasant climate, the living spaces in the vernacular

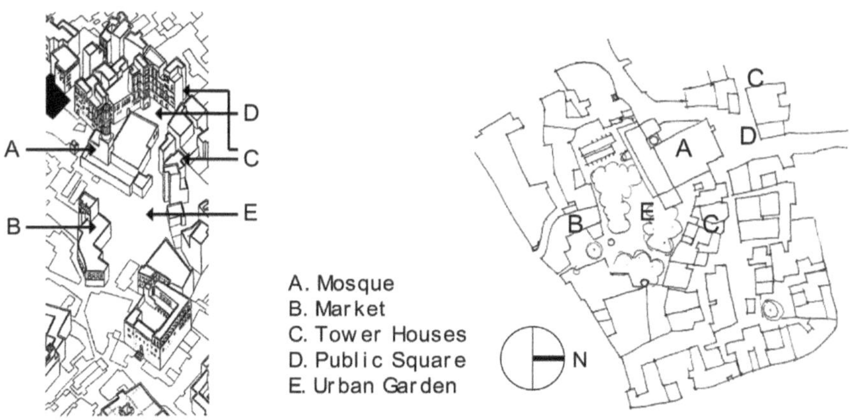

FIGURE 10.4 Isometric and plan of the housing cluster case study. (*After: Lane, M. B. [4]*)

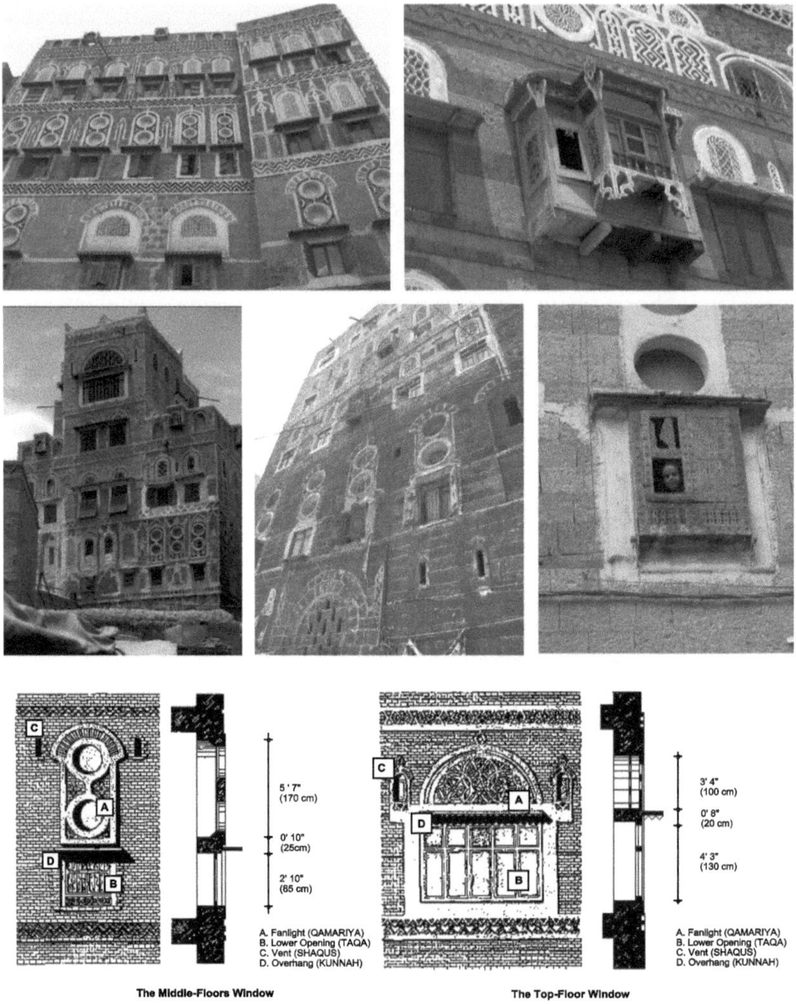

FIGURE 10.5 Images of Sana'a traditional window types, top. *(© Khaled Al-Sallal).* **Drawings of two types of windows showing their components.** *(Adapted from: Hirschi, S. and M. Hirschi [6]).*

house are oriented to south to benefit from the winter sun. This arrangement also helps occupants to escape the summer afternoon sun that strikes north facades (usually built with limited openings); this is a phenomenon that occurs only in the summer time on locations near the equator, such as Sana'a (latitude of 15° 31′) where the sun moves toward north of the equator. Sana'a's tower houses generally do not experience low altitude sun coming from east or west orientations, because their east and west sides are usually protected by attachment or proximity to other neighboring buildings. The windows in this type of house are usually small in the lower floors, middle size in the middle floors, and relatively large in the higher floors (Figure 10.5). The largest windows in the house can be found at the top floor level, where a small square room, named *Tairamana* in the Yemeni dialect, is used as an observatory space.

The indigenous window of Sana'a has four major parts: the upper section, the lower opening, the vents, and the overhang (Figure 10.5). Each part has a different function. One window can consist of some or all the parts, depending on its main function and its location on the facade as well as the plan of the house. Al-Sallal and Cook [5], documented the indigenous window of Sana'a and analyzed its design variations in terms of size, shape, and number, based on climatic influences and functional requirements. These authors categorized the window design according to its components, and described the geometry and function of each component. They concluded that the breakdown of the window into components improves integration of the different functions of the window and provides high flexibility in controlling each function.

10.2.3 Thermal Performance of the Vernacular House

Al-Sallal [7,8] investigated the role of indigenous fenestration in reducing energy requirements in the Sana'a house using computer simulation. The results showed that window shutters could save 7% of the annual heating requirements and overhang and drapery could save 10% and 33%, respectively, of the annual cooling requirements. Another study showed that the vernacular house in Sana'a provided better thermal comfort than the modern house [9]; yet very few of these buildings have been built in the last two or three decades. The dominance of imported design ideas and cheap building materials unsuitable for the climate or function had a great impact in limiting the building of new tower houses. Al-Sallal [10], investigated the balanced synthesis of form and space in the vernacular house of Sana'a, and how this was used as the chief means of mitigating the effect of the climate, and of providing functional requirements. The approach depended on a number of methods, including thermal performance analysis, solar access and shading study, and observation and analysis of common architectural practices and cultural aspects. A summary of the study's outcomes provided some rules of thumb that can be applied in design. Al-Sallal [11] employed three-dimensional computer modeling for visualizing solar access and shading in a traditional housing cluster in the historical city. The study found that the houses' urban pattern, orientation and vertical forms helped to maximize solar access in the winter from the south, while avoiding it in the summer.

Ayssa [12] evaluated the impact of traditional window shutters in reducing the cooling loads required by a typical seven-storey vernacular house sandwiched between two vernacular four-storey houses to due north and south. The results showed a 45% reduction in the cooling load due to the use of the traditional window shutters. The results also showed that the attachment of the houses to the neighboring buildings had the potential to reduce the heating load, but did not show significant effects in reducing the cooling load. In another study, Ayssa [13] investigated the vernacular houses of the old city of Sana'a with regard to improving thermal comfort and internal air movement.

Al-Motawakel et al. [14] developed a simple, steady-state, mathematical model which described the average total daily rate of heat loss through the walls, windows and flat roof of a generalized Yemeni building. The model can be used by designers to predict approximately the transient rate of heat loss via traditionally employed combinations of indigenous materials, as used in the walls and roof. The principal variable is the ratio of the total glazed area for each storey to the corresponding sum of the surface areas of the solid walls and roof in contact with the ambient environment. The predictions, expressed graphically, enable designers to select the most suitable combination of locally available, indigenous building materials, so that more energy-effective dwellings can be built.

To provide thermal comfort, the Sana'a vernacular house design is based on the principle of direct-gain solar system. Solar radiation enters the interior of the building through glazed surfaces on the walls. The sunlight is converted to heat when it strikes absorbing surfaces within the building, causing the temperature of these absorbing materials (thermal storage mass) to rise. The heat is gradually dispersed throughout the space from the directly radiated absorbing surfaces to the surrounding surfaces (walls, floor and ceiling) and to the room contents by convection, conduction and radiation. At night, heat is released by convection and radiation from surfaces in the room that have stored up heat energy during the day. For a passive building to have good interior comfort control, the following conditions must exist:

1. The structure must be well insulated and have a low rate of air infiltration.
2. The solar collection area provided for the building must be adequate in terms of size and orientation.
3. The thermal storage must be distributed in the building in such a way that it can absorb the energy of the solar radiation entering the building without the air temperature in the building rising to an uncomfortable extent.
4. The building should be provided with shading devices for the solar collection area, for use when energy collection is not needed, and with a method to exhaust excess energy already collected as warm air.

The way that the first condition was satisfied in the Sana'a house was discussed in a previous study by Al-Sallal [15]. Although this house is satisfactorily energy efficient, its thermal performance could be improved if more energy conservation measures are considered. The same study found that above 30% of the daily heat load for a one-storey house built from traditional construction materials was due to air infiltration. Insulation of walls and roofs depends on the use of locally available materials, such as brick and earth, which have relatively high thermal resistance. Stone, a material with poor thermal resistance, is limited to construction of the lower floors, which usually do not include residential quarters. Also, the same study found that the windows' exterior shutters, when closed during the night-time, had the potential to reduce the annual heating requirements by 7%. The second condition seems to be satisfied appropriately by the Sana'a house with regard to the appropriate orientation of the solar collection area according to the well known and widely applied building practice that emphasizes the importance of the southerly orientation; however, more investigation needs to

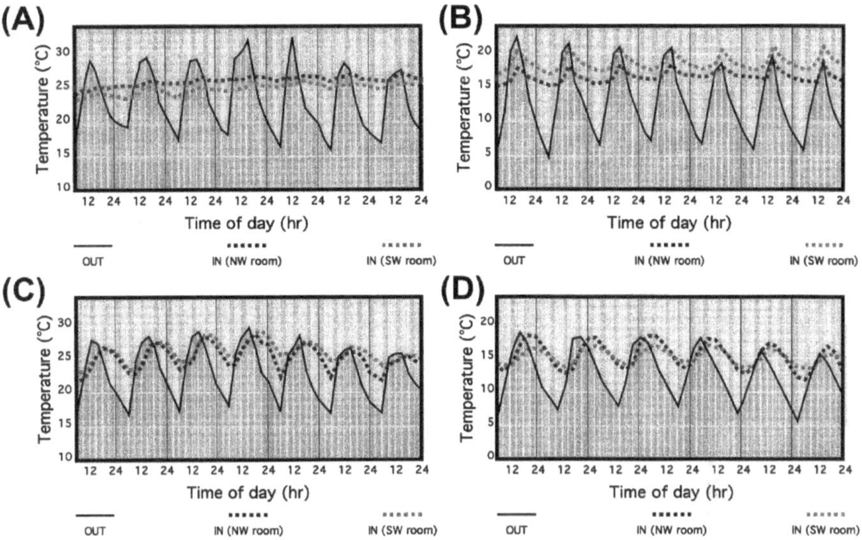

FIGURE 10.6 Indoor temperature profile over one-week period in the vernacular house in July (A) and December (B); and in the modern house in July (C) and December (D).

be carried out to find out the optimum size for solar windows. The size of the windows is typically determined by cultural and functional aspects.

Satisfying the third condition in the Sana'a house was discussed in a previous study [9], which showed how the massive construction materials of the vernacular house had a very important role in providing steady internal temperatures in both winter and summer when compared with the modern house (see Figure 10.6.) In Sana'a's winter, one can experience minor problems of thermal discomfort, especially at night; the average minimum outdoor temperature is 3°C and the average maximum is 24°C. In the face of this large variation in external temperature, the vernacular house provides very steady internal temperatures; as shown in Figure 10.6. This figure also shows the significance of the building's orientation in improving thermal comfort. The study showed that the internal temperature range of the vernacular tower house over a one-week period was 25–26°C in July and 16–20°C in December.

Satisfying the fourth condition in the Sana'a vernacular house is achieved by appropriate ventilation and shading. These houses have small vents integrated with the window design to overcome overheating problems; though such problems are rarely experienced [5]. As shown in Figure 10.6, the summer time in the vernacular house does not create a problem of overheating and provides better thermal comfort than the modern house. From the preceding discussion, it can be concluded that providing thermal comfort to the Sana'a house depends firstly on providing appropriate solar access; then, on improving its thermal performance based on the conditions explained previously. Exploring issues of solar zoning, solar shading, solar accessibility and the effect of the solar component on the building form is necessary for developing appropriate building codes and regulations.

10.3 THEORETICAL MODEL FOR SUSTAINABLE ARCHITECTURE

Vernacular architecture can be viewed as a creation produced through natural dynamic interactions among people's socio-cultural needs, ecological impacts, and technological solutions (see Figure 10.7). The creation of vernacular architecture is a very complex system involving many forces that are not always in agreement. It went through a long learning process that took many decades and centuries during which people through consecutive generations developed knowledge on how to operate interactions among multitude of forces, frequently not in agreement, and optimize architectural solutions governed by a system of generated rules of thumb (or traditional conventions). The cultural aspect, or the local way of practicing human activities, is an important component in the theoretical framework. The environmental component and the extent to which vernacular architecture was able to integrate between climatic conditions and functional needs is another important component that should be taken into

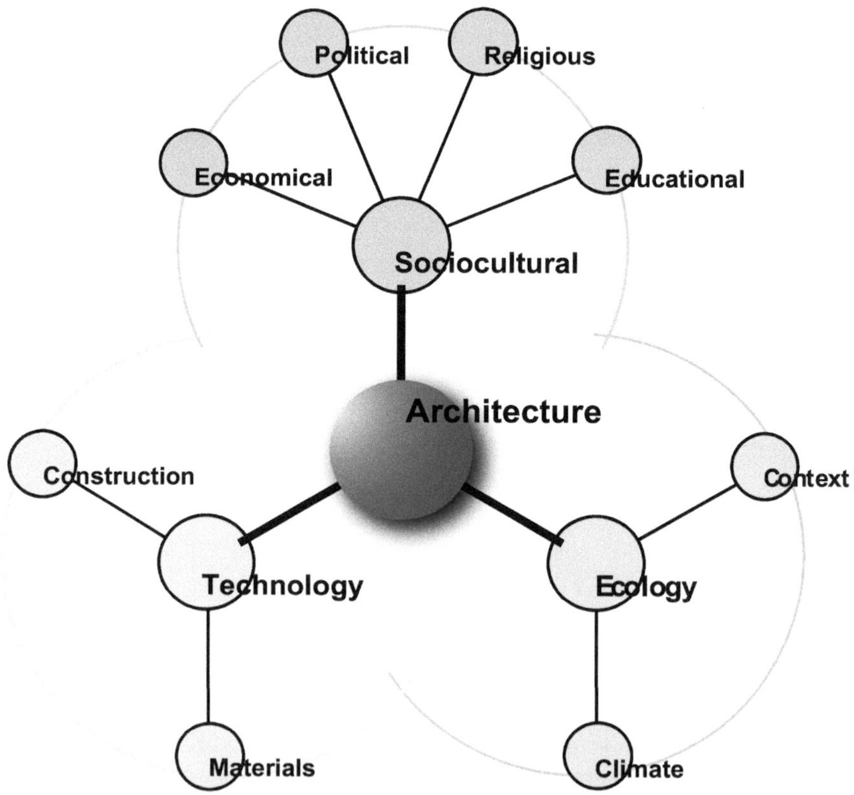

FIGURE 10.7 **Sustainable architecture as a result of interactions among socio-cultural, ecological and technological factors.**

consideration, especially in the case of Yemeni architecture where these values were very evident. The proposed model focuses on extracting abstract concepts (e.g., geometrical proportion or functional relations) from traditional environments and presenting methods to re-apply them in contemporary architecture; rather than to give in to the temptation to copy façades or architectural features.

The theoretical framework of the study comprises two main activities: design analysis and performance evaluation. In the design analysis activity, the architecture of a place is analyzed in terms of form-space relationships using the suggested model, and based on identification of levels of relationships and specifying certain settings. Then for each setting, an evaluation is conducted to check the ability of a physical form-space relationship to satisfy certain requirements (environmental and socio-cultural) and to test the level of agreement between both types of requirements. The form-space relationship model works here as a testing tool that provides the means to propose and develop solutions for use in contemporary architecture and as a methodology for comparison of solutions.

10.3.1 Form-Space Relationship Model

Without a doubt, form and space are inseparable and neither of them can exist without the other. The visual properties of form are shape, size, color, texture, position, orientation and visual inertia. Space can be understood as the opposite of form or the negative element that provides a background for the form. The elements of form and space together form the reality of architecture. In spite of their unity, both form and space can be seen as two different outcomes that result from different levels of functional and environmental influences. The term 'function' and 'functional' are used in this context to describe socio-cultural activities and standards. Typically, environment has a direct effect on form, while function has a direct effect on space. Yet, because they are inseparable, one should also look at how the chosen form will affect the function. In other words, although the function has a direct influence on space, space response to function comes via the proper form.

The form-space balance design approach can be defined as the search for a solution that harmonizes environmental and functional (socio-cultural) requirements. Environmental requirements can be defined as those factors mainly affecting the design of the building form, in order to maintain a comfortable and sustainable built environment. On the other hand, functional requirements can be defined as those factors mainly affecting the design of the space (via the appropriate form) in order to accommodate certain social or cultural activities.

The form-space relationship model is abstract and can be applied at several levels and scales of design. Yet, before applying the model, there is a need to identify the scope of its application, which is based on two types of limits:

1. Design Level – there is a need to specify the level at which analysis of form-space relationship depends on. Generally, one can define four levels of form-space relationships. These are the urban level, the housing cluster level, the building level and the building components level.

2. Form-Space Setting – at each level, the built environment (e.g., a building or cluster of buildings) should be divided into physical form-space settings, and each setting should be analyzed separately using the model.

The advantages of using this analysis model are as follows:

1. It helps make the complex relationships of environmental and socio-cultural variables more visible and design oriented, while keeping them coherent – this is necessary because many cultural values cannot be realized unless they are seen collectively. Dealing with issues in their entirety is necessary because it helps keep integration among cultural values, especially in traditional societies. This is also a well-recognized approach in architectural design.
2. Identifying different levels and dividing the built environment into physical settings helps limit the scope of the analysis and create a focus of attention on the most important issues.
3. Using one model to analyze several settings helps to improve consistency among different parts of the same problem, and creates opportunities for a fair comparison of solutions.

10.4 ANALYSIS

The factors that formed the traditional architecture can be divided into two groups; these are socio-cultural and environmental factors. Privacy and safety of the house's occupants, and sociability with the outside and within the house are examples of important cultural requirements that have a great impact on the architecture of Sana'a. Solar access/shading and ventilation are examples of environmental factors that played significant roles in forming the climate-responsive architecture of Sana'a. To respond to the influence of these factors, numerous innovative architectural solutions have been developed throughout history; such as the housing cluster, the tower form and the Multi-component window. These features have given the architecture of Sana'a a unique identity and attractive character.

10.4.1 The Urban Garden and the Social Square Setting

One of the important considerations in planning a housing cluster is locating buildings appropriately, based on their forms and functions. Creating the urban spaces (i.e. the urban garden and the social square) is achieved by creating the buildings around them. Appendix A shows the climatic and functional influence on form and space of the urban garden and the social square. Without rules that govern the making of these buildings' forms and their relationships to the outdoor spaces, the following problems could emerge:

1. Lack of solar access could produce uncomfortable, cold houses and an unhealthy outdoor environment for the humans, plants and animals that use theses spaces.

2. Inadequate space size for the number of people and their activities to be housed to practice daily activities, such as cultivation, or seasonal public activities such as feast celebrations.
3. Inadequate space for privacy and segregation of men and women.

Environmental factors – Buildings that are most demanding of solar access, such as houses, are placed along the north-northeast boundary of the urban garden or the social square with minimum heights of four floors (Appendix A). Buildings that are least demanding of solar access, such as the marketplace, are placed along the south-southwest boundary of the urban garden or the social square with maximum heights of two floors. Solar access should be ensured for the buildings that need it. Solar access is more important for houses than other types of buildings. This is related to the concept of solar zoning. The size and shape proportion of the urban garden (or the social square) should permit appropriate solar access to the housing sections located around its boundaries. This can be achieved if the size of the urban garden (or the social square) in the north-south direction exceeds the minimum required distance, 6 m, recommended by Al-Sallal [8]. Another important consideration is the shape ratio of the east-west (E-W) to north-south (N-S) dimensions of the urban garden (or the social square). It should maximize southerly solar access to the largest number of housing buildings. In a previous study by Al-Sallal [7] a ratio of 1.66:1 (in terms of the proportion of E-W to N-S dimensions) seemed to work well.

Socio-cultural factors – Because the activities of the social square are public, the privacy of the entrances and ground floor openings of the houses located around the social square should be maintained (Appendix A). This is an effect of the outdoor space function on building form. Privacy is ensured by designing indirect entrances and minimizing the area of openings on the ground floor. In many instances, the social square is directly connected to a mosque that is needed for the daily prayers, as well as to support certain social events happening in the social square, such as celebrations, feasts and weddings. The size, shape and proportion of the outdoor space are the design parameters that determine whether it is adequate for supporting certain functions. Hence, the space parameters of the social square should permit many people, usually males, to gather to celebrate religious feasts and social ceremonies; and the space parameters of the urban garden should be appropriate for cultivation activities.

10.4.2 Tower House Setting

Environmental factors – The climatic and functional influence on the form and space of the tower house setting is illustrated in Appendix A. The long axis of the tower house form runs east-west, which makes the long side of the building face north and south. This allows the majority of the windows to be placed in the north and south facing walls to exploit solar heat gain for heating in the winter and to reduce it in the summer, accordingly. The horizontal aspect ratio of the built form (i.e., length to width) falls between 1:1.6 and 1:3, which matches the general rule of thumb for temperate to arid climatic zones [15]. The verticality of the tower house helps improve solar access to the higher floors which are usually used as living spaces by households, and include the best rooms. Locating the main activities' spaces on higher floors maximizes the chance of solar access and consequently promotes solar gain; lower floors are largely shaded due to the tall neighboring buildings and their existence on narrow streets. In addition, higher floors are usually warmer because of the movement of air currents by convection; hot air layers rise up while cold air layers descend. From personal experience, the problem of overheating in summer is rarely felt, even on higher floors; however, there is a need to investigate the indoor temperature profile on different floors. A house can be made more energy efficient simply by designing the plan so that the order of rooms in which the normal daily sequence of activities occurs, 'follows' the path of the sun [16]. Specific rooms or functions in the Sana'a house are planned to coincide with the solar orientation. This strategy becomes most effective when it is combined with partitioning the interior into separate heating and cooling zones. By relating zones to the sun's movement, solar energy can be put to use when it is most available by direct orientation. The southerly orientation of the traditional Sana'a house is reserved for the best rooms. Less important rooms are oriented to the east or west; and service spaces are oriented to the north. There is a proverb in Sana'a that says 'a southern room is worth of one room, a western or eastern room is worth half a room, and a northern room is worth no room'. People change their diurnal and seasonal patterns of interior use to benefit from the climate's assets or to escape its liabilities. The flexible use of living space and the absence of cumbersome furniture help the normal daily sequence of activities to follow the path of the sun. Low-use spaces, storage, and utility areas are located toward north façades or on the ground floor to provide climatic buffers.

The vertical form of the traditional tower house provides significant bioclimatic advantages over a horizontal form that is equal to its volume and floor area, as is present in the modern house design [10]. The change of the solar altitude from high angles in the summer to low angles in the winter makes the vertical form better than a horizontal one, for both summer and winter in Sana'a. In winter, the vertical form allows greater exposure to the sun through its large vertical surfaces (mainly through its southern facade, because the eastern and western facades are assumed to be blocked by attached buildings). This can be seen in Appendix B, Table 1 for the case of December, 21. This higher solar exposure offered by the vertical form results from the relatively low solar altitude angles of the winter that allow the sun to reach more vertical surfaces (i.e., walls) than horizontal ones (i.e., roofs). Furthermore, the vertical form in summer allows a lower exposure to the sun through its vertical surfaces than the horizontal form, as seen in Appendix B, Table 1 for the case of June, 21. The lower solar exposure offered by the vertical form results from the relatively high solar altitude angles of the summer that allow the sun to reach fewer vertical surfaces than horizontal ones. Accordingly, if the form can provide smaller areas of horizontal surface (and larger areas of vertical ones), this limits the solar heat gain.

In fact, the horizontal form is not appropriate at all, because it will greatly increase the solar heat gain through its large roof area. Also, the tremendous reduction in the number of floors in the horizontal form results in exposing a higher proportion of the building's indoor spaces to direct contact with the roof. This is usually hot in the summer, because of the high solar heat gain. Appendix B, Table 2 shows two examples for comparison; a vertical form, 12 m * 12 m * 30 m (i.e., 4320 m^3), which represents a typical eight-storey tower house building, and a horizontal form, 32 m * 18 m * 7.5 m, which represents a two-storey imaginary building. The vertical form would experience solar heat gain through a roof area of 144 m^2 while the horizontal form building would experience solar heat gain through a roof area of 576 m^2, which is four times that of the vertical form. Consequently, the vertical form is better than the horizontal one during the winter because it helps in promoting solar heat gain, and is also better during the summer because it helps in reducing the amount of solar heat gain.

Socio-cultural factors – Sana'a's tower houses are surrounded by beautiful views of green gardens and a skyline of decorated towers' façades and minarets, all with a background of remote mountains and blue sky. These fascinating views, combined with the mild climate, encouraged the creation of extrovert architecture (open to the outside), as opposed to the introvert architecture (courtyard configurations) that can be found in other parts of the Arabian Peninsula and Middle East, which are characterized by harsher environments. The Yemeni people succeeded in choosing an appropriate form to provide views to all rooms (Appendix A). This is the tower form that is emphasized by its verticality and characterized by its small floor area. The verticality of the form provides better view angles and multiple view ranges from different heights (i.e., distant views of the mountains and sky and closer views of Sana'a's gardens and architectural facades). As opposed to the vernacular house, the visual link in the modern house to outside views is very restricted, especially on the ground floor, by the existence of high fence walls [16]. Spaces that need high levels of privacy are located on high floors that are usually used by households. The concept of the multistory house helped to provide several levels of seclusion (see Appendix A); the higher the floor is, the more secluded it becomes. The modern house on the other hand is isolated from the community context around it by the existence of high fence walls, constructed for privacy and safety reasons [16]. Utility spaces such as storage areas are usually located on the ground floor. The suitability of lower floors to house certain activities make the application of this design practice more common and useful. Storage spaces are usually located on, or at least nearby, the ground floor which facilitates carrying and handling of things.

10.4.3 Building Skin and Multi-Component Window Setting

Environmental factors – The climatic and functional influence on form and space of the building skin and window design setting is illustrated in Appendix A. The variation in the horizontal arrangement of windows (shapes, sizes and number) is a result of climatic influences, such as solar motion and wind direction. Southern facades have larger windows with more components and controls than northern facades, which helps to exploit solar energy for heating. The windows have two shading devices for protection from the sun's heat: the overhang and the shutters. The multi-component configuration of this window type provides high flexibility to admit air for ventilation or capture daylight at any time, even when the shutters of its main part are closed.

Socio-cultural factors – the variation in the vertical arrangement of the windows on the building facade comes from socio-cultural influences derived mainly from people's activities and traditions. While the higher floors have large windows to maximize visual connection to outside views, lower floors have smaller windows for privacy. The shutters of the multi-component window provide high levels of privacy and safety, while other functions such as ventilation or admitting daylight are still provided. However, when the shutters are closed, the view is restricted. The windows of the upper floors of the tower house are made wide enough at sight level to provide maximum connection with the outside views.

The window designs found in modern houses, unlike the traditional ones, are designed without any rules or conventions and show no logical order. In most cases, the designs do not reflect climatic or cultural needs. Also, their variation within the same building does not make any sense either with regard to outside factors, such as climate or view, or inside factors such as their arrangement in the indoor spaces. While windows were the most beautiful elements in the traditional house of Sana'a, they have now lost their unique architectural identity.

10.4.4 Relationships Between Factors

The way that the traditional architecture has responded to socio-cultural and environmental factors by innovative concepts has been investigated and described. Table 10.1 provides a matrix that illustrates how some traditional design concepts (or elements), such as the tower form, and the multi-component window, responds to the environmental and socio-cultural factors. This matrix can be a useful tool, not only to realize the function of each architectural element, but also to realize the complex relationship between each architectural element and the factors that influence its design and operation.

To understand these complex relationships, another matrix has also been devised, as shown in Table 10.2. This helps in the analysis of the binary

TABLE 10.1 Responsiveness of Traditional Tower Form and Multi-Component Window to the Complex System of Environmental and Socio-Cultural Factors

		Tower Form	Multi-Component Window
Environmental Factors	Shading/overshadowing	The tower form helps provide shading for itself during the summer season by limiting heat transfer by radiation through its limited roof. It also overshadows other neighboring buildings around it when the sun comes from the east and west.	The multi-component window has two devices for providing shading: the overhang and the shutters. This is a very flexible design because when the shutters are closed, other parts can still provide other required needs such as daylighting and/or ventilation.
	Passive Cooling	*Comfort ventilation* – The tower form acts as a wind generating tool in the house, hot air ascends and cooler air replaces it from the surrounding outdoor spaces such as the garden, the public square, and the roads. *Evaporative cooling* – The vegetation in the garden creates an evaporative cooling effect that lowers the air's temperature and increases its humidity.	The use of separate vents makes this window useful for ventilation at any time, even when the shutters of its main part are closed.

Continued

TABLE 10.1 Responsiveness of Traditional Tower Form and Multi-Component Window to the Complex System of Environmental and Socio-Cultural Factors—cont'd

		Tower Form	Multi-Component Window
Socio-cultural Factors	Privacy	The tower form helps create three levels of privacy. The most private zone is located near the top. The intermediate private zone is in the middle of the tower. The least private zone is located near the street level.	The shutters of the multi-component window can be closed at any time for privacy, while other functions such as ventilation or daylighting are still provided. However, when the shutters are closed, the view is restricted.
	Safety	The tower form keeps the most important rooms in the safest zones; i.e., far from the street level.	The shutters of the multi-component window provide high levels of safety while other functions such as ventilation or daylighting are still provided.
	Views Connection	The tower form provides a variety of view angles and ranges, that maximizes visual connection to outside views.	The large windows of the upper floors of the tower house maximizes visual connection to outside views. Only when the shutters are closed, the view is restricted.

TABLE 10.2 Analysis of the Level of Agreement between Environmental and Socio-Cultural Factors for each of the Tower Form and the Multi-Component Window

Tower Form	Multi-Component Window
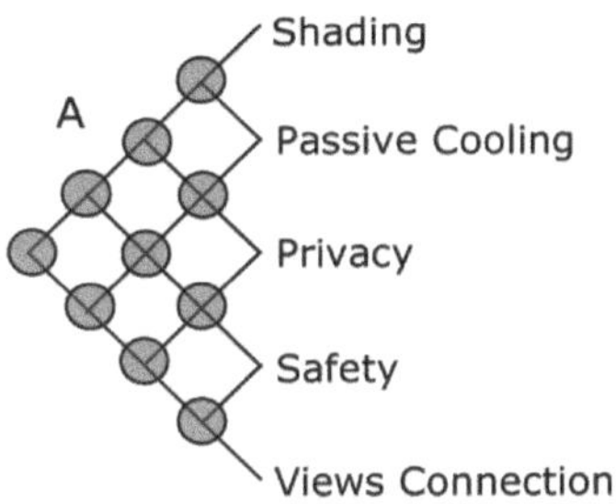	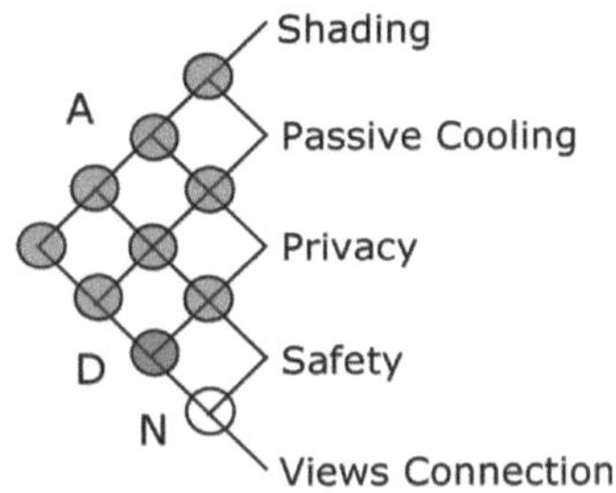

Three types of relations have been identified; agreement (noted as A), disagreement (noted as D) and no relation (noted as N).

relationship between any two factors regardless of its category (i.e., socio-cultural or environmental). Three types of relationships have been identified; agreement (noted as A), disagreement (noted as D) and no relation (noted as N). For example, the tower form provides shading and passive cooling for most of the day, while maintaining the privacy of the house's occupants. Therefore, in the tower form, there is a high degree of agreement between privacy and shading, and also between privacy and passive cooling. On the contrary, in the multi-component window case, there is a high degree of agreement between passive cooling and privacy, while there is a disagreement relationship between privacy and view. That is because the multi-component window promotes cooling by ventilation using its separate vents, while keeping the privacy of the inner rooms by shutters. When closed, these shutters totally prevent any chance of a view. This is a good example of matching between two factors (i.e., passive cooling and privacy) at the expense of a third one (i.e., sacrificing view.)

10.4.5 Influence on Form and Space Design

The traditional tower house was constructed according to certain conventions; some are stringent, to respond to the socio-cultural and environmental needs. Its design is constrained by a number of rules that specify its form proportion, orientation, zoning, and spatial arrangement. After identifying the environmental and socio-cultural factors that affect the tower house, it is necessary to separate those factors according to their impacts on form, space, skin and window design (see Appendix A). Visual presentation of

TABLE 10.3 Design Issues, Guidelines, and Design/Performance Variables

Design Issues	Sub-Issues	Design Guidelines	Design/Performance Variables
Built-form configuration	Orientation of long axis	• The long axis of the tower form is oriented east-west	• Controlled by site characteristics • Affects solar exposure
	Horizontal aspect ratio	• The horizontal aspect ratio of the built form (i.e., length to width) falls within 1:1.6 to 1:2	• Controlled by site characteristics • Affects solar exposure, visual connection
	Vertical aspect ratio	• The vertical aspect ratio (i.e., length to height) falls within 1:1.5 to 1:2:5	• Controlled by cost and family size • Affects solar exposure, solar access, visual connection
Arrangement of indoor spaces	Horizontal arrangement of spaces	• South for distinctive and high occupancy living spaces • East and west for average and less occupancy living rooms • North for only service spaces	• Controlled by room's frequency of use, type of user, and type of activity. • Affects solar orientation and visual connection
	Vertical arrangement of spaces	• Spaces that require more privacy are located in higher floors • Spaces that are higher connected to the public (or street) are placed in lower levels	• Controlled by room's frequency of use, type of user, and type of activity. • Affects privacy and visual connection

Façade design	Horizontal arrangement of windows	• South façades have larger windows with more components and controls to exploit solar energy for heating • North façades have smaller windows to avoid heat loss.	• Controlled by climatic influences such as solar motion and wind direction • Affects solar heat gain
	Vertical arrangement of windows	• Higher floors have large windows to maximize visual connection to outside views • Lower floors have smaller windows for privacy.	• Controlled by socio-cultural influences such as the people's activities and traditions • Affects visual connection and privacy
Window design	Size, shape, and number of windows	• Large windows mainly for view • Small windows mainly for ventilation	• Controlled by space type or activity • Affects daylighting, ventilation, and view
	Design of window components	• Main part for daylighting, ventilation, and view • Upper component (al-qamarya) for providing soft daylighting • Vents (shaqus) for ventilation only • Overhang for shading	• Controlled by the window's main function and its location on the façade as well as the plan of the house • Affects daylighting, ventilation, and view

this analysis would be most effective and useful to designers by showing the potential interaction among these factors and how they relate to the components' design. It is also a useful format for design application in contemporary architecture; they are presented here as basis for sustainable design (see Table 10.3 and Appendix A).

10.5 CONCLUSION

There is a need to understand how sustainable vernacular architecture succeeded in keeping a balance among the complex relationships between environmental and socio-cultural requirements. The vernacular tower architecture of Yemen is known for its strong character, remarkable identity, deep cultural meaning, and ecological harmony. The study proposes a model that helps to explore abstract concepts from traditional environments, and present methods to re-apply them in contemporary architecture. The model is based on form-space relationships and can be applied at several levels of design. It depends mainly on abstracting complex architectural environments into more simplified, yet more comprehensive and meaningful relationships. It helps in realizing the complex relationships between environmental and socio-cultural variables in a more visual depiction and design-oriented approach, while still keeping them coherent. The socio-cultural and environmental impacts on three settings from the vernacular architecture of Sana'a have been analyzed using the proposed theoretical model, to illustrate its operation and to pave the way for the analysis of other traditional urban and architectural settings in the future. It is believed that the same model can also be applied to the analysis of other settings or design proposals, for the purpose of understanding how they perform or comparing design alternatives. The study concludes by presenting the outcomes in a form of design guidelines which can be used as a basis for sustainable design.

REFERENCES

[1] M. Al-Oulfi, Characteristics of Yemeni Architecture: Forms and Directions of Development, Unpublished PhD Thesis, Shanghai Tongji University, China, 2000.

[2] R. Lewcock, The Old Walled City of Sana'a, UNESCO, Paris, 1986.

[3] H. Al-Sabahi, Sana'a Yemen Tradition and Modernity in Sanani Architecture, Sana'a, Sana'a Modern Offset Printing Press, 1996.

[4] M.B. Lane, Sana'a: Pilot Restoration Projects, UNDP/UNESCO YEM/88/006, 1988.

[5] K.A. Al-Sallal, J. Cook, Sana'a historical windows: integration between comfort and aesthetics, in: S. Burley, M.E. Arden (Eds.), Proceedings of 17th National Passive Conference of American Solar Energy Society, vol. 17, American Solar Energy Society, Boulder, CO, 1992, pp. 197–202.

[6] S. Hirschi, M. Hirschi, L'Architecture Au Yemen Du Nord, Berger-Levrault, Paris, 1983.

[7] K.A. Al-Sallal, The role of indigenous fenestration to reduce energy requirements in the Sana'a house, Proceedings of Fourth Arab International Solar Energy Conference. vol. 2, Royal Scientific Society, Amman, Amman, 1993, pp. 875–888.

[8] K.A. Al-Sallal, Traditional methods in new forms: insight to achieve energy conservation in the modern Sana'a house, in: A. Sayigh (Ed.), Proceedings of the World Renewable Energy Congress, vol. 3, Oxford: Pergamon-Elsevier Science, Denver, CO, 1996b, pp. 1438–1441.
[9] K.A. Al-Sallal, A.Z. Ayssa, H.A. Al-Sabahi, Thermal performance and energy analysis for Sana'a vernacular house, in: S. Ben Ghadi (Ed.), Proceedings of Applications on Renewable Energy in Yemen Workshop, Aden University, Yemen, 1995.
[10] K.A. Al-Sallal, The balanced synthesis of form and space in the vernacular house of Sana'a: bioclimatic and functional analysis, Architectural Science Review 44 (4) (2001) 419–428.
[11] K.A. Al-Sallal, Solar access/shading and building form: geometrical study of the traditional housing cluster in Sana'a, in: A. Sayigh (Ed.), Proceedings of the World Renewable Energy Congress, vol. 3, Pergamon-Elsevier Science, Oxford, 1996a, pp. 331–334. Denver, CO.
[12] A.Z. Ayssa, The impact of the traditional window shutters and attachment as comfort producing procedures in the vernacular houses of the old city of Sana'a, Proceedings of the World Renewable Energy Congress, Reading, UK, 1990.
[13] A.Z. Ayssa, The vernacular tower house of the old city of Sana'a, Yemen – an investigation of its internal air movement, in: A. Sayigh (Ed.), Proceedings of the World Renewable Energy Congress, Pergamon-Elsevier, Oxford, 1994.
[14] M.K. Al-Motawakel, S.D. Probert, B. Norton, Thermal behaviors of vernacular buildings in the Yemen Arab Republic, Applied Energy 24 (4) (1986) 245–276.
[15] K. Yeang, The Green Skyscraper: The basis for Designing Sustainable Intensive Buildings, Prestel Verlag Munich, Germany, 1999.
[16] K.A. Al-Sallal, Sana'a: Transformation of the Old City and the Impacts of the Modern Era', in: Y. Elsheshtawy (Ed.), Planning Middle Eastern Cities: An Urban Kaledioscope in a Globalizing World, Routledge, London, 2004.
[17] D. Watson, Climatic Building Design, McGraw-Hill Book Company, New York, 1983.

Appendix A

Sustainable Design Guidelines Derived From Response of Architectural Form and Space to Climatic and Functional Factors

Urban Garden and Social Square Setting

FORM response to CLIMATE

Form configuration

Buildings that are most demanding of solar access, such as houses, should be placed along the north-northeast boundary of the urban garden.

Buildings that are least demanding of solar access, such as the marketplace, should be placed along the south-southwest boundary of the urban garden

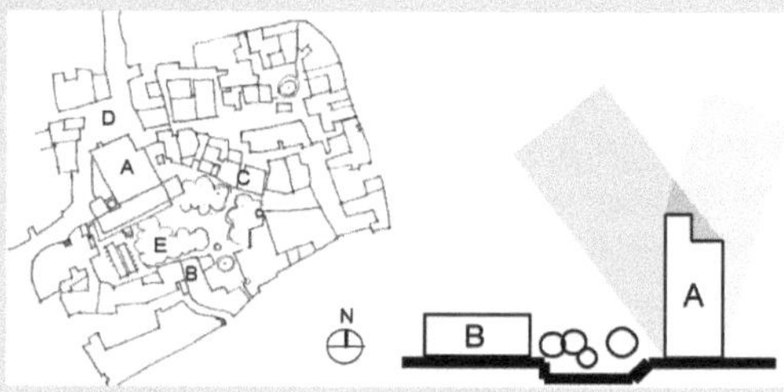

A: Mosque B: Market C: Houses
D: Social Square E: Urban Garden

FORM response to FUNCTION

Privacy of the houses located around the social square should be maintained by indirect entrances (a) and minimum area of openings in the ground floor (b).

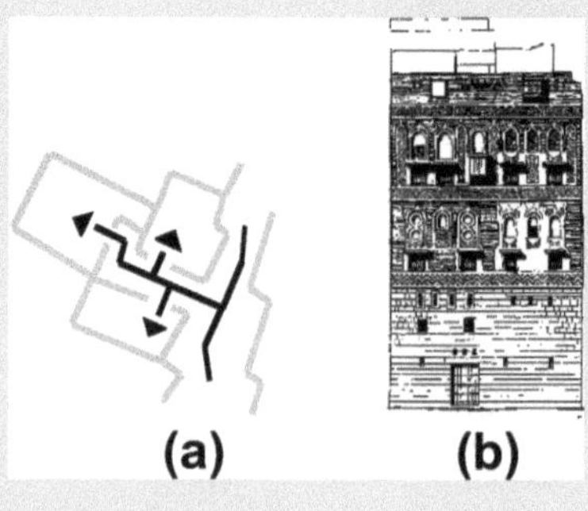

SPACE response to CLIMATE

Space Size and Shape Proportion

To permit appropriate solar access to the houses located around the urban garden (or the social square), the size of the space at north-south direction should not be less than the minimum required distance, 6m; and the shape proportion, 1:1.66, of east-west side to north-south side of the urban garden (or the social square), was found to be appropriate for south solar accessibility [10].

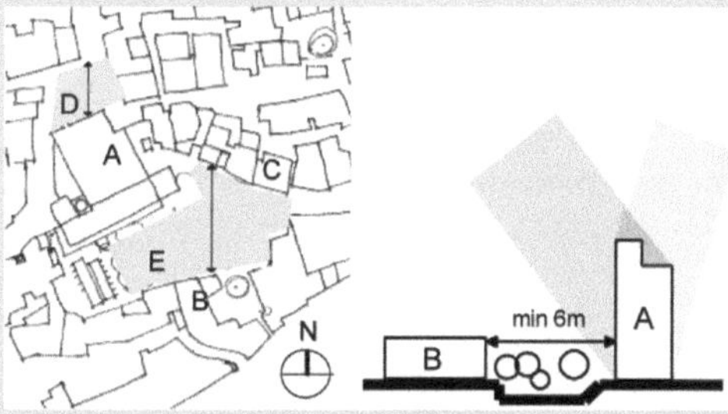

A: Mosque B: Market C: Houses
D: Social Square E: Urban Garden

SPACE response to FUNCTION

Space Size and Shape Proportion

The space of the social square should permit gathering of many people, usually males, to celebrate religious feasts and social ceremonies.

The space of the urban garden should be appropriate for practicing the cultivation activities.

A: Mosque B: Market C: Houses
D: Social Square E: Urban Garden

Tower House Setting

FORM response to CLIMATE

By selecting proper Orientation

The long axis of the tower house form runs east-west which means the long side of the building faces north and south.

This allows to place the majority of the windows into the north and south walls and accordingly to exploit solar heat gain for heating in the winter and to reduce it in the summer.

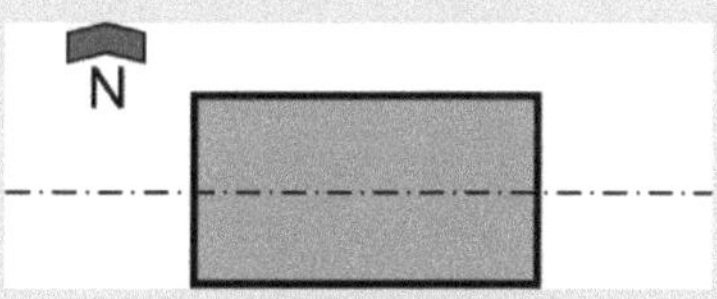

FORM response to CLIMATE

By selecting appropriate Horizontal Aspect Ratio of form

The horizontal aspect ratio of the built form (i.e., length to width) falls within 1:1.6 to 1:3.

This matches the general rule of thumb for temperate to arid climatic zones [15].

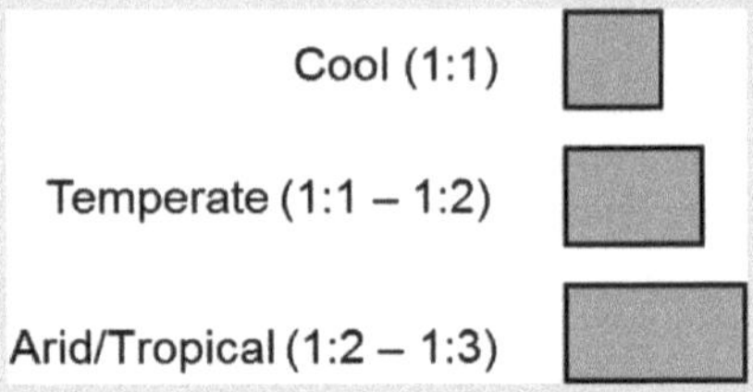

FORM response to CLIMATE

By selecting appropriate Vertical Aspect Ratio of form

The verticality of the form proportion provides several advantages. Because of change of solar ALT angles between seasons, walls receive most of the solar radiation in winter while roofs do that in summer. Accordingly, form A (the tower form) is better than form B for the following reasons:

- It has a larger wall area which promotes solar heat gain in winter.
- It has a smaller roof area which reduces solar heat gain in summer.
- It provides appropriate solar access to higher floors in winter.

Vertical Aspect Ratio = 1:1.5 – 1:4

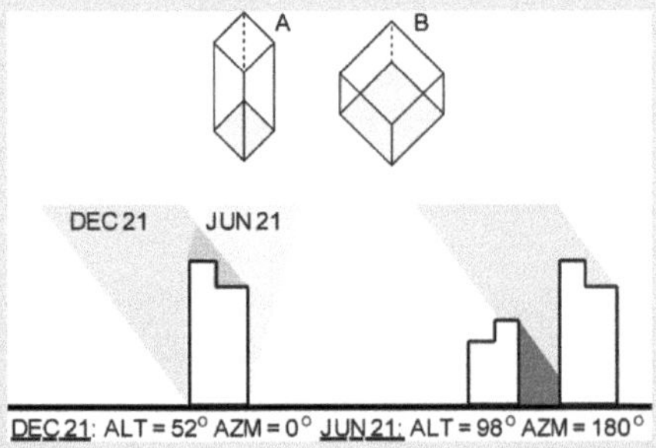

FORM response to CLIMATE

By selecting appropriate Vertical Aspect Ratio of form

The Sana'a house provides warmer upper floors than lower floors as a result of convective airflow currents; i.e., hot air moves to higher strata while cold air moves to lower ones.

The form of the Sana'a traditional house improves this effect by its vertical proportion and employs it to heating living spaces located in higher floors while cooling storage and service spaces located in the lower floors.

Vertical Aspect Ratio = 1:1.5 – 1:4

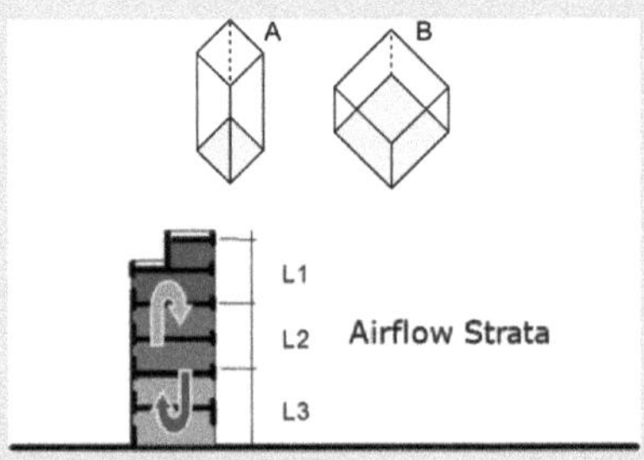

SPACE response to FUNCTION (via proper form)

By selecting appropriate Vertical Aspect Ratio of form that enables proper Space Arrangement

The vertical form allows the concept of a multistory house that provides different levels of privacy:

- L1: Spaces that need high level of privacy are located on highest floors.
- L2: Spaces that need lower level of privacy are located on middle floors.
- L3: Spaces that need lowest level of privacy are located on lowest floors.

Vertical Aspect Ratio = 1:1.5 – 1:4

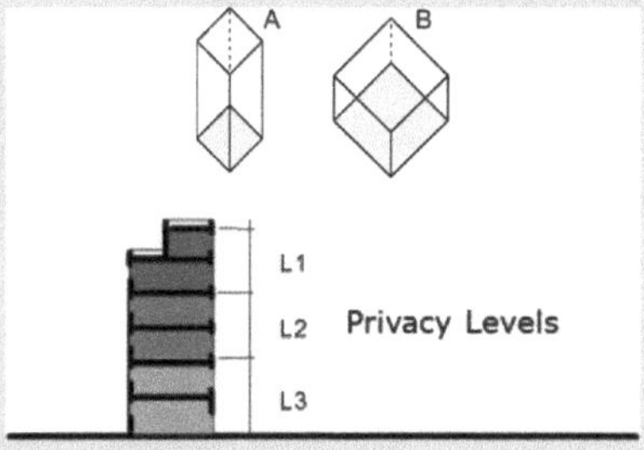

SPACE response to FUNCTION (via proper form)

By selecting appropriate Vertical Aspect Ratio of form that enables proper Space Arrangement

Form A (the tower form) is better than form B for the following reasons:

- It has larger wall area which promotes higher connection to the outdoor views.
- It has smaller core space that does not have connection to the outdoor views anyway.
- It is taller which provides more view ranges.

R1: cluster garden (form A, form B)
R2: nearby gardens (form A, form B less)
R3: city surroundings (form A only)

Vertical Aspect Ratio = 1:1.5 – 1:4

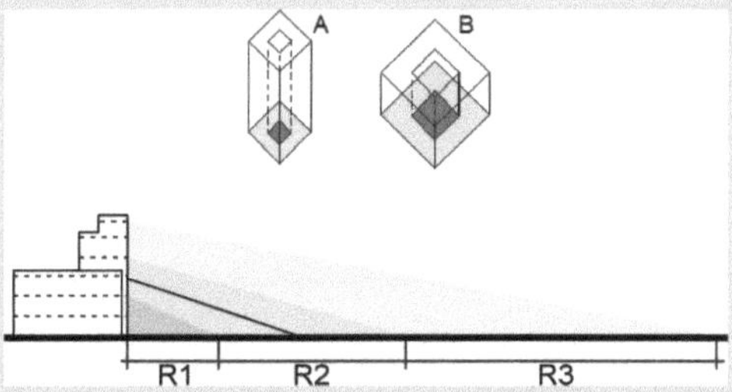

SPACE response to CLIMATE

By matching between Thermal Zones and Rooms' Ranks

1. Thermal zones are ranked according to their adequacy to provide thermal comfort. South is the best; east or west is average; north is the worst orientation.
2. Rooms are ranked according to their importance in terms of frequency of use, type of user, and type of activity.
3. Matching between ranks of zones and ranks of rooms.

A house can be made more energy efficient simply by designing the plan so that the order of rooms in which the normal daily sequence of activities occurs 'follows' the path of the sun [17]

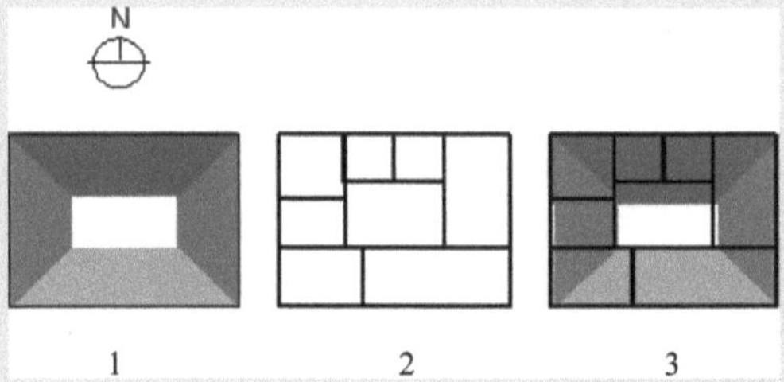

Building Skin and Multi-Component Window Settings

SKIN response to FUNCTION

By proper Windows' Vertical Arrangement

While higher floors have large windows to maximize visual connection to outside views, lower floors have smaller windows for privacy and security.

The vertical variation of windows' size and arrangement responds first to socio-cultural influences such as the people's activities and traditions then to climatic influences.

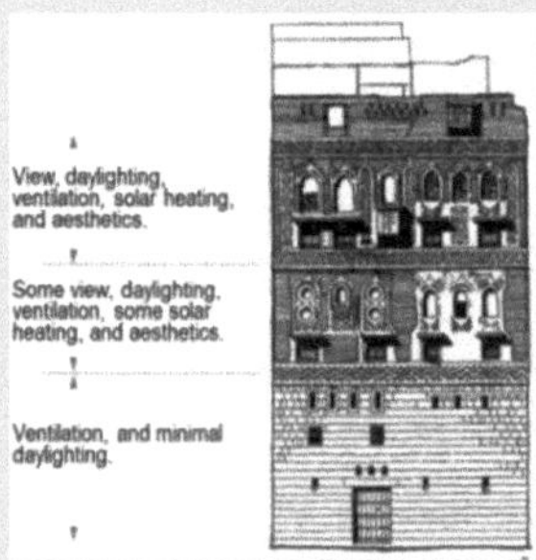

SKIN response to CLIMATE

By proper Windows' Horizontal Arrangement

The variation in the horizontal arrangement of windows (shapes, sizes and number) is a result of climatic influences such as solar motion and wind direction.

S facades have larger windows with more components and controls than N facades in order to exploit solar energy for heating.

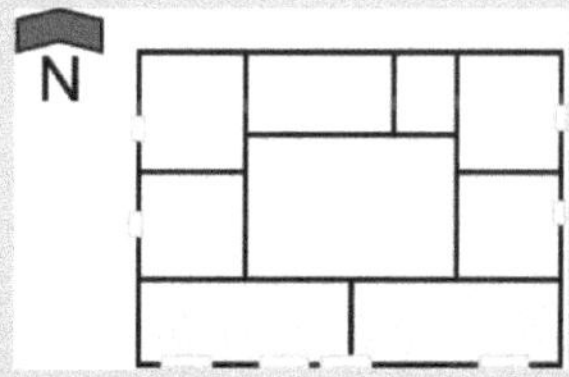

WINDOWS response to CLIMATE

By selecting proper Window' Size and Components

The indigenous window has four major parts to control the indoor environment: the upper section, the lower opening, the vents and the overhang.

Each part has a different environmental function. One window can consist of some or all the parts, depending on the space requirements and its location on the façade as well as the plan of the house.

Integration of the different functions of a window could be performed more efficiently by separating the window into components; each was responsible for 1 or 2 main functions [5].

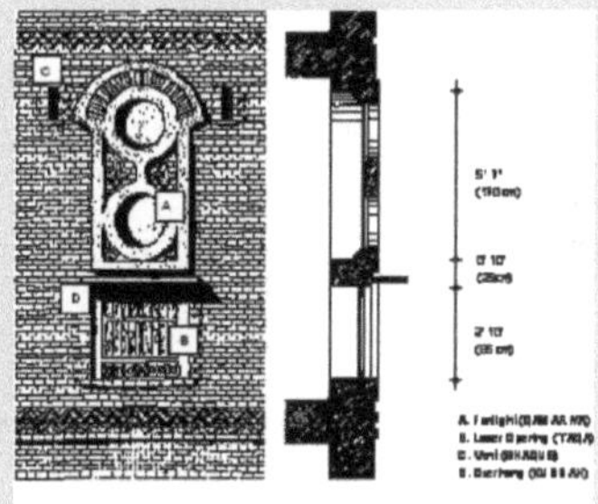

The Middle-Floors Window

WINDOWS response to FUNCTION

By selecting proper Window' Size and Components

The lower opening of the traditional window is sized according to the degree of privacy and connection to views. In this case, this window is large to maximize connection to views.

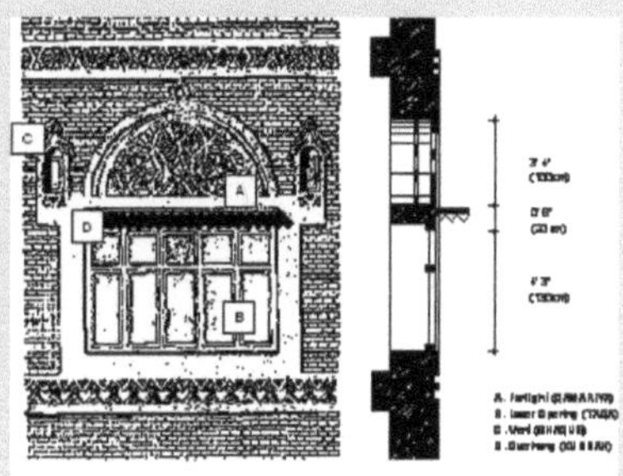

The Top-Floor Window

A Comparison between a Vertical Form and a Horizontal Form

APPENDIX B, TABLE 1 Perspective Views of Both Forms as Seen by an Imaginative Eye Located in the Center of the Sun During Different Hours of the Solar Solstices (i.e., Dec 21 and Jun 21) and Sol Solar Equinoxes Days (i.e., Sep 21 and Mar 21) [10]

	Dec, 21	Sep, 21/Mar, 21	Jun, 21
8:00 AM	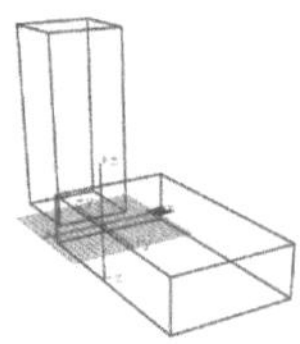	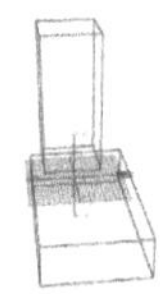	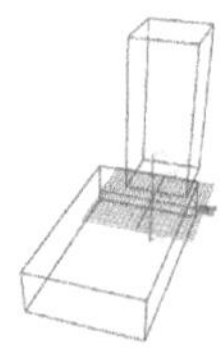
10:00 AM	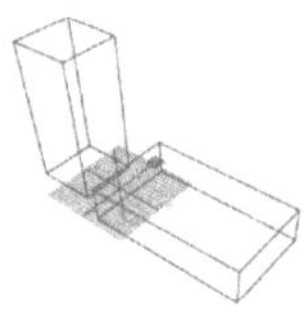	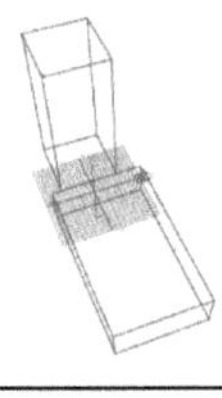	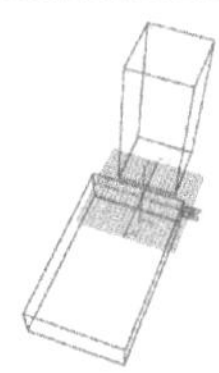
12:00 N	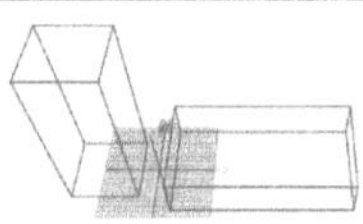	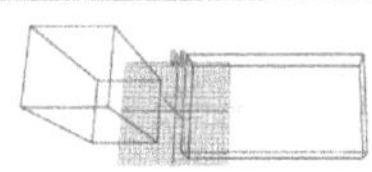	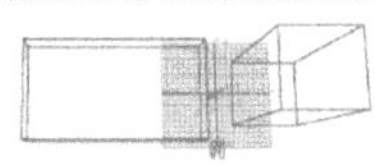
2:00 PM	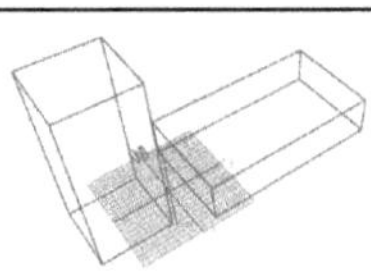	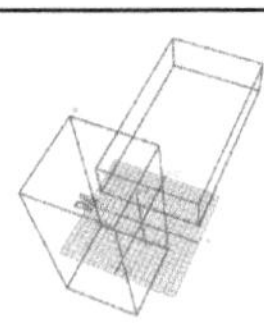	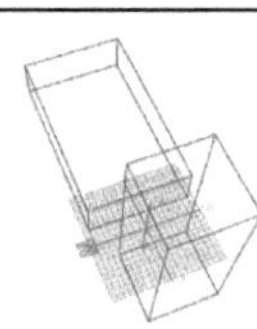
4:00 PM	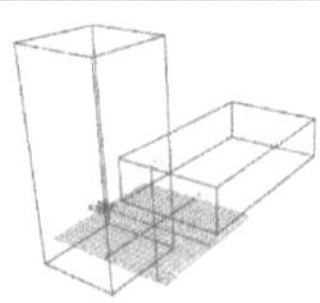	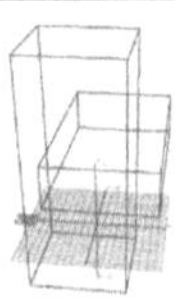	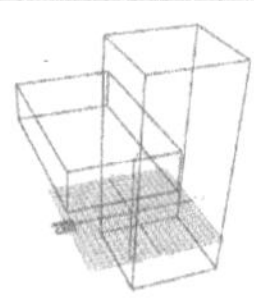

APPENDIX B, TABLE 2 **A Comparison between a Vertical Form, which Represent a Vernacular Tower House of Sana'a, and Another Imaginary Horizontal Form, Assumed to have the Same Total Area of Floors [10]**

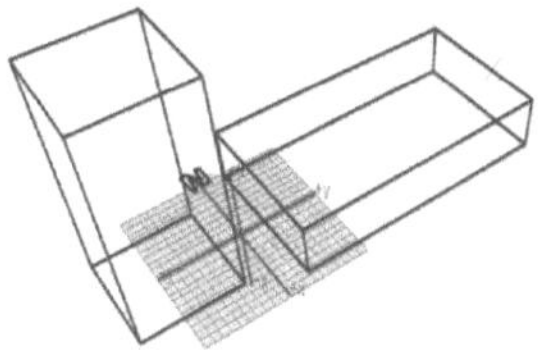

Perspective view of both forms as seen by an imaginative eye located in the center of the sun on Dec 21, 2:00 PM

Physical Aspects	Vertical Form	Horizontal Form	Remarks
Total area of floors	1152 m^2	1152 m^2	Assumed to be equal
Number of floors	8	2	Number of floors of vertical form = 4 times that of horizontal form
Roof area (also floor area)	144 m^2	576 m^2	Roof area of horizontal form = 4 times that of vertical form
Floor height	3.75 m	3.75 m	Floor height of both forms are equal
Total height	30 m	7.5 m	Total height of vertical form = 4 times that of horizontal form
Volume	4320 m^3	4320 m^3	Volume of both forms are equal
N or S façade area	360 m^2	240 m^2	N or S façade area of vertical form = 1.5 times that of horizontal form
E or W façade area	360 m^2	135 m^2	E or W façade area of vertical form = 2.7 times that of horizontal form

Chapter 11

Sustainable Buildings in Mediterranean Area

Marco Sala and Alessandra Carta
Università degli studi di Firenze, Firenze, Italy

Chapter Outline

Sustainability, Energy and Architecture. ISBN: 9780123972699

11.1 ABITARE MEDITERRANEO PROJECT

Abitare Mediterraneo is an applied research project, sponsored by the Tuscany Region under POR CREO FESR 2007–2013, and developed by the University of Florence in synergy with some construction companies.

The Project aims to realize an 'Open System' to promote technological innovation and architectural quality in the construction process, in order to encourage the development of building initiatives focused on high energy efficacy; the catalogue of the 'Open System' is a flexible tool dedicated to help enterprises to promote innovative products in Mediterranean areas. The research proposes a formula to create a synergy between research and production in the building sector, and also involves a 'construction model' to adopt for architecture in this kind of climate.

It is dedicated to draftsmen and enterprises as a promotion tool supporting design and planning in the Mediterranean climate. Its purpose is to transform Tuscany into an International Laboratory for Research into high quality living in the Mediterranean area; to develop the analysis of case studies in order to promote the future of Environmental Sustainable Buildings, designed in the context of history, culture and the Mediterranean climate.

One key objective fosters the creation of a 'Center for Technological Competence' as a benchmark for research, innovation and implementation of environmental sustainability, eco-efficiency, quality and livability.

Innovative results:

- Test Cell – outdoor laboratory for looking at the thermal dynamic behavior of façade components

Prototypes:

- MIA – temporary living module specific to the Mediterranean climate
- Domino – façade system that guarantees significant energy savings
- AIW – façade system with an integrated heat exchanger
- Shading Screen – innovative ventilated wall

Experimentation:

- Lorenzana/Rispescia – two innovative residential houses designed within a sustainable approach

11.2 EULEB

The EULEB project – 'European high quality low energy buildings' provides information on existing, public, non-residential, high quality and low energy buildings from all over Europe.

This project was realized within two years (2005–2006) and has been partially funded by the 'Intelligent Energy Europe' program of the European Commission.

The following case study analysis is taken from EULEB (www.euleb.info) and from Abitare Mediterraneo research (www.abitaremediterraneo.eu), with the aim of providing information on high-quality public buildings in the Mediterranean area of Europe.

11.2.1 Location

Europe contains a large variety of differences in buildings' boundary conditions. The map shows the location and climatic conditions of buildings, classified according to the KOEPPEN classification system.

This system uses letter codes to identify the major climate zones: (A) tropical forest, (B) dry forest, (C) warm temperate rainy, (D) cold forest and (E) polar regions. Further subdivision according to temperature, rainfall and seasonal variations is described through sub-codes:

BSk: Mid-latitude steppe. Semiarid, cool or cold.
Csa: Interior Mediterranean. Mild winter and dry hot summer.
Csb: Coastal Mediterranean. Mild winter and dry, short, warm summer.
Cfa: Humid subtropical. Mild winter and moist in all seasons.
Cfb: Marine. Mild winter and moist in all seasons. Warm summer.
Cfc: Marine. Mild winter and moist in all seasons. Short cool summer.
Dfb: Humid continental. Severe winter, moist in all seasons with a short warm summer.
Dfc: Subarctic. Severe winter, moist in all seasons with a short, cool summer.
ET: Tundra. Very short summer.

11.2.2 Building Classification

Three building types are considered:

Education:

This group contains buildings which are dedicated to education on different levels. School and university buildings can be found here.

Office:

Public office buildings are mainly administrative buildings of the national or regional government, but research institutes also form part of this group.

Leisure facilities:

This group consists of a selection of museums and exhibition centers.

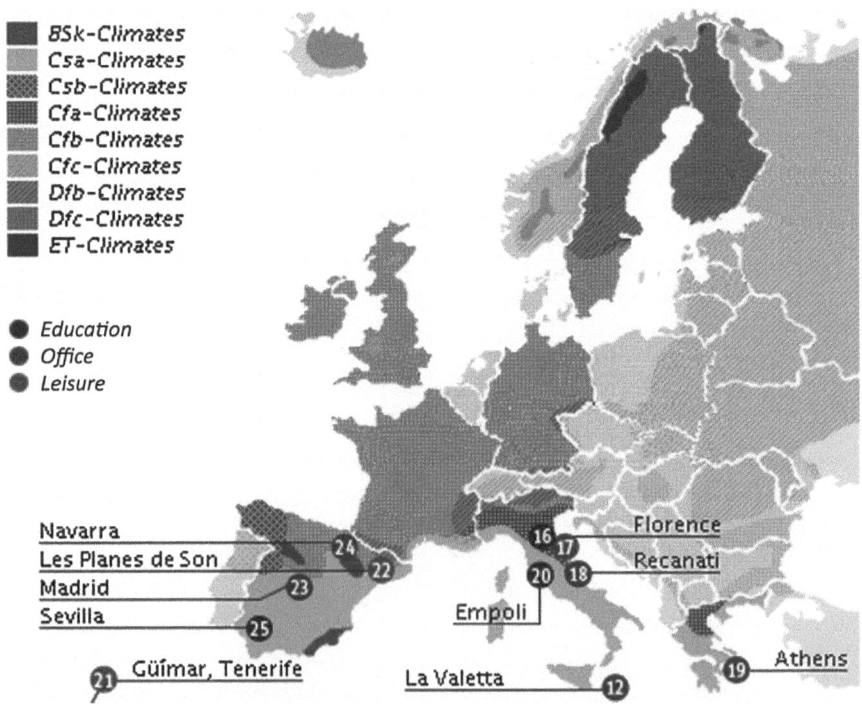

11.3 TECHNOLOGICAL AND BUSINESS INCUBATOR – LUCCA, ITALY

11.3.1 Identification

Name: Technological and business incubator
Owner: Camera di commercio Industria ed Artigianato Lucca
Country: Italy
City: Lucca
Street: Via della Chiesa in Sorbano del Giudice
Occupant(s) of Building: Arbitral Chamber of Lucca Headquarters
Typical days/hours of use: 9 h.
Designers: District of Lucca
Engineers: District of Lucca
Energy consultants: Centro Interuniversitario ABITA Prof. Marco Sala
Contractors: Skills center and ICT
Energy sources:
  Solar tubes – Greenhouse
  Photovoltaic – Daylighting strategies
  Natural ventilation – Radiant panels
Year of completion: 2011

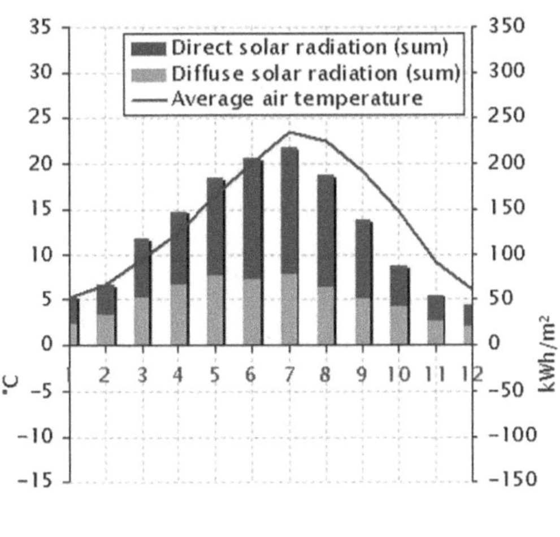

11.3.2 General Data

Number of floors above ground: 3
Number of floors below ground: 0
Heating or cooling gross floor area: 17.092 sqm
Usable floor area: 3.813 sqm
Building envelope area: 7.312 sqm
Average number of occupants: 300

11.3.3 Outdoor and Indoor Climate

Microclimate: urban
ASHRAE degree days heating/cooling: 2060 / 1917 Kd
Outdoor design temperatures/humidities: –2°C; RH 50%
Indoor design temperatures/humidities: 20°C
Design ventilation rates: 25 air changes of outside air per hour
Design illuminance levels:
Offices: 500 lux
Labs: 500 lux

The project's goal was to create a real technological pole including a group of buildings featuring environmentally friendly solutions. Natural solutions are developed according to the most modern standards of sustainable housing and energy saving so as to really serve innovation.

Strategy of the project:

- Choice of orientation and shape of the building;
- Highly efficient envelope integrated;
- Roof garden;

- Active integrated system for the exploitation of solar energy (solar and photovoltaics);
- Use of geothermal heat pumps and combined with trigeneration;
- Use of natural ventilation;
- Analysis of natural lighting of offices and labs;
- Use of innovative intelligent windows system;
- Sliding shutters for vertical and horizontal surfaces;
- Use of radiant floors and roofs;
- Use of 'natural' materials (mortar, plaster, stone and wood coatings, natural insulating).

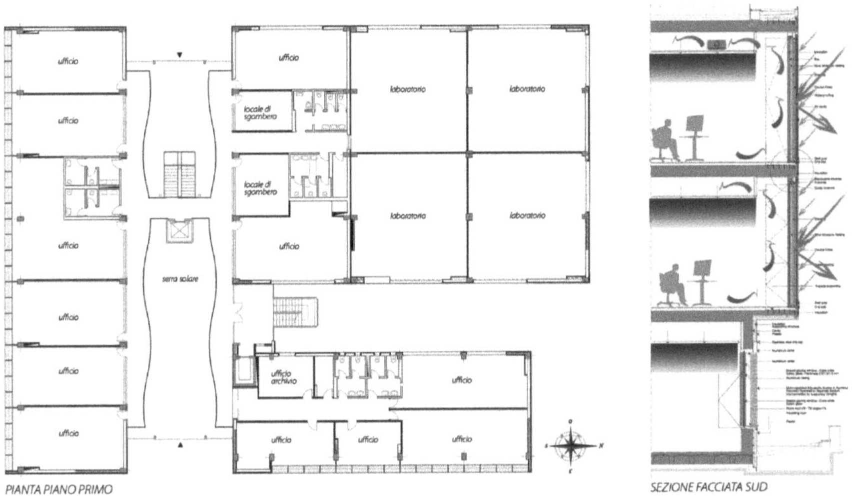

11.3.4 Insulation

Efficient thermal insulation, realized with a polystyrene panel, a fiberglass mattress and a steam barrier contribute to the energy-saving design of the building. The external walls of the offices and laboratories are designed as thermo-ventilated walls, faced with aluminum panels. The thermo-ventilated wall is characterized by the presence of a thermal insulation layer and an air gap, which give the envelope high thermal performance during the whole year, with a U-value of 0.29 W/mqK. They are composed of the following layers: lightweight plaster, blockwork, polystyrene insulation (6 cm), ventilated cavity (4 cm), aluminum panels.

11.3.5 Solar Control

11.3.5.1 Windows and Shading

Dynamic and innovative elements were designed for the south-west facing façades to improve the summer and winter performance of the building, containing integrated photovoltaic panels. The façade was designed to be a sunbreaker system, with grilles and openings to satisfy the various climatic needs. The philosophy which inspired this is that we should not be affected by the climatic conditions, but rather to exploit them.

Windows: south and east façade: clear inner glass: Uw = 0.9 W/m^{2}K, matt inner glass: Uw = 0.9 W/m^{2}K, vertical shielding, mosquito nets frame, glass and external photovoltaic panel; total UW = 1.1 W/m^{2}K.

11.3.5.2 Lighting

Large windows that connect the interiors, greenhouse and skylights on the roof are designed to deliver optimum levels of natural light in the work spaces and other open spaces, in order to minimize the use of artificial lighting. The openings and shields have been designed to prevent direct glare, ensuring a good distribution of natural light into the space. All installed lamps are high efficiency installations, and the total annual electricity demand is 12.3 kWh/m^2. These give an energy saving of about 35% compared to the energy demand of a similar building without these features.

Sun pipes and light ducts are used to improve daylight in laboratories; sun pipes are installed in the roof garden, and they contribute to the daylight levels in the labs. The use of sun pipes allows the artificial lights to be switched off during the morning throughout the year.

11.3.6 Cooling

Radiant panels for cooling and heating were installed in the offices; radiant solutions for cooling allow transmit physiologic wellbeing free of the costs and damage caused by air conditioning systems. Air conditioning is often the cause of cooling. Health problems, easier diffusion of germs, discomfort, sharpened by poor management and also involving high costs. The radiant system prevents all these drawbacks, especially in terms of cooling.

11.3.7 Ventilation

The decision to fit the façade systems with domestic systems to enable natural ventilation during the intermediate seasons and summer has produced a further 30% reduction in energy needs compared to a traditional building. It was also decided to equip the outer components (doors and windows) with heat reducers. These further cut down on energy consumption in winter by recovering the heat from the outgoing exhaust air and reusing it to heat the incoming air. Some protected openings are specifically dedicated to ventilation, either automatically or semi-automatically controlled. These allow the formation of air currents inside the building, thus guaranteeing appropriate natural ventilation.

11.3.8 Energy Performance

The annual estimated specific energy consumption is 75 kWh/m^2 per year. The specific values shown in the figure include the mechanical installation for cooling/heating systems, which is 50 kWh/m^2, and the other electrical installation for lighting, which is 25 kWh/m^2.

Estimated heating energy consumption during the winter is 25 kWh/m^2 per year.

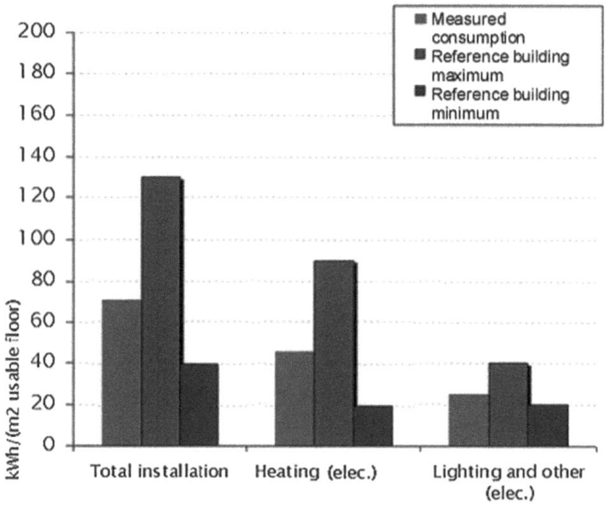

Annual primary energy consumption per m^2 usable floor area.

11.3.9 Monitored Comfort

Location: Office; the view is of south facing façade
System: Low emission glazing
Sky condition: Sunny day with direct sun on façade
Description: The luminance picture shows a good distribution of luminance on surfaces, also into the depth of the rooms

11.3.10 User Acceptance

Users of this building reacted well to the project specifications since the inauguration on 2012, but it wasn't possible to measure user acceptance so far.

11.3.11 Financial Data

Building:

The total investment cost of the building per m^2 of gross floor area is 1637,00 €.
The total cost is divided into:

Planning and safety: 340,00 €/sqm
Construction: 710,00 €/sqm
Technical systems: 586 €/sqm divided into:
- Electrical systems: 129 €/sqm
- Thermal systems: 180 €/sqm
- Fixture: 29 €/sqm
- Special systems: 248€/sqm

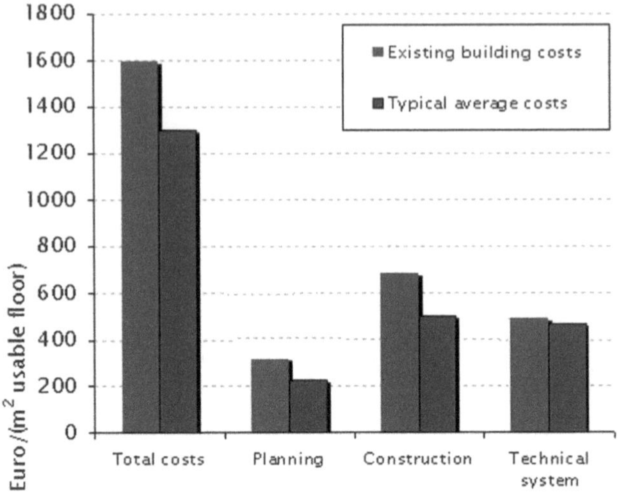

11.4 BARDINI MUSEUM – FLORENCE, ITALY

11.4.1 General Data

Number of floors above ground: 3
Number of floors below ground: 0
Heating or cooling gross floor area: 3,200 sqm
Heated or cooled volume: 76,784.4 m^3
Building envelope area: 3,431 sqm
Average number of occupants: 150

11.4.2 Identification

Name: Bardini Museum
Owner: Municipality of Florence
Country: Italy
City: Florence
Street: Piazza dei Mozzi 1
Occupant(s) of Building: Bardini Museum
Primary Use: Museum
Typical days/hours of use: From Monday to Saturday 9.30 am to 7.00 pm, except during summer
Designers: Arch. Lombardi
Engineers: Structural Eng. Giancarlo De Renzis; Electromechanical Eng. Roberto Innocenti; Electromechanical Eng. Raffaele Viscomi
Energy consultants: Centro Interuniversitario ABITA Prof. Marco Sala, Prof. Paola Gallo
Contractors: Municipality of Florence
Centro Interuniversitario ABITA
Energy sources: Daylighting strategies
Natural ventilation
Year of completion: 2003

11.4.3 Outdoor and Indoor Climate

Microclimate: urban
ASHRAE degree days heating/cooling: 2060 / 1917 Kd
Outdoor design temperatures/humidities: −2°C; RH 50%
Indoor design temperatures/humidities: 20°C
Design ventilation rates: 16 changes of outside air per hour
Design illuminance levels: 200 lux on oil paintings, 50 lux for watercolors

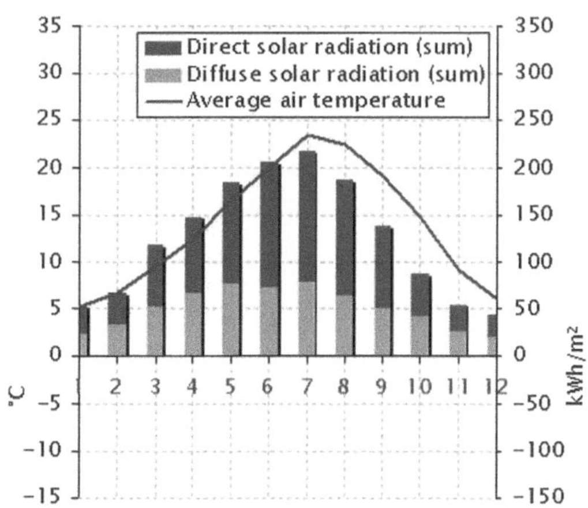

The building achieves specific energy targets with very substantial energy reductions, more than 30% according to measurements. It is estimated that application of the proposed measures to the studied building may easily achieve:

- up to 30% reduction of CO2 emissions;
- up to 25% reduction of heating loads;
- up to 30% reduction of electrical loads.

The set targets have been achieved with acceptable investment and running costs; in fact the application of the total package will be achieved within an amortization period of close to 15 years. In order to reach the proposed standards, the project works on three levels of intervention:

1. Thermal comfort by the installation of an HVAC system.
2. Energy saving supplementing passive strategies and the use of low efficiency appliances.
3. Daylighting comfort through the design of skylights and the installation of new lamps.

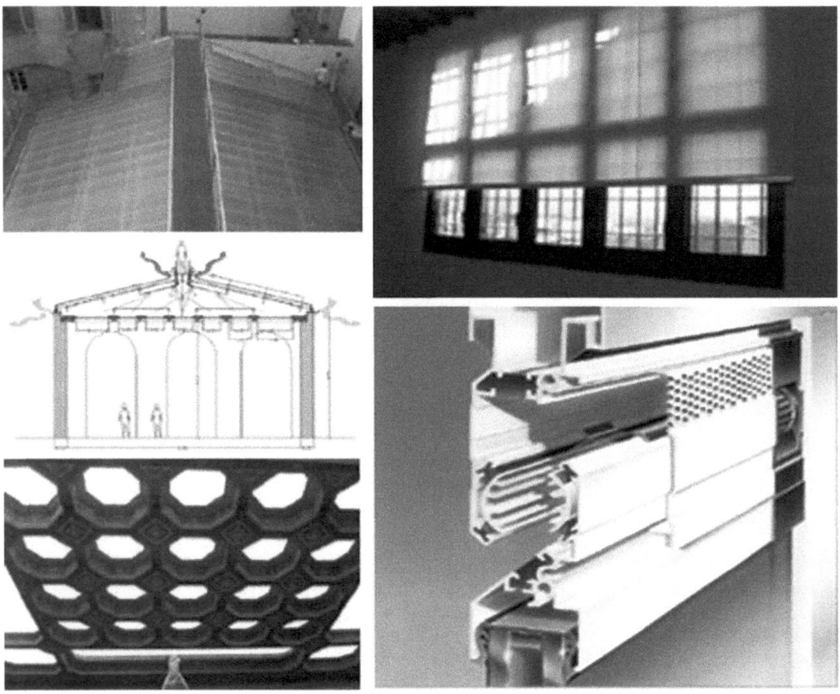

11.4.4 Solar Control

Top floor windows are fitted with system METRA NC 65 STH frames, in painted aluminum 6060 according to UNI EN 573 UNI EN 755-5 to the physical state T5 according to UNI EN 515. For solar control, the design foresees double glazing (6/7+gap 12+8/9) with UV film applied in order to reduce the glare effect on the exhibited materials. For the reduction of heat gains and direct sunlight ingress, new windows have been equipped with a special UV absorbing glass in order to reduce ultraviolet radiation to 75 microwatts per lumen, and screens have been placed in front of the windows.

11.4.5 Lighting

The high electrical consumption of the Bardini Museum was caused by excessive power, low control flexibility and inefficient lamps.

The first action was to reduce the installed power by replacing the existing lamps with high efficiency ones. This action has not only drastically reduced the energy consumption, but has also increased visual comfort inside the exhibition halls. This is due to devices equipped with special reflecting floodlights, which can direct more light into the exhibition space, reducing glare and ensuring optimal illuminance levels while using only half the number of lamps used previously.

To optimize daylighting, after simulations had been performed using specific software tools (Radiance), the glazed elements of the central skylight were entirely removed and replaced, together with the overhanging transparent roofing.

11.4.5.1 First Floor

Instead of the heavy glazed roofing, Thermoclear luminaries were used: these are composed of a transparent, 30 mm, twin-welled, polycarbonate panel with a special reflector which can reduce glare and increase illumination levels. With regard to the existing wooden false ceiling, all the bullet-proof glass will be replaced with special high transmittance diffuser components (Barrisol system).

11.4.5.2 Ground Floor

With regard to different types of exhibitions, new lamps have been installed in order to ensure the right illuminance levels as follows (combining daylight and electric light):

- 200 lux for oil paintings
- 50 lux for watercolors

11.4.6 Cooling

The installed centralized heat pump system (two pipes) has a reversible direct expansion cycle with a variable cooling volume, of a modular sort, and is divided into four zones. The system uses ecological cooling gases such as R407C. A cooling/heating controlled air system has also been installed that, combined with passive natural cooling strategies, allows optimal temperature and relative humidity parameters to be achieved.

11.4.7 Ventilation

A ventilated roof system was designed to carry out two different kinds of 'air circulation' in the roof:

- The first micro-ventilation under the tiles is essential to prevent stagnation of humidity;
- The second macro-ventilation under the roof, activated between the tiles and the insulating layer.

During summer season, heat stored thanks to the envelope's thermal mass, is transferred to internal spaces during the night; night cooling is increased by the opening of special grids designed into the window frames. Based on this principle, thermal exchanges during the night are planned from 10.00 pm to 8.00 am. The combined effects of these strategies decreases maximum internal temperature of 1–2°C during the day, with a significant energy saving.

The intelligent windows allow different interconnected functions such as ventilation, solar and anti-glare protection, heating, cooling and sound insulation for summer and winter on a largely individual basis without losing the psychological effect of an opening.

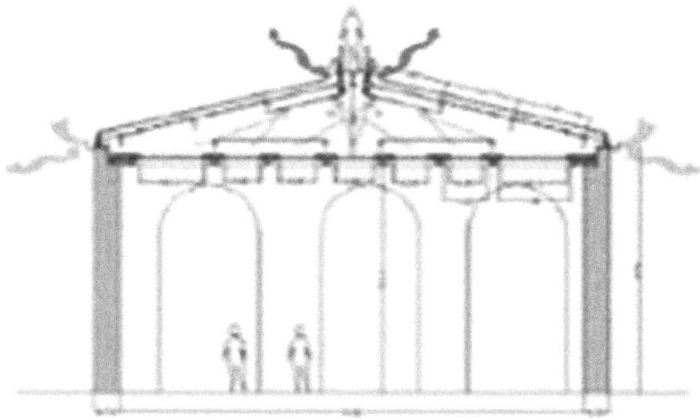

11.4.8 Energy Performance

The project increased the roof insulation levels, and installed a cooling/heating controlled air system and passive natural cooling strategies: these allowed optimal temperature and relative humidity parameters to be achieved (to suit both staff and exhibits). The increase in insulation has produced a reduction in the roof's U-value from 1.90 W/m^2°C to U = 0.36 W/ m^2°C. The insulating panels (10 cm) are made of natural cork, installed without artificial additives and adhesives.

In a museum, the difficulty of maintaining set environmental conditions without increasing the energy consumption needs advanced control systems to optimize the final energy balance. These intelligent systems, formed from three basic elements (sensor, controller, actuator), can manage a large number of sensors (such as fire-alarms, smoke-alarms, ventilation, security and air treatment plants) according to the various required internal comfort levels. In the Bardini Museum project, BMS has been used for thermal-hygrometric and daylighting control of the rooms using humidity, temperature, occupancy and illuminance sensors.

The chart shows the difference between thermal and electrical consumption calculated in kWh/m^2 per year, in standard buildings and in the revamped museum. The overally savings are: heating 48%, cooling 48%, ventilation 26%, lighting 53%, other 20%; total energy saving 48%.

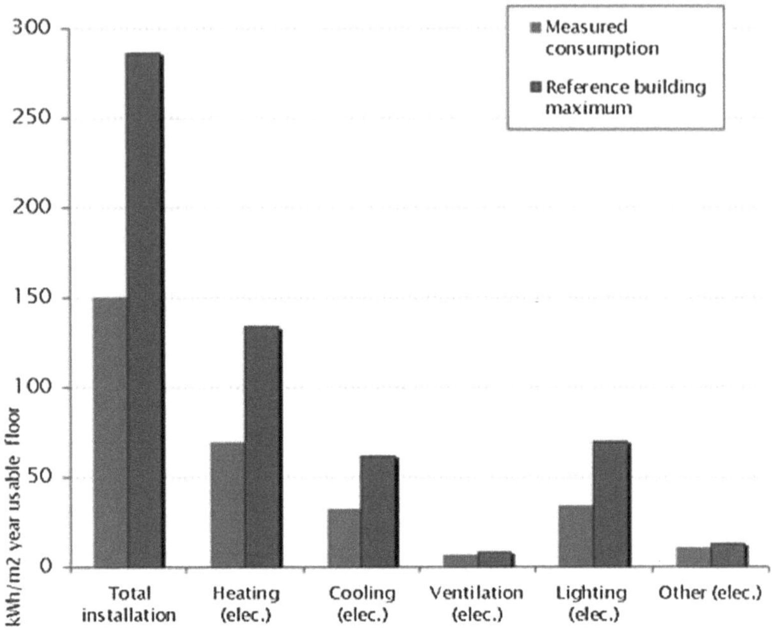

11.4.9 Monitored Comfort

The as-is state (picture on the left) shows the surface at the back of the image with an illuminance value that is over the maximum admitted by the laws. To reduce the illuminance to around 200 lux, the project (picture on the right) foresees the introduction of a glass film and of a translucent diffuser that is set inside the window panes and can be lowered or raised, as a flowing curtain, in order to exploit the possible penetration of natural light during periods of low luminance, while always trying to avoid the entry of direct solar rays.

Thanks to this screening system the illuminance of the wall complies with the limits set by law.

11.4.10 User Acceptance

Questionnaires have shown a general increase in perceived comfort, both by staff and visitors. Employees have reported a great improvement in thermal comfort during both winter and summer. This is probably because they become accustomed to working in unheated/cooled spaces. They also noted not just that the temperature had increased in winter and decreased in summer, but that they really felt good, being comfortable throughout the year.

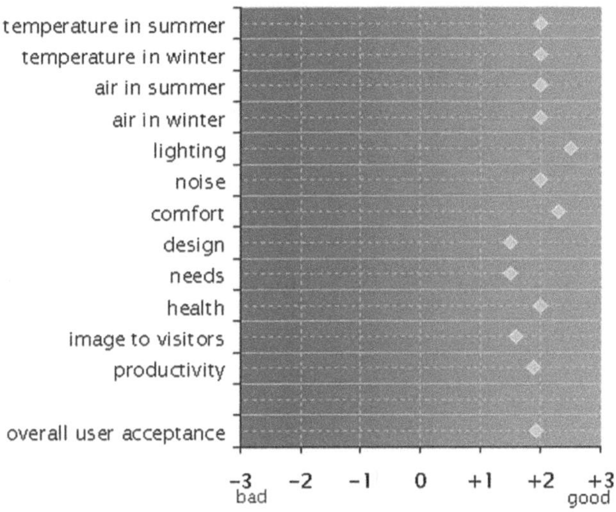

11.4.11 Financial Data

Architectural design: € 42,000
Energy and environmental design: € 47,000
Construction: € 29,000
Monitoring: € 27,000
TOTAL: € 145,000

The chart shows that immediate cost savings are mostly achieved in the building construction, planning and technical systems, which impact the total cost. But we must always remember that this type of building requires lower maintenance, so reducing future general management costs, which will have a higher impact in the future and shorten the payback time.

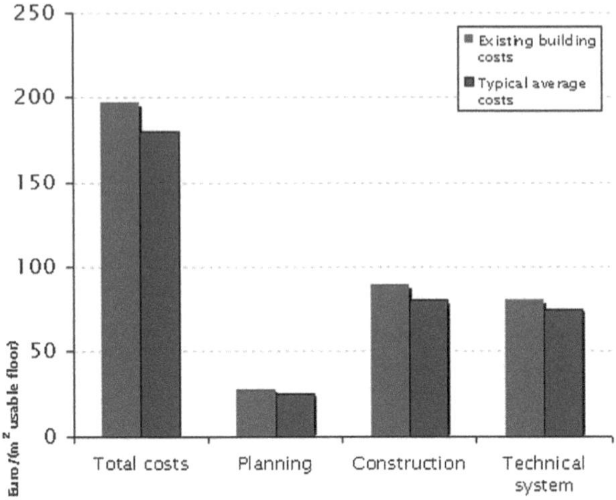

11.5 NEW MEYER HOSPITAL – FLORENCE, ITALY

11.5.1 General Data

Number of floors above ground: 2
Number of floors below ground: 1
Heating or cooling gross floor area: 21,600 sqm
Usable floor area: 15,000 sqm
Heated or cooled volume: 60,238 m^3
Building envelope area: 32,671 sqm
Average number of occupants: 130 patients + 35 day patients

11.5.2 Identification

Name: Meyer Children's Hospital
Owner: Fondazione Meyer
Country: Italy
City: Florence
Street: Careggi
Occupant(s) of Building: Meyer Children's Hospital Staff
Use: Hospital
Typical days/hours of use: 24h
Designers: Studio Cspe-Anshen Dyer – Studio Chiarugi
Engineers: A6l Ingegneri Associati – CMZ – Studio Lombardini Engineering
Contractors: First site (central pavillion and technological platform area): Grassetto e Gemmo, Second site (lateral parts and greenhouse): Cogepa
Parking: Montinaro
Energy sources: Solar tubes greenhouse, lighting strategies, thermal mass, photovoltaic
Natural ventilation
Year of completion: 2006

11.5.3 Outdoor and Indoor Climate

Microclimate: urban, ASHRAE degree days heating/cooling: 2060 / 1917 Kd
Outdoor design temperatures/humidities: minimum 10.6°C maximum 40.2°C
Indoor design temperatures/humidities: 22°C
Design ventilation rates: 35 air changes of outside air per hour
Design illuminance levels:
  Offices: 420 lux
  Atrium: 280 lux
  Circulation space: 150 lux
  Patient rooms: 350 lux

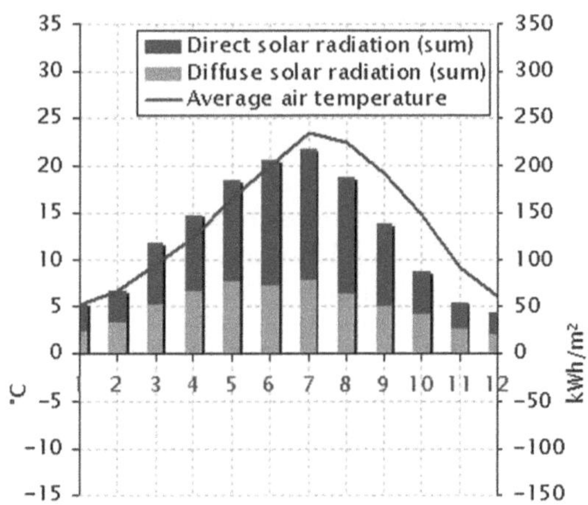

Innovative technologies in Meyer Children's Hospital include:

- Sun pipes and light ducts: daylight inside the hospital will not only be a good solution for energy saving, but will impact on good spirits.
- Green roof: the green roof has a strong character in this project.

The idea that the hospital is a place in which psychological aspects are very important for children and parents suggested the creation of green roof terracing with gardens.

- Buffer space on north façade: this will be used during rainy days in winter as a hall. The particular section and orientation of the buffer space will contribute to solar gains during winter. In summer, it is partly openable to reduce overheating.
- Optimum insulation inside walls: simulations have been done to find the optimum cavity insulation.
- Insulation material used on the first and second floors is recycled material.
- Radiant panels: to achieve a better and uniform temperature in patient rooms.
- Condensing combi boilers for a high efficiency heating system.
- Shading for patient rooms and halls.

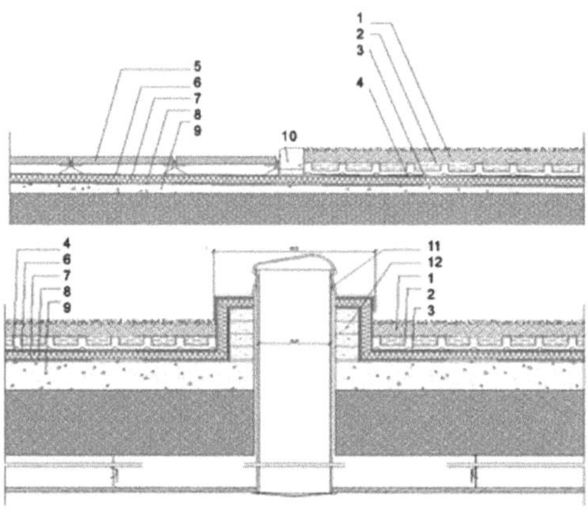

Transmission losses are stated in terms of the heat flow through the envelope: that is, the quantity of energy which passes through the envelope per unit of time. These losses depend mainly on the temperature difference between the inside and outside faces of the envelope and the thermal resistance of the material – or combination of materials – of which the envelope is made. These losses take place through conduction, convection and radiation. One method of reducing them is to prevent heat conduction by adding thermal insulation to the envelope in order to increase its thermal resistance.

The patient room of the Meyer Hospital has been carefully studied: note the two drawings of the external cavity wall with insulation inside. The description is of the wall used for the hospital building, with 4 cm of thermal insulation and a U-value of 0.37 W/m^{2}K. The use of this improved wall insulation on surfaces that are directly exposed to external climatic conditions (19 sqm for the chosen patient room) reduces annual energy consumption for heating, giving an annual energy saving of 12%.

11.5.4 Green Roof

To reduce transmission losses as much as possible it is necessary to insulate all the opaque elements in the building, not just the walls.

Investing in green roof technologies helps to diminish the environmental impact on our communities while providing a fresh approach with visually appealing organic architecture.

Benefits:

- Lower environmental impact.
- The grass absorbs the solar radiation and the evaporation process reduces air temperature and delta T.
- Transmission loss reduction.

The green roof package produced for the Meyer Hospital has a U-value of 0.79 W/m^{2}K compared to a traditional flat roof with a U-value of 1.16 W/m^{2}K.

The proposed solution increases the insulation material in the cavity wall, and the contemporary use of a green roof reduces the annual energy demand for heating by 36% per patient room.

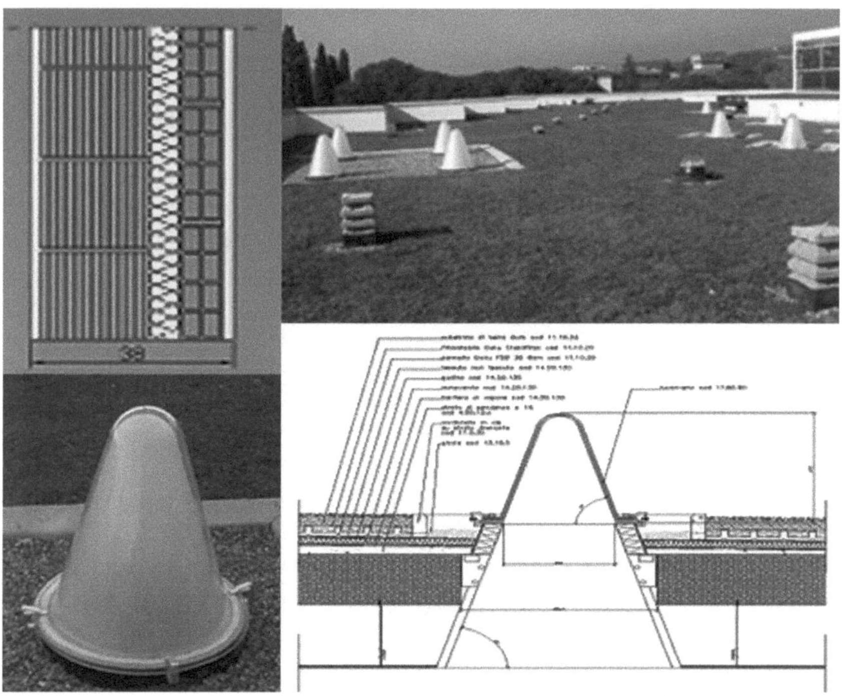

11.5.5 Solar Control

11.5.5.1 Windows and Shading

The windows are constructed with wooden frames. Patient rooms are protected from direct sunlight by an overhanging structure externally covered with copper plates, to reduce the visual impact of the building in the park, with an internal surface covered in wood. The greenhouse is shaded by internal white blinds, which are adjusted by an automatic control system. This shading device is a system of sails.

11.5.5.2 Lighting, Sun Pipes and Light Ducts

Sun pipes and light ducts are used to improve daylight levels in corridors and halls. Sun pipes are installed in corridors in front of patient room windows: they add just a small contribution in terms of daylight factor in the rooms themselves, but they have a positive impact on patients psychologically. The use of sun pipes allows artificial light in corridors to be switched off during the morning, throughout the year. Sun pipes send sufficient daylight into patient rooms every day of the year.

The best possible energy saving is around 60%, but this will depend on the efficiency of the energy facility manager of the hospital and on the system being used sensibly.

Positive surroundings, with plenty of daylight and high thermal comfort levels, are an important aspect of patients' wellbeing. Sun pipes are installed to achieve a good illuminance value in patient rooms. Solar-tubes and roof-lights in corridors and halls give a good level of daylight. During an overcast day, a DF of 2.5% in the principal corridors and 1.5% in the others without windows can be reached; this means that it should not be necessary to use artificial light in several spaces during the morning.

For daylight calculations we have to specify an overcast sky, but in the climatic area analyzed there are fewer cloudy days than sunny days.

All installed lamps are high efficiency and the total annual electricity demand is 12.3 kWh/m^2. Compared to the energy demand in which all these features are not applied, the energy saving is of about 35%.

11.5.6 Heating

Heat pumps are used to generate heating and cooling. These are appropriate where both summer cooling and winter heating are required. Radiant panels and high efficiency boilers are used for the heating system. Radiant floor heating panels are installed in patient rooms, in which we want to achieve a good level of thermal comfort at a low energy cost.

For winter heating and DHW generation there are two boilers: they are condensing combi boilers with an efficiency of about 106%. The boilers use gas and not electricity. Another, conventional boiler is also installed.

11.5.7 Cooling

For summer cooling there are two electrical chillers. A third chiller is of the water/water type: the heat generated from this last machine is used for DHW.

In the hospital there are also two heat-pumps to be used in case of emergency (i.e., if the gas does not reach the hospital because of gas supply problems). They are also used in summer, but only when necessary. A thermostatic valve inside the patient room area will measure temperature and relative humidity; when the temperature falls below 21°C (in winter) or above 27°C (in summer) the heating/cooling system will be switched on.

The installation of a control system to set temperature, relative humidity and air velocity, together with the clothing level value and metabolic rate, will give a predicted PPD (predicted percentage dissatisfied) of below 6%. This is the percentage of patients who are dissatisfied (uncomfortable) in their patient room.

11.5.8 Ventilation

Ventilation is done by manually opening windows, which move up and down.

A combination of shading and ventilation systems can keep the indoor temperature 10°C below the outside temperature. To save on cooling energy, passive cooling and ventilation techniques are used as much as possible, with air conditioning operating only where necessary. A sun space functions as a buffer area for the building. The heated air is used to create solar draughts, thus providing a natural air flow through the building. A centralized energy management system selects the best operational strategy in each case.

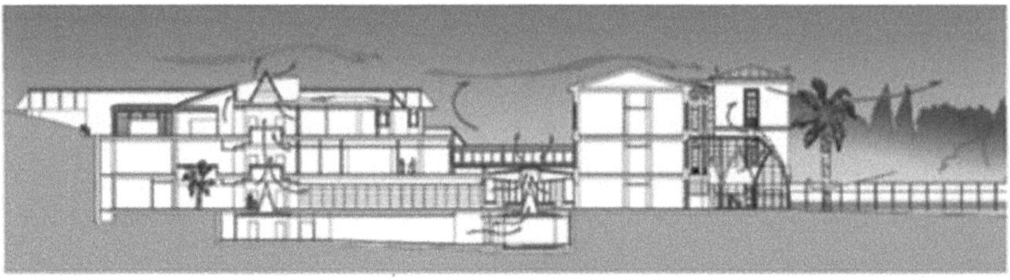

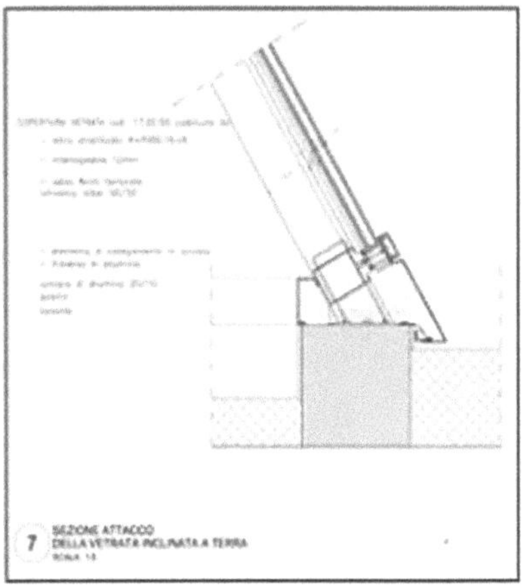

11.5.9 Renewable Energy

The Meyer's photovoltaic greenhouse is a south facing structure with unobstructed solar access to the main solar glazing in order to maximize the collection of winter sunshine; it is not only a particular type of structure but also, and more importantly, a particular kind of space. The design objective not only considered energy and environmental aspects but also its social impact. The primary objective was to create a pleasant 'socializing' space which can be used for semi-outdoor activities during most of the year without any extra energy needed; a social space well-integrated into the adjacent green park. PV installation integrated into building greenhouse façades allows energy production to be combined with other functions of the building envelope, such as shading, weather shielding and heat production. Cost savings through these combined functions can be substantial, e.g. in expensive façade systems, the cladding costs may equal the costs of the PV modules. Additionally, no high-value land is required and no separate support structure is necessary. Electricity is generated at the point of use. This avoids transmission and distribution losses and reduces the utility company's capital and maintenance costs. The photovoltaic system is 30 kWp and uses glass/glass PV modules.

11.5.10 Co-Generation

The co-generation plant is a gas turbine, with an electrical power of 7.5 Mwe (ISO), which allows self-produced energy to be used in the hospital complex. The annual electrical efficiency of the turbine is 29.9%. The annual thermal efficiency of the turbine – calculated as the ratio between produced and used thermal energy and the amount of thermal energy supplied by the methane's combustion – is 40.5%. On the basis of these values, it is also possible to calculate the total efficiency with respect to the supplied thermal energy, which is 70.4%. The obtainable annual energy saving as equivalent energy is equal to around 4.400 Tep.

11.5.11 Energy Performance

The performance objective was to achieve a 40% reduction in consumed energy. Results of energy consumption are discussed in this section, and are derived by simulation, calculation and monitoring. Consumptions taken refer to lighting, heating and cooling. Sun pipes and roof-light in corridors and halls provide a good level of daylight and help to reduce the energy consumption for lighting. Furthermore all installed lamps are high efficiency.

The total annual electricity demand is 12.3 kWh/m^2, giving an energy saving of 35% compared to the energy demand of a building without all these features. The heating and cooling internal temperatures and relative humidities measured during the monitoring phase are in accordance with simulations. The insulation used in the walls and roof gives an energy saving of 35% for heating and cooling. The annual heating demand is 73.4 kWh/m^2. The annual cooling demand is 87.3 kWh/m^2. During the summer period, two chiller machines are used for cooling the hospital. The heat produced is used for DHW. The annual heating for DHW demand is 13% less than in a conventional Italian hospital.

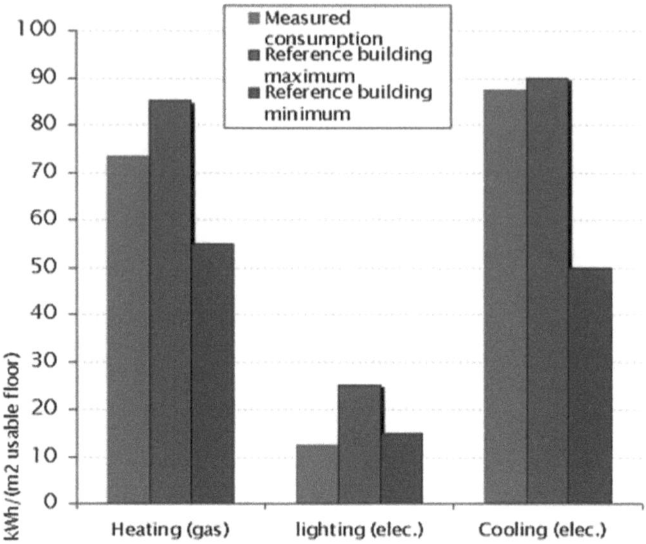

11.5.12 Monitored Comfort

Location: Children's playroom, 2nd floor, south facing façade
System: External overhang roof and low emission glazing
Sky condition: Sunny day with direct sun on façade
Description: The luminance picture shows a good distribution of luminance on surfaces, also into the depth of the room. Lighting is also maximized by the presence of two light ducts.

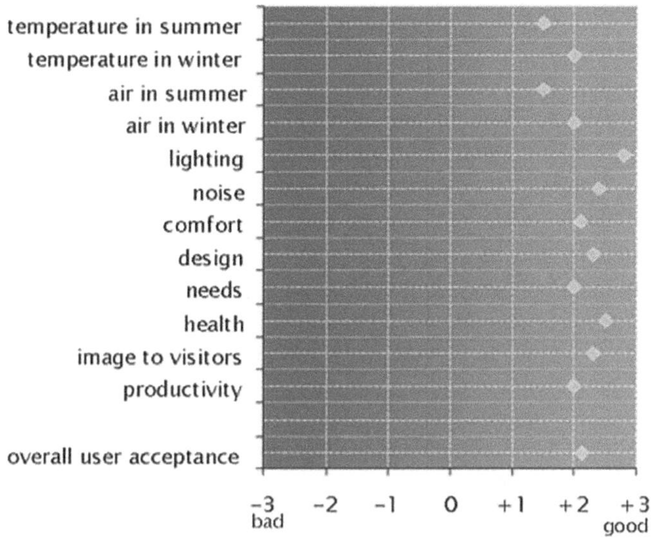

11.5.13 User Acceptance

The graph shows that the children's hospital achieved an excellent user rating. In fact all the values are above +2 on a seven point scale which goes from −3 (bad) to +3 (good), with an overall user acceptance of +2.23.

11.5.14 Financial Data

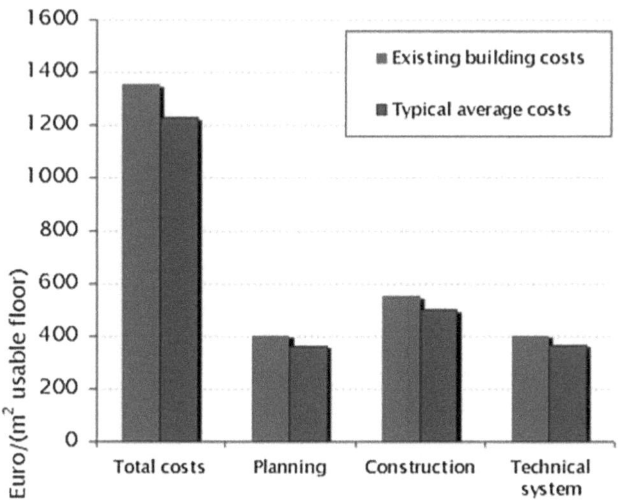

Total cost: € 1,160,000 of which € 960,000 was for the construction and € 200,000 was for furnishing. The co-generation plant was not included in these figures, because it has yet to be completed and will be financed by the district.

The costs of the energy saved or produced annually is € 181,679.

11.6 PRIMARY SCHOOL – EMPOLI, ITALY

11.6.1 General Data

Number of floors above ground: 1
Number of floors below ground: 0
Heating or cooling gross floor area: 1800 sqm
Usable floor area: 1800 sqm
Heated or cooled volume: 7500 m^3
Building envelope area: 600 sqm
Average number of occupants: 70

11.6.2 Identification

Name: Ponzano Primary and Nursery School
Owner: Empoli municipality
Country: Italy
City: Empoli
Street: Via di Ponzano
Occupant(s) of building: Ponzano Primary and Nursery School

Primary use: School
Typical days/hours of use: 8.00 am– 5.00 pm
Designers: Marco Sala Associati
Engineer: Ing. Luigi Campa – Structures CMZ – Plant Enginering
Contractors: Muicipality of Empoli
Manufacturers of energy saving products: Consage
Energy sources: Radiant panels, condensing combi boiler
Year of completion: 2001

11.6.3 Outdoor and Indoor

Microclimate: urban, ASHRAE degree days heating/cooling: 2060 / 1917 Kd
Outdoor design temperatures/humidities: –2°C; RH 50%
Indoor design temperatures/humidities: 20°C
Design ventilation rates: 30/h
Design illuminance levels:
Classrooms: 500 lux
Circulation space: 150 lux

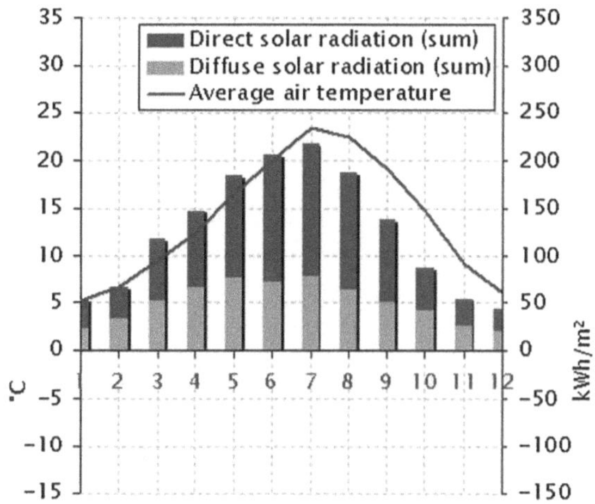

The building was constructed with a reinforced concrete skeleton system and an innovative insulation system. Window types were employed depending on orientation and function.

A high thermal floor insulated with radiant heating was planned. Shading devices were used for the southern side of the building.

The windows have double pane thermal protective glazing and timber-PVC frames, with internal air exchangers to regulate the entry of air. A double ventilated roof guarantees good insulation and building ventilation.

BUILDING COMPONENT	U-VALUE (W/M²K)
External wall with bricks	0.28
Window	1.7
Ventilated roof	0.28
Floor: store and technic	0.50
Average U-value	0.43

11.6.4 Insulation

Properly sealed, moisture-protected, insulated walls help increase comfort, reduce noise and reduce energy costs. The walls are however the most complex component of the building envelope. The precast insulation wall is an innovative system which uses concrete walls able to combine mechanical resistance and insulation. The precast element is in casing form, having two polyester panels (EPS) facing each other and connected by a separator which creates a cavity between the two surfaces. This insulating system is able to ensure a trasmittance U-value of 0.15 [W/m²K].

11.6.5 Solar Control

Well-designed sun control and shading devices dramatically reduce building peak heat gain and cooling requirements, and improve the natural lighting quality of the interiors. The school reduces the amount of annual cooling energy consumption from 5% to 15%. Sun control and shading devices also improve user visual comfort by controlling glare and reducing contrast ratios. This often leads to increased satisfaction and comfort. Shading devices offer the opportunity of differentiating one building façade from another which can add interest and human scale to an otherwise undistinguished design.

The use of sun control and shading devices is an important aspect of this energy efficient building; particularly in employing passive solar heating and daylighting through sun control.

During cooling seasons, external window shading is an excellent way to prevent unwanted solar heat gain from entering an air-conditioned space. Shading is provided by movable aluminum overhangs.

Exterior shading devices are particularly efficient in conjunction with the clear glass façades of the 'intelligent windows'.

Lighting is controlled by the 'intelligent window', essentially a façade device which acts as an intelligent interface between inside and outside, as it is installed on the 'skin' of the building. It provides the appropriate thermal insulation and air exchanges necessary for improving indoor conditions. Its use parameters may be described as: solar energy control, daylight and ventilation control, building façade aesthetics, cost saving in heating or air conditioning and automatic adjustment through neural network.

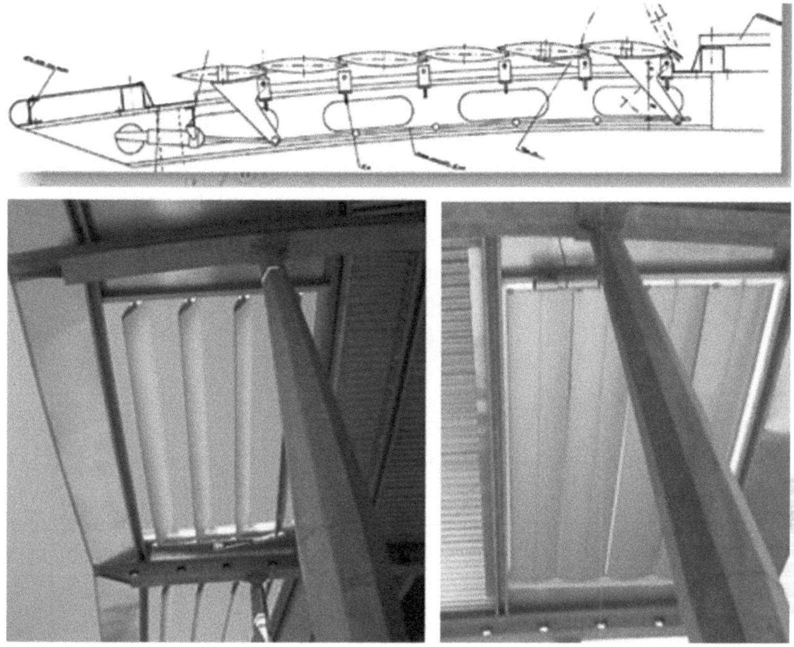

11.6.6 Lighting

Lighting is controlled by the 'intelligent window', essentially a façade device which acts as an intelligent interface between inside and outside, as it is installed on the 'skin' of the building. It provides the appropriate thermal insulation and air exchanges necessary for improving indoor conditions. Its use parameters may be described as: solar energy control, daylight and ventilation control, building façade aesthetics, cost saving in heating or air conditioning and automatic adjustment through neural network. The creation of an intelligent interface between indoor and outdoor conditions remains the primary objective. The window includes a set of elements, each with a specific or variable function, depending on outdoor conditions. The elements are contained in two main sections: the upper section contains glazing panels which in turn enclose variable transparency film operated on a roller system. The employed materials are PVC frame, temperate glass, low emissive glass, a roller system and air flow control.

11.6.7 Heating

The best high efficiency boilers can operate with seasonal efficiencies in excess of 90%, by recovering and using heat that would otherwise be lost in the flue. Recovering heat from the flue reduces the temperature of the flue gases to a point where the water vapor produced during combustion 'condenses out'. Thus the name: high efficiency condensing boiler. A side effect is that this 'condensed out' water, known as condensate, which is usually acidic, has to be piped away to a drain or soakaway. All condensing boilers will produce 'pluming' from the flue terminal which appears as steam. This pluming can drift into neighboring property, causing annoyance and possible condensation on window glass or frames, therefore careful consideration should be given to the positioning of the flue terminal, especially if it can affect neighboring property.

A condensing boiler is a highly efficient modern boiler that incorporates either a larger or even a second heat exchanger. It produces lower flue gas temperatures, lower flue gas emissions and reduced fuel consumption. It typically converts more than 88% of the fuel used into useful heat, compared to, typically, the 78% of modern conventional types.

11.6.8 Natural Ventilation

Natural ventilation is guaranteed by many incorporating south and north facing openings. During the daytime winter season 'intelligent windows' enact a natural ventilation control together with the air exchanger and window apertures. The roof is ventilated and double: one hollow block floor and a second roof realized with structural steelwork and aluminum panels. Air flows from north to south, also ventilating classrooms from the window.

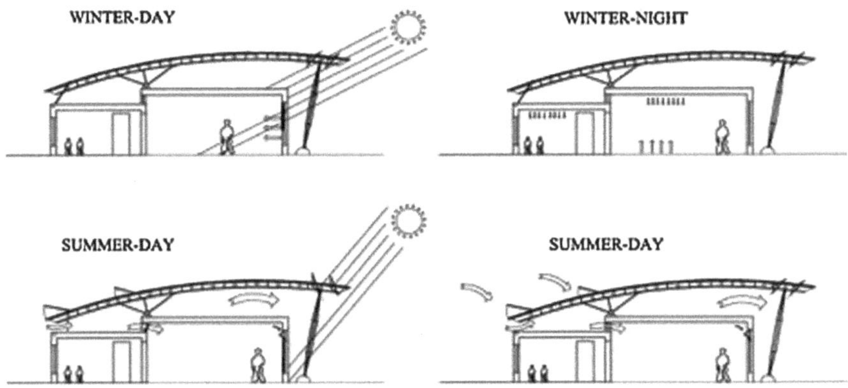

11.6.9 Cooling

Natural cooling is guaranteed by the presence of the 'intelligent window', formed by a number of elements each with a specific or variable function, depending on the outdoor conditions. The elements are contained in two main sections:

- The upper section contains glazing panels, which in turn enclose a variable transparency film on a roller system.

- The lower section is contained inside a compartment, internally clad with a filter panel and externally with an opaque glass panel.

Within these panels the heat exchanger and the upper and vertically mounted fans for air intake and exhaust are respectively located. The intelligent control system, sensors and the local control for different configurations, are also located within this section. The window heating strategy includes the concepts of solar collection, heat storage and heat distribution, while the cooling strategy comprises solar control, minimizing internal gain and heat dissipation. Shading component: roller shutter with 30% of radiation control; two air flow controls with T shape valve for reducing heat in summer and preheating air ventilation in winter.

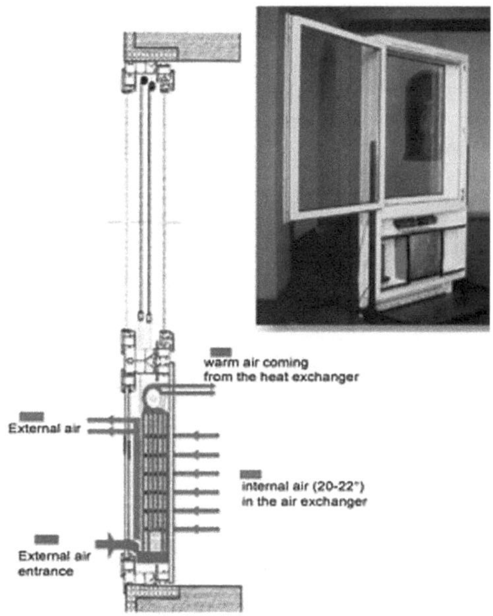

Section and picture of the intelligent window.

Effect:

- high solar gain
- active solar systems
- high ventilation rate
- control of solar radiation
- night cooling
- u-factor: U= 1.7W/m²K
- visible transmittance: U glass = 3.5 W/m²K
- admissible chance air: m = 50 m³/

11.6.10 Monitored Comfort

The total energy consumption was monitored for twelve months between October 2000 and September 2001, and was found to be 80 kWh/m^2 for the year. The diagram shows how the adopted passive systems reduce energy consumption for both mechanical and electrical installation.

Location:
Classroom on ground floor; the view is of south facing façade
System: Intelligent windows with inside louvers and curtaining
Sky condition: Sunny day with direct sun on façade
Description: The luminance picture shows a good shading distribution and a good luminance on surfaces, also into the depth of the room.

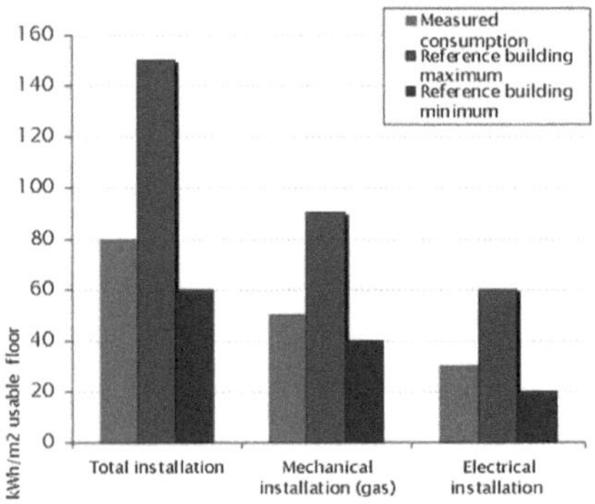

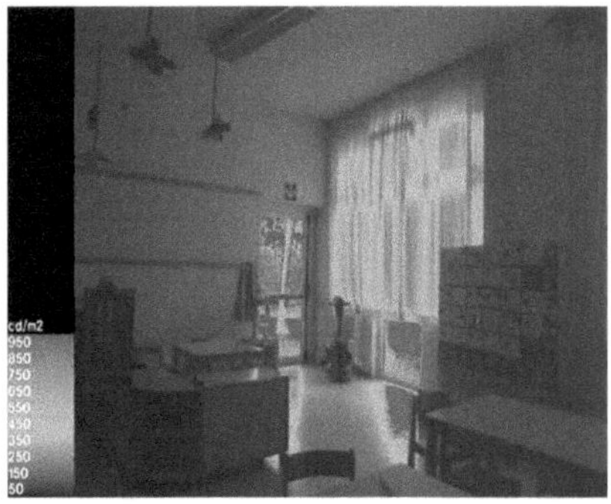

11.6.11 User Acceptance

The mean response of occupants voting on a seven-point scale from bad (–3) to good (+3), was 2.18 – which is a really good evaluation. The school building thus rates well on productivity, occupant control and air quality.

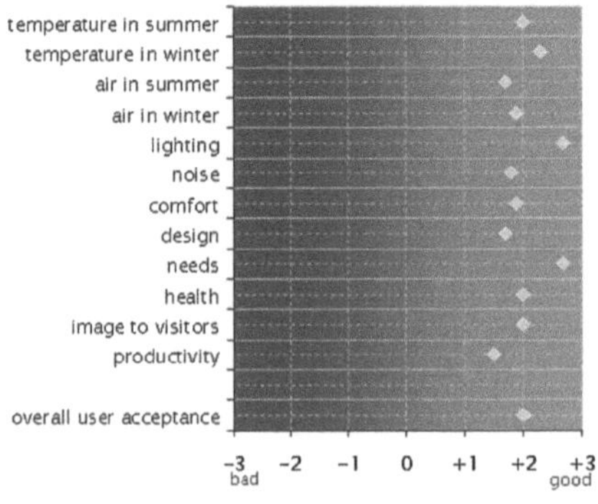

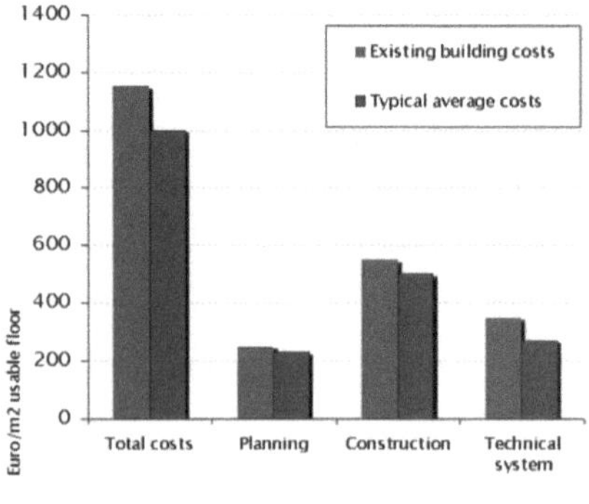

11.6.12 Financial Data

Total building cost: € 1,200,000

The chart shows that an immediate cost saving is mostly found in the building's construction (planning and technical systems remain the same) which impacts the total costs. However, we must remember that this type of building has lower maintenance costs, which will have a higher impact in the future and hence shorten the payback time.

11.7 MALTA STOCK EXCHANGE – LA VILLETTA, MALTA

11.7.1 General Data

Number of floors above ground: 3
Number of floors below ground: 3
Heating or cooling gross floor area: 1862 m^2
Usable floor area: 1271 m^2
Heated or cooled volume: 7600 m^3
Building envelope area: 2800 m^2
Average number of occupants: 45 occupants

11.7.2 Identification

Name: Malta Stock Exchange
Owner: State of Malta
Country: Malta
City: Valletta
Street: Garrison Chapel, Castille Place

Occupant(s) of building: Malta Stock Exchange
Primary use: Office
Typical days/hours of use: 7:00 am – 5:00 pm, Monday to Friday
Designers: Architecture Project, Brian Ford & Associates
Engineer: TBA, Frank Franjou, MTS
Contractors: Medairco, Vassallo Builders, Peter Cox
Manufacturers of energy saving products: Grundfos
Energy sources: Electricity
Year of completion: 2001

11.7.3 Outdoor and Indoor Climate

Microclimate: Mediterranean; site: exposed city center ASHRAE degree days
Heating/cooling: 832 / 3080 Kd
Outdoor design temperatures/humidities:
  T summer : 38°C, RH=30–80%
  T winter : 8°C, RH=30–80%
Indoor design temperatures/humidities:
  T summer : 24–25°C, RH=70–75%
  T winter : 21°C, RH= 70–75%
Design ventilation rates:
Design illuminance levels: 300–500 lux

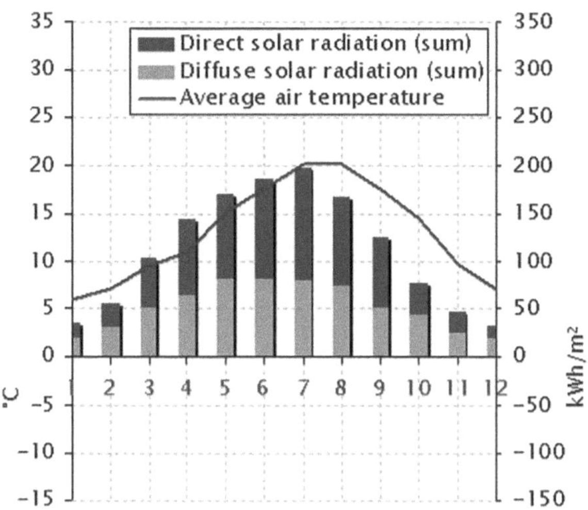

11.7.3.1 Identification

11.7.3.2 General Data

The project involved the refurbishment of an existing listed building. The converted building incorporates a five-storey atrium surrounded by perimeter cellular offices. A new metal inner structure divided by concrete slabs has been inserted into the original stone building envelope. The thickness of the stone walls is 60 cm. The glazing ratio is about 5%. The offices are separated from each other by gypsum board partitions and are bounded by glazed partitions facing towards the central atrium. The roof structure is of wood. The roof has a U-value of 2.5 W/m^{2}K.

11.7.3.3 Cooling

The cellular offices and the meeting rooms of the Malta Stock Exchange are air-conditioned. The central atrium is cooled by three complementary strategies, as follows:

Indirect cooling: two chilled water circuits serve cooling coils installed in the roof ridge. They are linked to the automatic vents and operate in conjunction with each other. The rising warm air passes over and through the coils resulting in a down draught of cooled air.
Direct cooling: the passive downdraught evaporative cooling system (PDEC) relies on hydraulic nozzles. The air is moisturized by the micronisers situated on the ridge. The downdraught process drives the airflow pattern inside the building, hence avoiding the need for fans, ductwork and a suspended ceiling.
Night time convective cooling: this comes into operation when the external temperature drops below 23°C. The movement of the air is driven by buoyancy forces reversing the daytime air movement pattern (also see 'Ventilation' below).

11.7.4 Ventilation

High level window vents in the ridge of the roof and low level vents on the lower ground floor mounted in the east, west and south walls provide ventilation if opened.

Night time convective ventilation comes into operation when the external temperature drops below 23°C. When this is initiated, all low and high level vents are fully opened. The movement of the air is driven by buoyancy forces reversing the daytime air movement pattern. Air enters below ground level and rises by a stack effect to exit via the roof-ridge vents. The thermal capacity of the masonry helps stabilize conditions within the building, while the timber louvers on the main window prevent solar gain.

11.7.5 Energy Performance

All conversions of delivered energy in primary energy were made according to the German Standard DIN V 18599-1:2005-07. Only the non-renewable energy part is considered.

Electricity-Mix: 2.7
Estimated cooling energy consumption: 113 kWh/ m^2/year

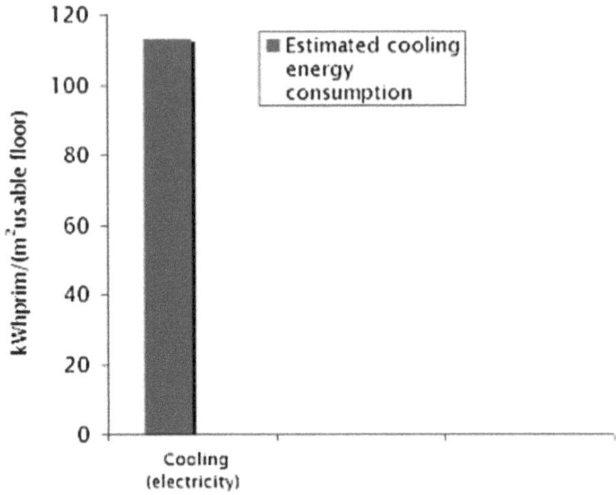

11.7.6 Monitored Comfort

Location: Office room, 1st floor, south facing façade
System: Folding shutters, completely closed
Sky condition: Clear with direct sun on façade
Description: The luminance picture shows an even distribution with low values due to the completely closed shutters on the outside. As there is only part of the space visible a more detailed description here is not possible.

11.7.7 User Acceptance

The figure shows the subjective perception of comfort in the building. The occupants' subjective evaluation is given as the arithmetic mean of the responses given by a population consisting of 23 persons. The occupants voted on a seven-point scale from bad (−3) to good (+3), as illustrated in the figure. The visitors' impression of the building is a relatively high rating.

The majority of the responses are values between −1 and +1. Some concerns have been raised regarding the air quality in winter and summer. The overall user acceptance and the comfort have a rating value close to 0.

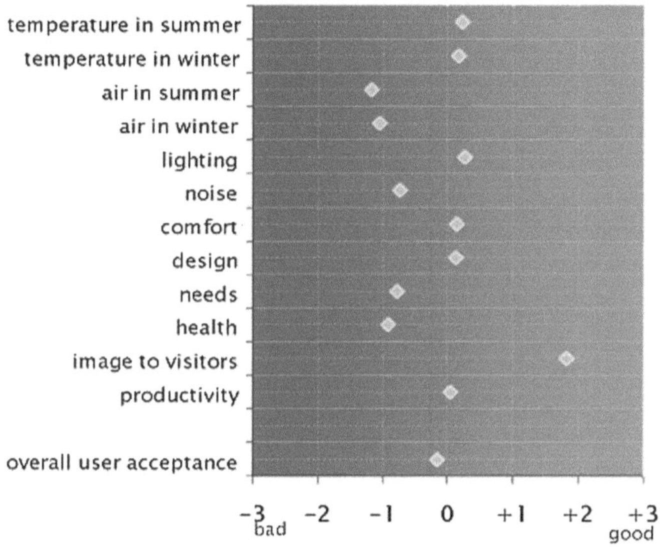

11.7.8 Financial Data

The total cost in 2001 is € 2,470 per/m^2 including taxes.

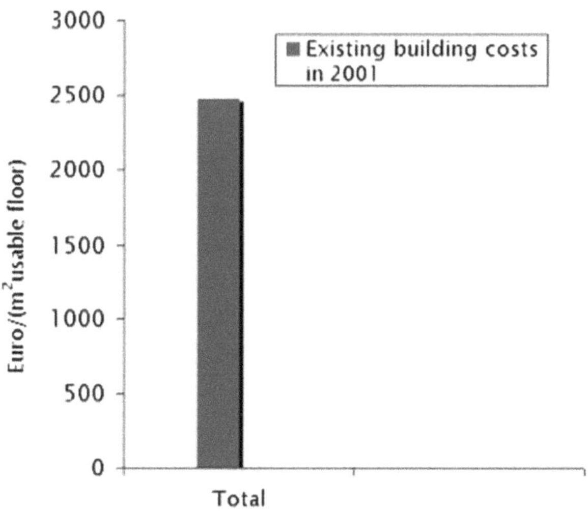

Chapter 12

A Low-Energy Building Project in Sweden – the Lindås Pilot Project

Fredrik Karlsson* and Bahram Moshfegh†
*Sweco Systems AB, Norrköping, Sweden, †Division of Energy Systems, IEI, Linköping University, Sweden

Chapter Outline

Sustainability, Energy and Architecture. ISBN: 9780123972699

12.1 INTRODUCTION

One of today's great challenges is to find opportunities to create and sustain a modern, comfortable existence while facing the growing threat of global warming and the future depletion of resources. A continuation of material and energy deployment and growth in waste products is not sustainable environmentally or ecologically. More than 50% of the world's population now lives in urban areas, and this figure is increasing. Thus, adaptation and development of urban as well as rural areas for tomorrow needs to be tailored to reduce environmental impact and use of finite energy resources, improve associated economic prosperity and create social stability. As a result, there is a need to develop a well-defined energy plan with clear energy and climate transition to sustainable societies.

The world is currently faced with many major challenges within the fields of energy, environment and climate, primarily due to the large-scale, globally increased use of fossil fuels. During the past 20 years, the total demand for energy in the world has doubled, while during the same time period the demand for electrical energy has tripled. The arguments for rational energy handling are both economic and environmental. The EU has a goal of reducing energy use and carbon dioxide emissions by 20% by the year 2020. One of the areas prioritized in order to reach the EU's stated goal is energy efficiency in housing construction, which is presented in EU Directive 2010/31. By 2050, Sweden has a national goal of reducing energy use by 80% compared with today's use, primarily through energy efficiency in the housing sector.

EU Directive 2010/31 is a guideline on how the member states must actively work on and promote improvements in the energy performance of buildings. The overall aim is to provide long-term sustainability and less dependence on energy use, which will create safety, security and a more stable economy. The Directive covers both new buildings and existing buildings when renovated. Special attention has been placed on buildings occupied by public authorities. These buildings should serve as inspiration and good example, as well as showcase the cutting edge of technology and its implementation. The consequence of EU Directive 2010/31 for Sweden has been analyzed and presented in a report from Swedish Parliament 2011 and Swedish National Board of Housing, Building and Planning (report N2010/1474/E), which is incorporated in Swedish building regulations and laws.

The building sector accounts for 35% of end use of energy and 15% of CO_2 emissions in Sweden (see Figure 12.1). The annual energy use in the building sector in year the 2009 was 149 TWh. Residential buildings and service and commercial buildings account for 87% of the total energy use in the building sector. The figures become even more significant in a global context, as these sectors are responsible for more than 40% each of primary energy use and CO_2 emissions. In order to reach the agreed long-term EU and national targets, energy use in buildings in urban areas must be sharply reduced. The challenge is to build new houses in a more energy-efficient and sustainable manner, but even

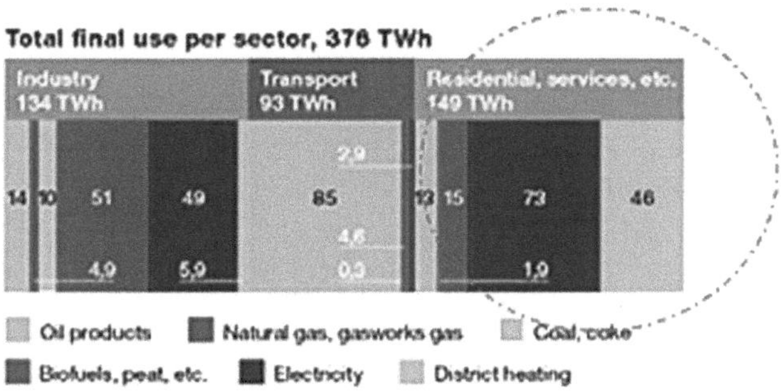

Source: Statistics Sweden and the Swedish Energy Agency.
1. These are large heat pumps in the energy sector. 2. Nuclear power is shown as gross power, i.e. as the nuclear fuel energy input, in accordance with the UN/ECE guidelines. 3. Net import of electricity is treated as supply.

FIGURE 12.1 **Total energy use per sector in 2009.**

greater challenges are to substantially reduce energy use in existing buildings and to find good solutions to cover the remaining energy needs with renewable energy. It is worth mentioning that it is also important that future building be adapted for a change in climate, such as changes in temperature and moisture content in the air. Making changes in construction practices is a laborious process, and in order to achieve such a radical reduction in energy use by 2050, implementation must start now.

Research and development as well as innovation are key elements to overcome these challenges. The contribution from public and private societies, industry and the research community to the development of energy-efficient products and services, of systems for electricity and heat production using renewable energy resources, and inventions and applications of conservation measures for rational use of energy are only a few examples that will help with resource efficiency and sustainability. These contributions will also lead to positive business opportunities that benefit both industry and society in general.

More and more energy-efficient buildings are being built in Sweden. There is still a need for greater knowledge, including systems thinking, a holistic view and a long-term perspective, along with the promotion of resource conservation, reduced environmental impact and an understanding of people's needs and behavior. Housing must be designed to have the lowest possible environmental impact, while remaining economically feasible to live in. In addition, the indoor climate should not adversely affect or compromise human health through the energy efficiency measures put in place. There are many goals to be met at once. Further research activities in the field of energy systems, energy technology, building physics, building materials, building service technology, life cycle assessment, environmental technology and applied psychology are needed in order to increase our knowledge of how to build in sustainable and energy-efficient ways.

Design of energy-efficient, environmentally sound buildings, renovating buildings to meet higher energy conservation and energy efficiency standards, energy-efficient installations and components, and effective energy distribution and supply systems with a variety of renewable energy sources are research areas in which industry is looking for technical solutions.

12.2 THE BUILDING'S ENERGY SYSTEMS AND BUILDINGS IN ENERGY SYSTEMS

In order to achieve the goal of more energy-efficient buildings that remain financially feasible and do not compromise human health or comfort, we need to look at buildings at different system levels. Figure 12.2 shows the various system levels of a building, where the room is the first level in which a good indoor climate is achieved in an energy-efficient manner. Beyond this level there is interaction between people and technology where the use of technology is the focus. The third level is the building as a whole, and the energy needed to provide the functions that should be in the building. The fourth level is the building and its relationship to the surrounding energy system. By applying a system perspective, one can demonstrate how efficiency measures at various levels will affect the other levels.

An important task is to convert the building energy systems into long-term sustainability, including reduction of CO_2 emissions. Buildings account for a large share of energy use in society, and are thus important in developing a

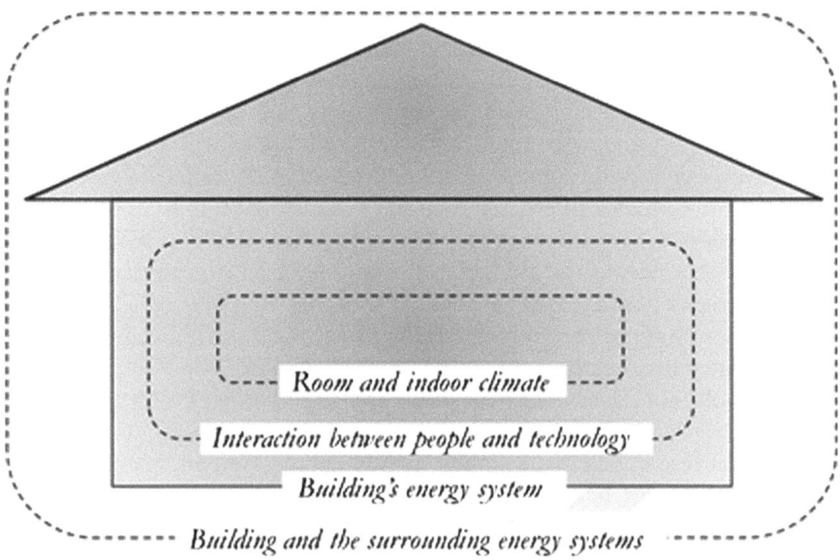

FIGURE 12.2 System levels of the buildings.

long-term energy-efficieny strategy. Measures of building stock can be performed in four principal ways, as shown in Figure 12.3:

(1) By increasing the efficiency of electricity and heat usage in buildings;
(2) Through the load management for efficient use of available resources; and
(3) and **(4)** By conversion to CO_2-neutral heat and power, including the use of bio-fuels both on the building energy system's level and on the surrounding energy system's level.

12.3 ENERGY USE IN SWEDISH BUILDING SECTOR

Energy is used in buildings to create appropriate thermal comfort and indoor air quality, domestic hot water production, lighting and for operation of household appliances. Today, buildings in Sweden account for about 35% of the country's total energy use (see Figure 12.4). Energy for heating, ventilation and domestic hot water production accounts for more than 70% of energy use in buildings. Thus, there is a great potential for energy conservation and energy efficiency measures.

Figure 12.4 shows how energy supplied to residential buildings in Sweden has remained fairly constant during the past 40 years. However, a major conversion has occurred from mainly oil products to electricity and district

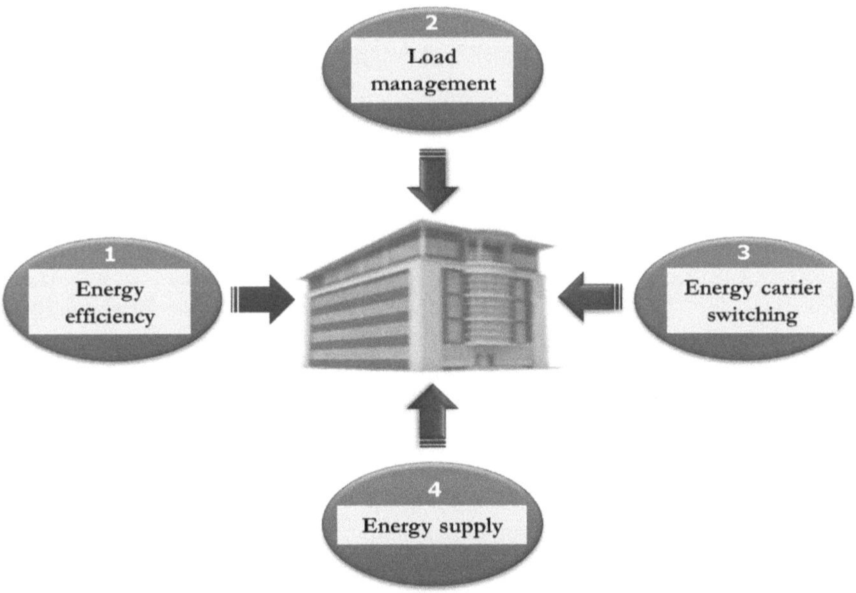

FIGURE 12.3 Opportunities for energy measures in buildings. Sustainable development in buildings requires management of materials and energy both during construction and during the actual use of the building. As previously mentioned, almost 40% of Sweden's total energy use is in buildings and thus a major part of Sweden's CO_2 emissions are attributable to buildings.

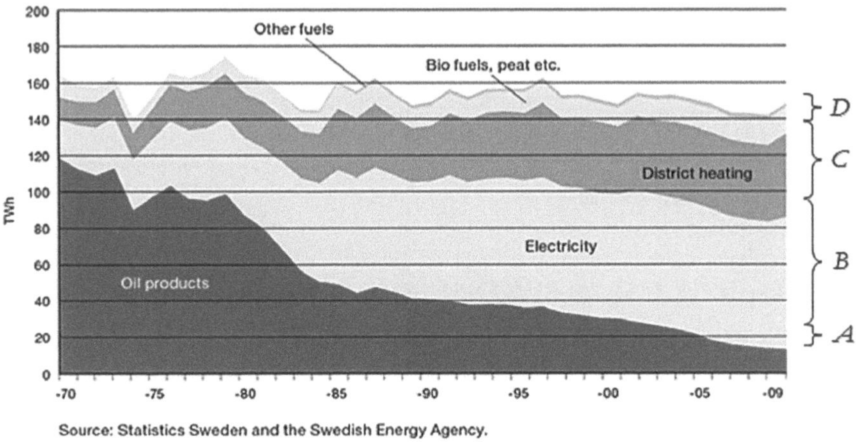

FIGURE 12.4 Final energy use in the residential and service sectors 1970–2009.

heating. The use of oil products as a primary energy source for the building sector has fallen sharply since the 1970s. According to Figure 12.4, oil products, electricity, district heating and bio-fuel account for roughly 9%, 50%, 30% and 11% of energy supply to building sector in 2008, respectively.

Electricity is the most important supplied energy source in residences, and district heating is the most common energy source for heating in high density residential areas such as down-town and neighborhoods.

12.4 ENERGY USE IN RESIDENTIAL BUILDINGS

The energy use in the building sector in Sweden normally falls into three main categories: houses using electricity for direct heating; houses using other energy forms or in combination with electricity for heating systems; and multifamily buildings. Electricity is the dominant energy form used in houses which in direct form stands for 31% and in combination with other energy forms for about 22%. Bio-fuels and oil products stand for 14% and 3% respectively. However, only 12% of Sweden's 1.6 million single-family houses have been connected to district heating, corresponding to a district heating load of approximately 5.4 TWh in 2008. Approximately 80% of all multifamily buildings and 65% of service and commercial buildings are connected to district heating, corresponding to a district heating load of 22.3 TWh and 14.8 TWh in 2008 [33]. Energy use in the building sector for different years is presented in Figure 12.4.

12.4.1 District Heating is an Efficient, Clean and Environmentally Sound Form of Heating

Swedish Combined Heat and Power (CHP) plants are a both competitive and resource-efficient energy supply system for heating, electricity and cooling.

Increased investment in CHP is a key measure in the transition to lower CO_2 emitting energy systems. In Sweden, the share of CHP electricity is low. Only 6% of Sweden's electricity generation was based on CHP in 2006, while the corresponding figure for Finland was 35%, and for Denmark it was 60%. The average throughout the EU was 10% of electricity from cogeneration in 2006. Under the EU CHP Directive, the percentage of electricity from cogeneration should increase in the EU to 18% and in Sweden to 14% in 2010. Here, the local authority and private energy companies play an important role. The Swedish municipalities with their local anchoring have the opportunity to invest in upgrading existing district heating plants to CHP-based production utilities, and to take steps that lead to the expanded use of district heating systems. They are thus better able to share CHP-based power generation in accordance with the EU CHP Directive.

Today, only part of the Swedish district heating system has the possibility for power production. We have therefore built district heating systems, but we have not fully taken the opportunity to produce electricity as well. According to the Swedish District Heating Association, the production of electricity from cogeneration plants will double, and about 35 new power-heating plants will be operational by 2015. Bio-fuels and waste will be the dominant fuels in the new plants.

District heating as an energy source affects the climate through its CO_2 emissions. But thanks to current and future conversions from fossil fuels to renewable energy, CO_2 emissions from heat production are expected to decline from 86 kg CO_2/MWh in 2007 to an estimated emission level of 46 kg CO_2/MWh by 2015. The total amount of natural gas, oil and coal used for heat production was 10% in Sweden in 2008 and is expected to be 5% of heat production in 2015 [1].

12.4.2 The Role of District Heating Systems as an Energy Source in Low-Energy Buildings

Buildings should also be seen in a broader perspective in order to achieve sustainable solutions. For example, it is important to get collaboration between district heating companies and property owners to continue the use of district heating in a cost-effective way for energy supply to buildings. A dialogue is needed on the way that district heating should be maintained as an energy source with low environmental impact, and how it can be profitable for a district when the need for the amount of purchased energy is greatly reduced. In low-energy buildings the supplied heat is mostly used for domestic hot water, the use of which is distributed throughout the year, and not connected to seasonal variations as it is with comfort heating.

Figure 12.4 shows how the energy supplied to residential buildings in Sweden has been converted from mainly oil products to electricity and district heating. According to this figure, oil products, electricity, district heating and bio-fuel account for roughly 9%, 50%, 30% and 11% of the energy supplied to the building sector in 2008, respectively. One important measure that is needed to achieve

2020 and 2050 targets is to significantly reduce energy use in the building sector; see the dashed line in Figure 12.4. One can draw the following conclusions from the figure: a reduction in the use of fossil fuels will reduce the environment impact (see notation *A*); reduced electricity consumption in buildings will be beneficial for the environment because the excess electricity can be exported to the continent to replace power plants using fossil fuels as their energy source (see notation *B*); the condition for a district heating system is not as simple as other supplied sources, because a reduction in district heating use is a waste of existing capital resources, and the district heating system in Sweden with its fuel mix is an efficient, clean and environmentally sound form of heating (see notation *C*); and finally, a reduction in bio-fuel use in buildings benefits the environment, because bio-fuel is a scarce resource and should be used with caution (see notation *D*).

Thus, in order to minimize the risk for sub-optimization, it is vital to distinguish between buildings within and outside the district heating systems. For the former, one needs to focus on reducing electricity consumption and minimizing use of resources. This could be achieved by the introduction of small scale renewable energy systems for electricity and heat production in district heating areas, including the allowance of third-party entry to the district heating systems, and the introduction of smart grids for district heating systems. Buildings outside the district heating should be constructed based on passive house standards, with proper selection of durable materials in both new construction and renovation, and the prioritization of sustainability rather than low-cost procurement.

Special attention should be paid to low-temperature heating systems that use the district heating system as supplied energy, which offer the opportunity to use combined heat and power more efficiently. Usually the focus is on reducing the building's energy demand, but it is equally important to consider the impact of supply and distribution systems. Otherwise, there is a risk that a delivery system designed for periods of greatest need will be too large for much of the year. The majority of the 'million program' buildings are connected to district heating systems. Special attention should be focused on the holistic approach of a systems perspective to analyze the interaction between systems for energy supply and energy efficiency measures in million program buildings, to avoid sub-optimization and to achieve a significant reduction in peak loads and carbon emissions, as well as longer usage time of the utilities.

12.5 NEW TECHNOLOGIES THAT MAKE BUILDINGS MORE ENERGY-EFFICIENT AND ENVIRONMENTALLY SOUND

Technologies that can be applied to buildings can be divided into three categories:

(1) *Demand* measures which decrease the demand for energy use;
(2) *Transmission* measures to increase the effectiveness in distributing energy throughout the building; and
(3) *Production* measures to convert useful energy in a more efficient way.

The energy demand in a building can be decreased by using better insulation materials with lower heat transmission, for example vacuum insulation panels. Windows with better insulation performance (which can change their solar protection characteristics with the need for heat inside the building) can decrease the need for heat during winter and cooling during summer. Highly efficient heat exchangers in the ventilation system will decrease heat losses through ventilation. Recently, fan motors have increased in efficiency, which has an impact on the amount of electricity needed to power them.

There is also a potential to recover heat losses through sewage. New, more reliable products are emerging which can be used more commonly in the future. Other technologies focus on demand control, such as demand-controlled ventilation systems that are used when needed instead of all the time. Sensors and control devices, as well as the comfort and health consequences of demand-controlled ventilation in residential buildings needs further research. Other measures to reduce energy use include decreasing the indoor air temperature, e.g., during the night or during the daytime when the tenants are at work.

Efficient distribution systems can decrease the need for energy. For example, more efficient control systems can be used to supply the building with accurate temperature and flow, depending on conditions inside and outside the building. Load management does not have a great impact on energy demand in a specific low-energy building, but it can be of interest on a larger scale to utilize production and distribution capacity more efficiently.

Lots of action is taking place in developing sustainable supply systems, including renewable solar (electricity and thermal), wind power and small scale CHP. Each of these common techniques needs to be further developed and integrated with building energy systems and fed into the building market. Increased use of small scale power production on site will further decrease the need for energy supply from surrounding systems. From this point of view, load management is of great interest, as the surrounding supply system then needs not be dimensioned to supply maximum power at all times. Here one has to consider the need for redundancy for different kinds of buildings.

12.6 ACTION PLANS AND ENERGY POLICIES TO ACHIEVE ENERGY-EFFICIENT BUILDINGS

To achieve a more sustainable building sector, efforts should be made to increase research and development on components as well as combinations of components in systems. There is no single solution for this issue, and a palette of both technical and socio-economical solutions needs to be put in place. The pilot project mentioned below is an example from Sweden that relies on common technologies. However, users do have trouble with the technology. Technology that is easy to handle is very important in achieving energy-efficient buildings.

Policies have an impact on the user's energy use, as well as on how the building itself performs. The user is using a product that has been designed based on policies, for example building rules, energy targets by the builder and knowledge about energy design in the design team. Having an energy policy that emphasizes the importance of low energy demand will act as a driving force for the building sector.

Another driving force is environmental systems, such as Leadership in Energy and Environmental Design (LEED) and BRE Environmental Assessment Method (BREEAM), which can be used to produce more sustainable buildings. Policy makers can use certified buildings as examples of sustainable building that will encourage others within the building sector.

Feed-in tariffs for energy-efficient technologies can be used to introduce new technology to a market. In Germany, this has been successfully used to introduce wind power and solar heating. Sweden has a tradition of large-scale solutions, which does not benefit small scale, renewable sources, such as wind or solar power. However, there are some subsidies that can be used for solar energy. However, the problem remains in Sweden, that if you produce more electricity than you need, you will not receive any financial benefit, and the energy produced will be available for free on the power network. Policy makers have an opportunity to increase small scale energy production by changing this.

12.7 BUILDING AND THE HEALTH OF OCCUPANTS

Issues of climate change and sustainable development are currently high on the political agenda, and the solutions will be found in people's living environments, their behavior and the local environment that surrounds them. Sustainable development in buildings requires management of materials and energy in existing buildings and measures to reduce the environmental impact of additional buildings' lifecycles. Sustainability includes a good indoor environment that offers sufficient comfort and is not harmful to the people who work or reside in the building.

Building must be designed with the lowest possible CO_2 impact, but at the same time should not adversely affect the indoor climate or compromise human health through its energy efficiency measures. Buildings are built to live and work in, not to save energy. As a result, energy saving is a secondary issue to accommodation, though it is by no means less important.

These important issues require systematic and dispersed interdisciplinary collaboration between different disciplines, such as energy, installation technology, building physics, building materials, environmental technology and applied psychology, to study the whole chain of people – technology. Human health, comfort, performance and cognition are the final and necessary criteria for indoor environments, both for residents and for workers. Therefore, indoor environmental quality, ventilation systems, the chemical composition of indoor air and the costs of energy are measured against the environment, and also the well-being and psychological functioning of the users. Poor indoor air quality in buildings costs society large sums of money annually in the form of health care, administration and lost productivity.

In the search for future near zero energy buildings, it is vital that the human body is thermally comfortable in its surroundings. Moisture safety, adequate ventilation and low emissions from materials are examples of important parameters that must be included in the energy-efficient solutions to be developed.

12.8 SOME EXAMPLES OF LOW-ENERGY BUILDINGS IN SWEDEN

The meaning of the term 'low-energy building' is somewhat diffuse. Sometimes it means a building that uses almost zero purchased energy; sometimes low-energy is related to the energy amount needed for heating. This criterion is changing over the decades. A definition of low-energy buildings is found in Abel [2]:

'A building that is used for developing and testing new technologies with focus not only on decreasing energy demand for space heating, but also to decrease the need for electricity.'

Another definition of a low-energy house is one that uses considerably less energy than a building corresponding to present building regulations and building traditions [3]. Other definitions or names used are advanced houses [4], high-performance buildings [5], houses without conventional heating [6] and passive houses [7]. Advanced houses are houses with low energy usage, low demand for natural resources and low environmental impact. The definition is thus broader than of a low-energy or an energy-efficient building. On the other hand, most low-energy buildings can also be defined as advanced houses. The term high-performance building is used to describe buildings that have ambitious energy saving goals, ranging from 40% better than regulations to a net zero energy performance, and which use thermal envelopes better than the present standard. Houses without conventional heating and passive houses are actually the same thing. Passive houses have very well-insulated building envelopes and windows that do not cause any downdraught. The heating demand is thus so low that a conventional heating system with radiators is unnecessary. Passive houses have the goal of decreasing their heating energy to zero, within economic limits. That is, such a house can be defined as a low-energy building in one sense but as an energy-efficient building regarding the economic limits.

The first generation low-energy houses were built in Sweden in the beginning of the 1980s. Three projects were evaluated ten years later by Berggren et al. [8]. The projects, named Sparsam, Lättbygg 85 and Rockwool, all used increased insulation to provide energy savings. In addition, the Sparsam and Rockwool projects involved an exhaust air heat pump. The Sparsam houses aimed to be air-tight and have a sun parlor to utilize passive solar irradiation [8]. Electrical heating was used in all of them. The energy use was measured immediately after the occupants moved in, and again about ten years later. A detached house was built in Kungälv, Sweden, in 1981–82, with the aim of constructing a low-energy building with high comfort standards but a competitive price.

The house uses a well insulated envelope, small window area (12% of floor area), triple glazed windows, air-to-air heat exchanger, and heat exchange of waste water by heat pump and heat exchanger. The total energy demand was measured at 75 kWh/m^2 [9].

A multifamily building was built in Lund, Sweden, named Jöns Ols. The aim was to minimize the energy purchased. For example, the insulation thickness was increased over that installed in most buildings, the number of thermal bridges was minimized, the heat from waste water was recovered, an exhaust air heat pump was used for heating purposes and solar power was used for heating water. The energy demand has been measured at 82 kWh/m^2 (60 kWh/m^2 electricity and 22 kWh/m^2 district heating), of which the energy demand for heating is 10 kWh/m^2, which is covered by the heat pump. The additional charge for the measures has a pay-off time of about ten years with current energy prices [10].

Two energy-efficient buildings from Malmö, Sweden, were monitored extensively for about one year in the beginning of the 2000s. The buildings are called energy-efficient and participate in a project sponsored by the Swedish Energy Agency (SEA). Among the restrictions for the buildings it can be found that the buildings should not use more than 100 kWh/m^2 per year (80 kWh/m^2 if using electric direct heating), they should have a good indoor climate, and the U-value of the windows should not exceed 1.0 W/m^2K [11]. The first house, built by Yxhult, had an area of 149 m^2 and included an exhaust-ventilated suspended foundation. It is air-tight and uses an air-to-air heat exchanger to reduce the heat losses through ventilation. District heating is used for space heating and hot water. The other house, built by LB-hus, is built on a concrete floor slab and uses an exhaust ventilation system connected to an air heat pump, which is used for space heating and hot water requirements. An auxiliary electric heater is used when the heat pump energy is not sufficient.

The Brogården, Alingsås project is an example of the refurbishment of multifamily houses to passive house standard which was started in 2008 (see Figure 12.5). The vision of the project was that refurbishment of municipally-owned buildings could be achieved because the reduced energy costs would pay for the restoration. The project was carried out by the municipal housing company Alingsåshem, and was targeted at people with limited incomes. The project foresees the renovation of 300 apartments in 16 buildings that were constructed in the early 1970s and are in urgent need of restoration.

The project considers sustainability in three aspects: economic, ecological and social. The final costs for the tenants should not differ much from their existing ones, since the reduced energy costs should pay for the renovation to passive house standard. The project is performed as a partnership development between the housing company, an architect specializing in passive houses and a building company whose aim is developing refurbishment processes for existing buildings. All participating contractors are briefed and educated in the requirements of passive house standards. Tenants living in the neighboring buildings are also

FIGURE 12.5 The buildings of Brogarden before (left) and after refurbishment (right).

involved, through briefings and regular information updates about the progress of the project and what they might expect after moving back into their apartments.

The refurbishment of the houses mainly comprises the replacement of the outer walls by new air-tight walls avoiding heat bridges. New balconies have also been added outside the wall structure. The heating system was exchanged from radiator heating based on low flow air heating system with heat recovery. Solar heating supports DHW (domestic hot water) production. For the first conversions, annual energy use for heating and hot water has been reduced by 100 kWh/m^2 to 55 kWh/m^2 (district heat) and annual electricity use has reduced by 15 kWh/m^2 to 44 kWh/m^2. The main R&D issues were air tightness and the air-borne low flow heating system. Another important issue was the feedback process regarding the innovations, in which tenants and entrepreneurs could respond on 'do' and 'don't' items.

The most well-known passive house project in Sweden comprises the terraced houses in Lindås, outside the city of Gothenburg, which were completed in 2001. The Lindås project will be presented thoroughly in Section 12.10.

12.9 ENERGY-EFFICIENT BUILDINGS AND CITIES – A STRATEGIC DIRECTION FOR URBAN POLICY MAKERS

The law on municipal energy planning came into effect in 1977. But this law gives no clear guidance on how to perform energy planning in practice, nor does the act include any penalties for municipalities that do not carry out an energy plan. As a result, up to 30% of Swedish municipalities have not formulated an energy plan. The European Energy Efficiency and Energy Services Directive (ESD) went into effect in 2006, and requires each Member State to reduce its energy consumption by 9% by 2016. It covers virtually all players except those involved in the EU emissions trading scheme. Each Member State is obliged to submit a National Energy Efficiency Action Plan (NEEAP), which should also include instruments for efficient end use. The Swedish NEEAP indicates that the public sector should set an example. Much more needs to be done in terms of research for the Swedish government to be able to concretize ESD through NEEAP for local stakeholders, such as municipalities. Important factors that promote and hinder effective energy planning should be identified, methods must be developed, and transparent evaluation methods must be constructed.

Despite major technological advances in construction, such as energy-efficient, CO_2 neutral and green buildings, which can realize very large reductions in energy usage and greenhouse gas emissions, the community's energy usage is in general not decreasing. There are examples of cities with successful energy plans and climate strategies, but these are few. The reason for this is that the process of achieving consensus on the optimal energy plan that follows the investment decision is very complex and time-consuming, and many stakeholders are involved. As a result, regulations and policies issues, decision-making process, transition management are much more important than introducing cutting edge technological innovations. While the latter has been examined in many studies, research on the former is scarce, underlining the need for such research. Thus, methods are needed to enable a strategy for urban planning, to treat overall long-term energy strategies, to develop overall energy (or climate change) plans for the proper implementation of strategies, and techniques for introducing successful municipal energy policies.

12.10 THE SWEDISH LINDÅS PILOT PROJECT – HOUSES WITHOUT HEATING SYSTEMS

Passive houses are houses that are built with the aim of significantly reducing the building's energy use, preferably to zero. The most publicized example of passive houses in Sweden is probably the houses built in the early 2000s in Lindås, south of Gothenburg. These houses have started a trend of passive house construction in Sweden that is still ongoing. They were built with the aim of showing that it is possible to maintain a good indoor climate in a house despite using less than half of the energy of ordinary buildings in cold climates. This is done by reducing heat losses to the point where a heating system is unnecessary, without affecting the indoor climate. Moreover, the ambition was that the occupants should live in the houses without any professional knowledge of the technical systems. The concept was based on experience from earlier passive house projects in Germany, participated in by the architect.

12.10.1 Description of the Lindås Buildings

The Lindås project consists of twenty terraced houses (see Figure 12.6), situated south of Gothenburg on the west coast of Sweden (a detailed description can be found in Karlsson [12]). In total, there are four groups of townhouses, two of

FIGURE 12.6 The situation of the four sets of buildings in Lindås (left), and one of the four blocks of buildings (right).

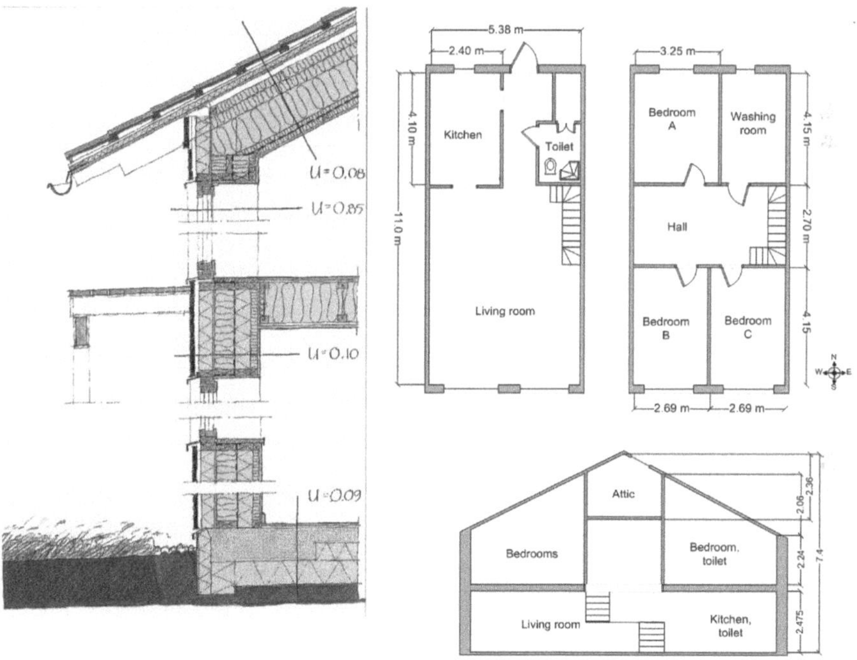

FIGURE 12.7 The building construction of the Lindås low-energy building (left, the drawing shows the north façade, Illustration EFEM Arkitekter), drawings of the terraced house (middle and right).

four houses and two of six houses, respectively. No advanced technical systems were used; instead common techniques and simple solutions were chosen.

The houses are heated mainly by the emissions from household appliances, the occupants themselves, and by solar irradiation. The building envelope is well insulated, with *U*-values presented in Figure 12.7 and Table 12.1. The window areas on the south and north sides of the houses are 13.6 m^2 and 3.5 m^2

TABLE 12.1 U-Values of the Building Elements in the Real Building

Structure	Area (m^2)	*U*-value (W/m^2K)
External walls	38	0.1
Roof	66	0.08
Floor	62	0.09
Windows (average for all windows)	18	0.85
Average *U*-Value		0.17

respectively. The large window area on the south façade allows solar radiation to heat the building. In addition, all the houses have a roof window, which makes efficient airing possible during the summer season. The windows used in the houses are triple-glazed with three 4-mm panes (*U*-value 0.85 W/m^2K). The space between the panes is filled with either argon or krypton. Three layers of low-emission coating have been applied on the panes in order to reduce their emissivity.

An air-to-air heat exchanger with a thermal efficiency of about 77% is used to minimize heat loss through ventilation. On cold days, an integrated electrical heater of 900 W can be used to heat the air, which is distributed through the ventilation system. The supply air terminal devices are situated in the three bedrooms and in the living room. The exhaust air terminal devices are situated in the toilets on both levels. The ventilation duct is mounted in the framing of joists, causing the supply air terminal devices on the upper floor (e.g. the bedrooms) to be placed on the floor, and near the ceiling on the bottom floor. Both floors have contact with the attic through the stairwell (see Figure 12.7). A 5 m^2 thermal solar collector combined with an auxiliary electric immersion heater used for DHW is included. All technical features are summarized in Figure 12.8.

The house consists of a ground floor, upper floor and attic. Table 12.2. shows some basic data for the house. On the ground floor there is a kitchen, living room and toilet (see Figure 12.7). On the upper floor there are three bedrooms, with rather high ceilings, and a bathroom. In addition there is a roof window, which makes effective natural ventilation possible during summertime (see Figure 12.8).

12.10.2 Energy Usage – Measurements and Building Energy Simulations

The energy used by the houses has been monitored for two years, between September 2001 and August 2003 – see Ruud and Lundin [13]. The average house uses about 8,100 kWh annually and the gable and middle houses use 8,950 and 7,500 kWh annually, respectively. For a middle house, this

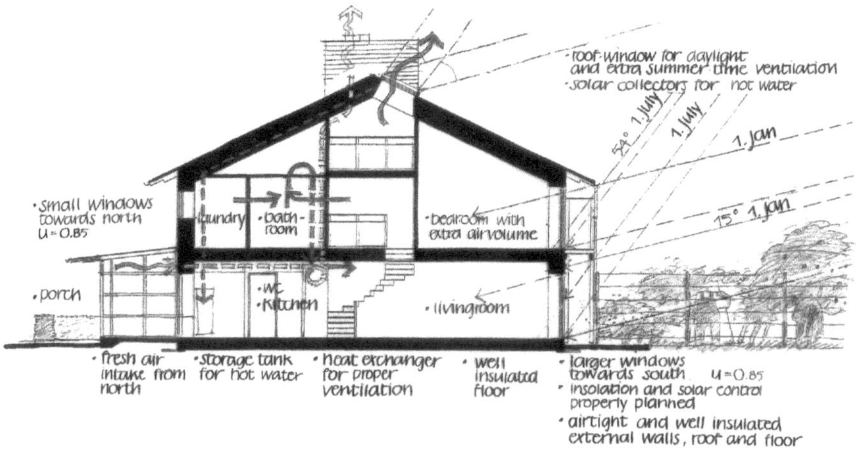

FIGURE 12.8 Schematic picture of the object of this study. Notice the shading of solar irradiation during summertime to prevent overheating. *(Illustration EFEM Arkitekter).*

TABLE 12.2 Some Basic Data for the Terraced Housing

Ground area	60 m^2
Total floor area	120 m^2
Total volume	340 m^3
Ceiling height, ground floor	2.5 m
Ceiling height, upper floor	2.2 to 4.3 m
Infiltration at a differential pressure of 50 Pa	0.2–0.4 l/s m^2
Geographical location	Lat.: 57.5°, Long.: 11.5°

represents about 62.5 kWh/m^2 annually (this figure does not include the heat provided by the solar panel), of which 12.8 kWh/m^2 is used for comfort heating purposes [14]. The maximum value was measured as 101.7 kWh/m^2 and the minimum value as 49.2 kWh/m^2. The difference between these figures shows that the activities of the occupants has a considerable impact on the resulting energy demand.

For six buildings, the measurements were divided into electricity for heating, fans, domestic water heating and household electricity. The average distribution is shown in Figure 12.9 and compared to a typical, comparable Swedish building. To investigate how different measures affect the energy

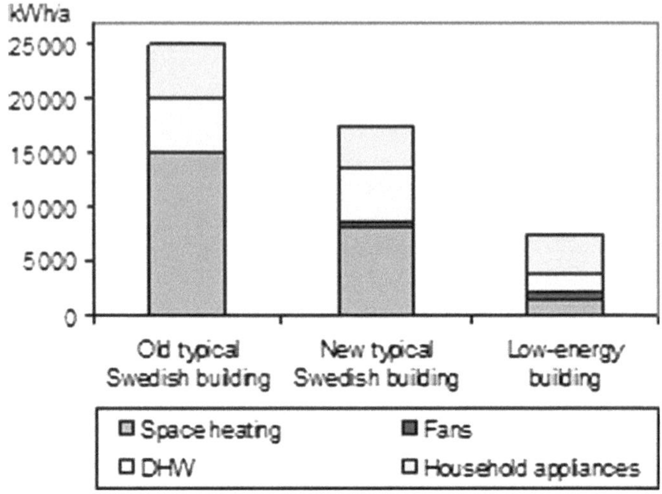

FIGURE 12.9 Energy usage pattern in an average household in the Lindås low-energy building, compared to an older and new typical, comparable Swedish building [27].

demand and the indoor climate, several energy simulations were performed, using the whole building energy simulation program ESP-r [15]. The energy demand and the PPD value (Predicted Percentage of Dissatisfied) with different temperature set-points illustrate that a heating set-point of 21–23°C gives the most satisfied occupants [16]. A load management case, in which the heating system was turned off during the daytime, shows that the saving potential, with daytime electricity price 1.5 times the night-time prices, was only 22 € annually. It was also shown in the simulations that the indoor temperature was relatively high during the summertime, sometimes above 30°C [16].

As shown in Figure 12.9, the energy demand of the houses in Lindås corresponds to roughly half the energy use of a corresponding 'normal house' [17,18]. The average total energy of all households was measured at just under 8,300 kWh, which for the most part can be considered household electricity (see Figure 12.10).

12.10.3 Indoor Environmental

Applying different measures to increase energy saving and energy efficiency are vital, but at the same time it is also important that the indoor climate does not deteriorate. Using air-distributed heating can affect the air distribution in ways that do not correspond to the desired ventilation strategy. The temperature and airflow pattern was investigated by using a full-scale simulation model of the studied object. The Computational Fluid Dynamics (CFD) code Fluent is used for this purpose.

Four cases were simulated, representing climatic conditions in winter, summer and autumn, the last at both low and high airflow rates. For detailed

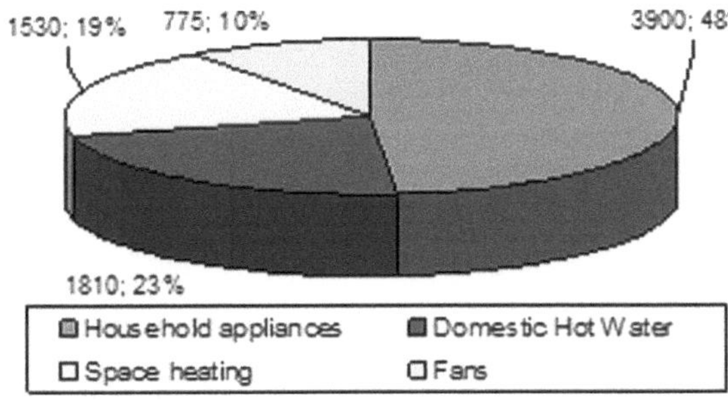

FIGURE 12.10 The average yearly energy demand for household appliances, domestic hot water, space heating and mechanical ventilation.

TABLE 12.3 Average Predicted Air Temperatures, T_{av}, on the Ground Floor and the Upper Floor

	T_{av} ground floor [°C]	T_{av} upper floor [°C]
Winter case	19.4	19.5
Summer case	23.7	22.6
Autumn case floor		
Low airflow rate	24.5	22.7
High airflow rate	23.6	21.9

information about the mesh generation, boundary conditions and numerical procedure, see [19].

The predicted values for the autumn case show temperatures within the building ranging from 19.1°C to 32.4°C. The average measured temperature is 25.4°C in the living room during the same period and 24.6°C in the kitchen. This can be compared to the values in Table 12.3 where the average predicted temperature ground floor for the autumn case (low airflow rate) is found to be 24.5°C. The predicted values are slightly lower than the measured values. Table 12.3 shows the average predicted indoor air temperatures.

The predicted air velocities for winter conditions in the door opening between room *B* and the upper hall have been compared to measured values and show a similar pattern. However, the measured velocity magnitudes are lower than predicted values, which can be explained by the high uncertainty in the measured values due to the low velocity magnitudes.

Figures 12.11 and 12.12 show the temperature and velocity contours for the summer and the winter cases in the mid-plane of the house. The summer case was simulated with a supply air temperature of 19°C, which represents evening

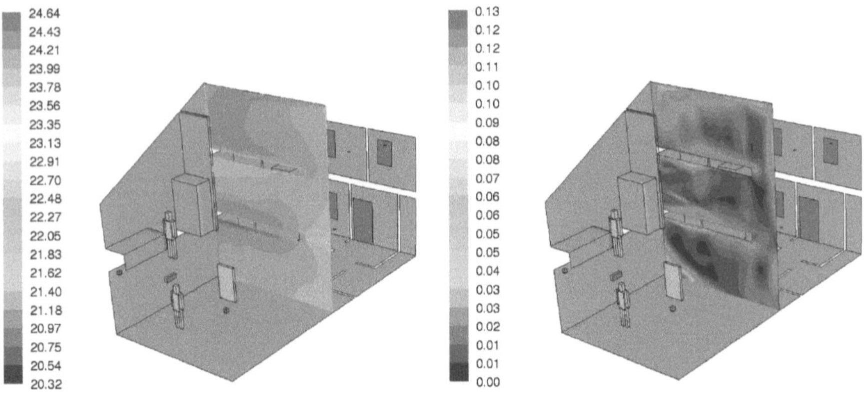

FIGURE 12.11 Contours of temperature (left) and velocity (right) at mid-plane for the summer case.

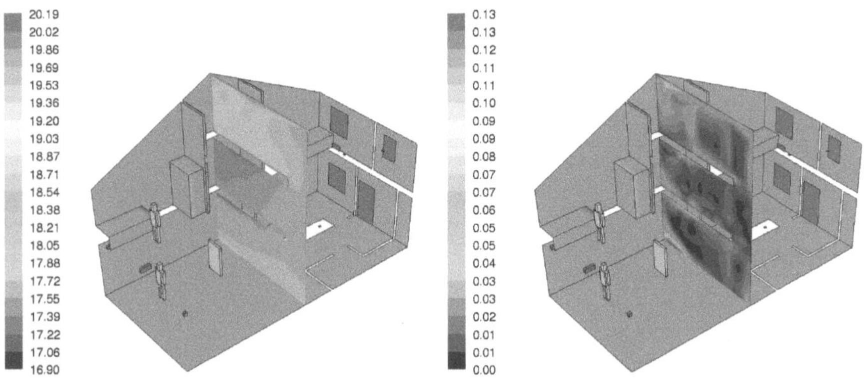

FIGURE 12.12 Contours of temperature (left) and velocity (right) at mid-plane for the winter case.

conditions. The temperatures in the real house will be higher during a hot summer day. An air circulation pattern is found in the attic and there air exchange occurs between the attic and the rest of the house. High air velocities are found on the ground floor near the kitchen and in the upper hall. The airflow in the upper hall near the floor and ceiling comes from the bedrooms *A* and *B*. Almost the same flow pattern and velocity magnitude have been observed for the winter case, but the temperature is approximately 3°C lower in the winter case than in the summer case. In the summer case, the average temperature on the upper floor is higher than on the ground floor, while for the winter case the situation is the opposite. Figures 12.11 and 12.12 show that there is airflow through the stairwell to the attic, which is in agreement with results from constant concentration tracer gas measurements that showed that the flow of fresh air to the attic was considerable (see [11,20]).

Figure 12.13 shows the contours of the velocity magnitude for both the summer and winter cases. The predicted airflow around and above the representative

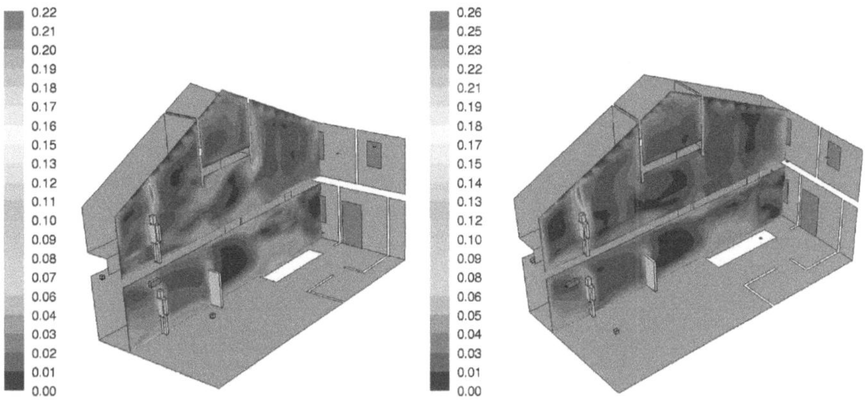

FIGURE 12.13 Contours of velocity magnitude in one plane for the summer case (left) and winter case (right).

human can be compared to calculations of plume velocities according to empirical findings for a heat source. The airflow to and from rooms *A* and *B* into the hall is also shown and is more pronounced in the summer case.

Special attention has been given to mapping the iso-velocity 0.15 m/s and 0.25 m/s contours for the summer and winter cases, respectively. These air velocities are the limit for avoiding discomfort due to draughts during the cold and warm seasons, respectively. Simulations reveal that the air velocity close to the supply device in rooms *A* and *C*, above the human body's head on the upper floor, in the kitchen and at the air gap below the toilet doors on both floors is greater than 0.15 m/s. The toilets are the location that the occupants of the low-energy buildings mention in connection with draughts in interviews about the indoor climate [21]. The downdraught from the supply devices in the living room is more pronounced in the summer case than in the winter case. The air velocity is in general lower or equal to 0.15 m/s in occupied zones in all simulated cases, and consequently the risk of local discomfort due to draughts is very low.

12.10.4 Environmental Performance and Embodied Energy

The environmental performance of the building was exemplified by calculating the CO_2 emissions for different energy systems. The composition of the surrounding energy system is an essential factor. The result will be different if the surrounding system is regarded as national or larger, i.e., by using marginal or European average energy production values. The CO_2 emissions per kWh electricity depend on the surrounding energy system. Either the national average supply mix can be used to calculate the CO_2 emissions from electricity production or a marginal production method where the CO_2 emissions from the marginal production plant are used [22].

Sweden is connected to the other Nordic countries through the Nordic market, NordPool, which also has connections to the European continent.

Hence, average European emissions can be used in the accounting method. The energy production mix in those three areas is very different. In Sweden, nuclear and hydroelectric power are the main production methods, accounting for more than 50% of the production mix. The European mix is more diversified, using mainly nuclear, coal, hydro and natural gas-fired power production [23].

Table 12.4 shows the CO_2 emissions for different energy systems, and the final CO_2 emissions from different energy systems for the object of this study are shown in Figure 12.14.

TABLE 12.4 CO_2 Emissions from Different Energy Systems

Accounting method	CO_2 emission [kg/MWh]
Swedish average electricity production [22]	11
Average Nordic electricity production [22]	101
Marginal electricity production [22]	933
European average electricity production [28]	475
Swedish average for district heating [29]	89
Pellets [29]	4

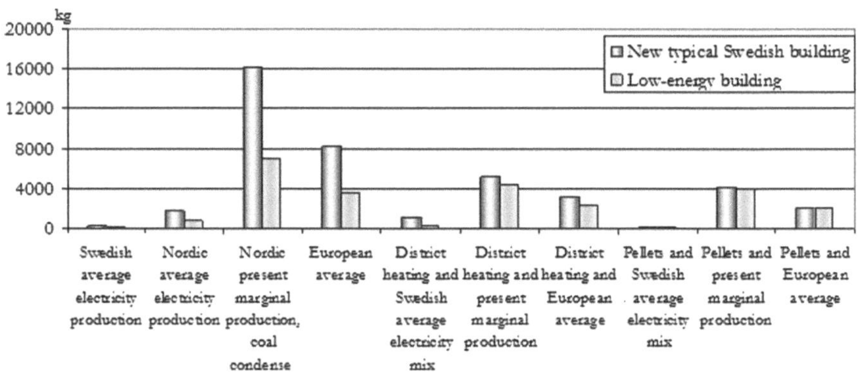

FIGURE 12.14 CO_2 emissions from the object of this study and from a typical Swedish building. The rates of emissions from different sources are shown in Table 12.4. In the cases using district heating or pellets the demand for comfort heating and DHW is considered to be converted from electricity to district heating or pellets, respectively.

It is considered possible to convert the energy used for comfort heating and DHW to any waterborne system, in this case conversion to district heating or pellets. The emissions arising from the district heating system are average figures for Sweden. If the district heating system includes combined heat and power production, an increased use of heat will imply a possible increase in electricity production as well, but this is not considered in the present calculations. The CO_2 emissions according to a Swedish average accounting scheme are 190 kg annually for a normal Swedish building, compared to 83 kg for a low-energy building. This implies a reduction of about 56% for the low-energy building. The largest reduction in CO_2 emissions is obtained if district heating is used in a Swedish energy system context, as shown in the fifth pair of columns in Figure 12.14. The lowest absolute value is 60 kg annually, which occurs when pellets are used with a Swedish average accounting scheme.

The result is shown in Figure 12.14. In the first four cases the reduction is about 60%. If heating and domestic hot water are provided by district heating, the reduction in CO_2 emissions is 93%, if the Swedish average electricity mix is used as accounting method. This reduction decreases to 33% if present marginal production, coal condensing, is used instead. The same picture appears if pellets are used; the reduction is 47% with Swedish average electricity mix and 24% with marginal production, respectively. On the other hand one can argue that domestic heating, if produced in a cogeneration plant, gives rise to electricity production that can replace other production methods. In that case the marginal production will be replaced first if the demand is assumed constant regardless of the electricity price [24,25].

The energy requirements of a building do not only relate to the energy used in the building during the time it is occupied. Building construction also involves an energy demand and subsequent environmental impact. Figure 12.15 shows that the total energy use for a typical Swedish house during its life cycle is about 15% in production and about 85% during the operational phase. Figure 12.15 also shows that the total energy used is less for the Lindås houses, but at the same time, the split between the production and operational phases changed to 38% and 62%, respectively. Therefore, the energy use associated with production is more important for low-energy buildings than for typical ones.

The environmental impacts of different energy sources differ, and different forms of energy require different conditions and are used for different purposes. Making a difference in the various energy forms and their applications can highlight new opportunities to improve energy efficiency even further. Exergy is a term that denotes the amount of energy that can be converted to work, and thus provides a measure of energy quality. In an exergy analysis of a building, efficiency opportunities may emerge that are not seen as clearly in an energy analysis.

A holistic approach means that the surrounding energy systems are taken into account when a building's performance is assessed [12,17,26]. For example, the CO_2 emissions related to energy use could be different depending on whether the house uses direct electricity or district heating for heating. Figures 12.16 and 12.17 display energy uses in a number of different types of houses, and the

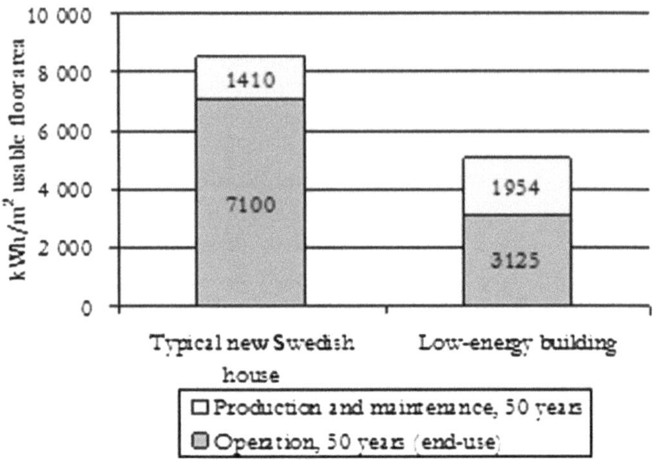

FIGURE 12.15 Embodied and operational energy in a normal building [30] and the object of this study. The difference in embodied energy mainly originates from the higher insulation levels in the low-energy building.

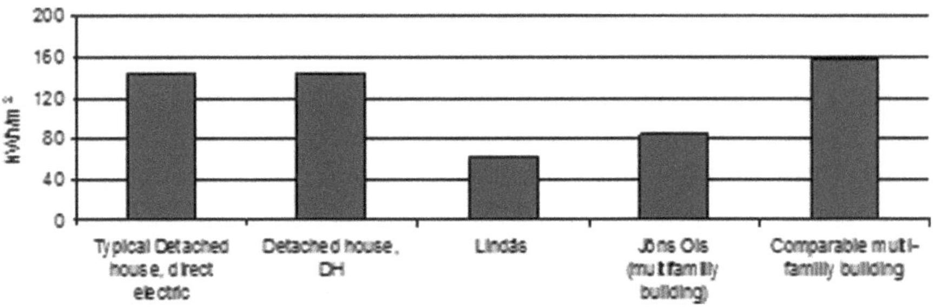

FIGURE 12.16 Energy use in various buildings.

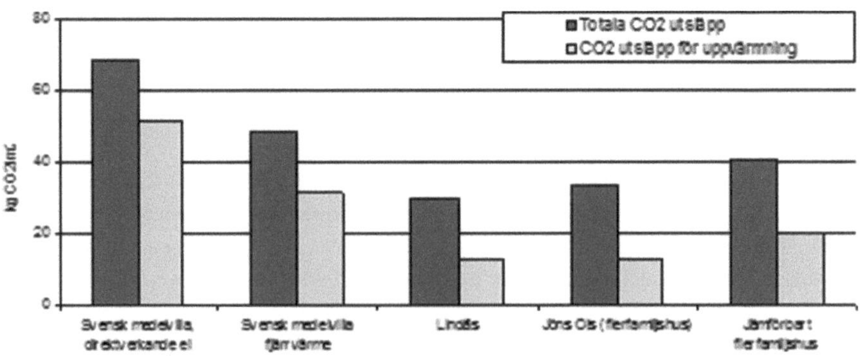

FIGURE 12.17 Carbon dioxide emissions in various buildings.

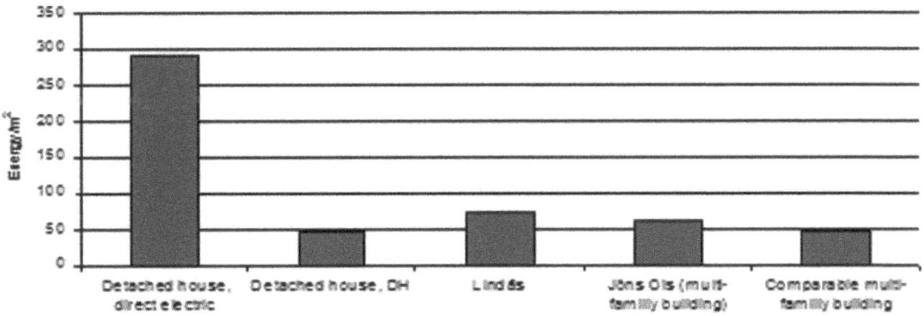

FIGURE 12.18 **The energy consumption of the heating system in the various buildings.**

related CO_2 emissions, depending on the heating system used. Five buildings have been compared; the first two are detached villas [27] with direct electricity and heat, the third is a house in Lindås [16], the fourth an energy-efficient apartment building in Lund called 'Jöns-Ols' [10] and the last one is a comparable apartment building with district heating. Jöns-Ols uses district heating and exhaust air heat pump for heating and hot water. The comparable multifamily house is supposed to only use heat for heating and hot water. CO_2 emissions are calculated with the European electricity mix and Swedish district heating system's mix.

Lindås uses the least energy per square meter, followed by Jöns-Ols. Energy use in the typical house does not depend on whether direct electricity or district heating is used. Carbon dioxide emissions however, differ markedly between the houses using direct electricity or district heating.

Although the houses in Lindås are regarded as low-energy, this does not take energy quality into account. As the house uses a small amount of energy, it is often most economical to use electricity for heating purposes, as direct electricity has low installation costs.

Taking into account energy quality, or how exergy consumption looks, involves different priorities. Exergy is the work a particular system can perform. Unlike energy, which cannot be consumed but only transformed, the exergy is consumed. The exergy, as opposed to the amount of energy, takes into account the environment the system is in. When we use electricity for heating, the house will consume exergy of high quality as opposed to when we use the waste heat from power production with temperatures around 60°C with a low exergy value.

Figure 12.18 shows a simplified calculation of exergy consumption of the heating system based on an exergy computational program developed within the framework of the accession countries before the IEA Annex 37 (http://lowex.org, 2009). Exergy demand, calculated by multiplying energy use by primary energy factor, describes the amount of primary energy used in the transformation process in the production system (e.g. boiler) and a quality factor based on the quality of the energy carrier. It is notable from the exergy analysis that a multifamily building that uses district heating consumes a minimum of exergy, and the relationship between the different types of buildings changes in surprising ways. Lindås still uses the least energy, but the picture is different for the exergy consumption, depending on the electricity consumption in the heating system.

Thus it is essential to achieve a precise balance between energy saving and energy at many different boundary conditions.

REFERENCES

[1] Swedish District Heating Association, Fjärrvärmen fortsätter växa, prognosrapport, 2010 (In Swedish).

[2] E. Abel, Low-energy buildings, Energy and Buildings 21 (1994) 169–174.

[3] G. Johannessen, Lågenergihus – nuläge och forskningsbehov, R31:1984, Byggforskningsrådet, Stockholm, 1984 (in Swedish).

[4] S. Carpenter, Learning from experiences with advanced houses of the world, CADDET analyses series no. 14, CADDET, Sittard, Netherlands, 1995.

[5] P.A. Torcellini, R. Judkoff, D.B. Crawley, High-performance buildings, ASHRAE Journal, September 2004.

[6] S.R. Hastings, Breaking the 'heating barrier' – Learning from the first houses without conventional heating, Energy and Buildings 36 (2004) 373–380.

[7] J. Schnieders, A. Hermelink, CEPHEUS results: measurements and occupant's satisfaction provide evidence for passive houses being an option for sustainable building, Energy Policy 34 (2006) 151–171.

[8] J. Berggren, Å Blomsterberg, F. Ekelund, M. Elfström, A. Elmroth, 1980-talets lågenergihus – hur fungerar de efter tio år, Rapport TVBH-3030, Department of Building Physics, Lund Institute of Technology, Lund, 1997 (In Swedish).

[9] L.-E. Larsson, A low energy house – Use of solar radiation and heat recycling, in: H. Erhorn (Ed.), Low energy buildings (2nd generation), CIB W 67 Annual meeting, CIB Proceedings Publication 130, Heidenheim, 1990.

[10] C. Warfvinge, Kv Jöns Ols i Lund – energisnålt och lönsamt flerfamiljshus med konventionell teknik, WSP Environmental Byggnadsfysik, 2005 (In Swedish).

[11] H. Bagge, A. Elmroth, L. Lindstrii, Energianvändning och inneklimat i två energieffektiva småhus i Västra Hamnen i Malmö, Rapport TVBH-3048, Department of Building Physics, Lund Institute of Technology, Lund, 2004 (In Swedish).

[12] F. Karlsson, B. Moshfegh, A comprehensive investigation of a low-energy building in Sweden, Renewable Energy 32 (11) (2006a) 1830–1841.

[13] S.H. Ruud, L. Lundin, Bostadshus utan traditionellt uppvärmningssystem – resultat från två års mätningar ('House without conventional heating – results from two years of measurements'), SP Rapport 2004:31, Swedish National Testing and Research Institute, Borås, 2004 (In Swedish).

[14] F. Karlsson, Varaktighetsdiagram för 'hus utan värmesystem' (Duration graphs for 'house without heating systems'), LiTH-IKP-R-1342, Division of Energy Systems, Department of Mechanical Engineering, Linköping University, Linköping, 2004 (In Swedish).

[15] J.A. Clarke, Energy Simulation in Building Design, second ed., Butterworth-Heinemann, Oxford, UK, 2001.

[16] F. Karlsson, B. Moshfegh, Energy usage and thermal environment in a low-energy building – Changed boundary conditions and control strategies, Energy and Buildings 38 (4) (2006b) 315–326.

[17] B. Moshfegh, F. Karlsson, An Example of a Low-Energy Building in Sweden, Proc. of World Renewable Energy Congress X and Exhibition, Glasgow, Scotland, UK. 19–25 July, 2008.
[18] Ruud and Karlsson, 'Husen utan värmesystem' halverar energianvändningen, artikel i VVS Teknik & Installation, oktober, 2004 (In Swedish).
[19] F. Karlsson, B. Moshfegh, Whole building CFD simulation of a Swedish low-energy building, Proc. of Clima2007, 9th REHVA World Congress - Wellbeing Indoors, Helsinki, Finland. 10–14 June, 2007.
[20] F. Karlsson, Experimental evaluation of ventilation rates in a low-energy building, Journal of Ventilation 5 (2) (2006c) 239–248.
[21] C. Isaksson, F. Karlsson, Indoor climate in low-energy houses – An interdisciplinary investigation, Building and Environment 41 (12) (2006) 1678–1690.
[22] J. Sjödin, S. Grönqvist, Emissions accounting for use and supply of electricity in the Nordic market, Energy Policy 32 (2004) 1555–1564.
[23] ÅF, Energi & Miljöfakta. Stockholm: ÅF Group. Available at: http://www.energiochmiljo.se, accessed 2006-09-06.
[24] B. Rolfsman, Interaction between energy systems of buildings in an ever-changing environment, Linköping Studies in Science and Technology, Dissertation No. 827, Linköping University, Linköping, 2003.
[25] J. Sjödin, Swedish District heating systems and a harmonised European energy market – means to reduce global carbon emissions, Linköping Studies in Science and Technology, Dissertation No. 795, Division of Energy Systems, Linköping University, Linköping, 2003.
[26] B. Moshfegh, F. Karlsson, Utveckling av lågenergihus mot en hållbar byggd miljö, artikel i Energi&Miljö NR 5, maj, 2009 (In Swedish).
[27] A. Persson, Energianvändning i bebyggelse ('Energy in buildings'), The Royal Swedish Academy of Engineering Sciences, Stockholm, 2002 (In Swedish).
[28] IVL (Swedish Environmental Research Institute), Energy database in LCAiT, IVL, Stockholm, 2006.
[29] SEA., Värme i Sverige 2002 ('Heat in Sweden 2002'), Swedish Energy Agency, Eskilstuna, 2003 (In Swedish).
[30] K. Adalberth, Energy use and environmental impact of new residential buildings, Report TVBH-1012 Department of Building Physics, Lund Institute of Technology, Lund, 2000.
[31] IEA ECBCS Annex 39 Lowex, http://lowex.org/guidebook/additional_information/index.htm, accessed 2009-04-20.
[32] F. Karlsson, Multi-dimensional approach used for energy and indoor climate evaluation applied to a low-energy building, Linköping Studies in Science and Technology . Dissertation no 1 065, Linköping University, Deparment of Mechanical Engineering. 91-85643-21-1, 2006d.
[33] Swedish Energy Agency, ET 2010:45, 2010 (In Swedish).
[34] Swedish National Board of Housing, Building and Planning, EU-direktivet om byggnaders energiprestanda – konsekvenser och behov av förändringar i det svenska regelverket, En redovisning av Boverkets regeringsuppdrag N2010/1474/E, 2010.
[35] Swedish Parliament, Hållabara städer – med focus på transporter, boende och grönområden, Rapporter från riksdagen 2010/11:RFR3, Civilutskottet, Trafikutskottet, Miljö– och jordbruksutskottet, 2011 (In Swedish).

Chapter 13

Key Characteristics of Top Performing Sustainable Buildings from the Perspective of the Users

George Baird
School of Architecture, Victoria University of Wellington, New Zealand

Chapter Outline

Sustainability, Energy and Architecture. ISBN: 9780123972699

13.1 INTRODUCTION

The overall mission of the author and his collaborators has been to provide an independent and unbiased evaluation of how the users perceive some of our commercial and institutional sustainable building developments. It is still surprising that building designers (with rare exceptions) do not systematically evaluate their projects, if only for the benefit of their own practices.

The overarching aim of the piece of research described in this chapter was to advance the practice of environmentally sustainable building design. A world-wide set of commercial and institutional buildings, all of which had well-recognized sustainability credentials or features, were sought out and evaluated. The intention was to find out the context for these projects, how they were designed, and most importantly, the users' perceptions of the performance of these buildings.

This chapter outlines the methodology used to determine the better performing buildings in terms of an overall 'Summary Index', which identifies their key features, and summarizes those features they had in common.

13.2 THE BUILDINGS AND THEIR USERS

The thirty-one buildings selected were all commercial or institutional in nature. Fifteen accommodated office activities predominantly, ten were tertiary-level academic teaching buildings, four housed laboratories or research organizations, and two contained a combination of light industrial and administrative functions. Virtually all were recipients of national awards for sustainable or low energy design, were highly rated in terms of their respective country's building sustainability rating tool, or in some way pioneered sustainable architecture.

13.3 SURVEY METHODOLOGY AND ANALYTICAL PROCEDURES

Generally speaking, these investigations involved undertaking several visits to each of the buildings to personally distribute and collect a questionnaire survey seeking the users' perceptions of a range of factors. The questionnaire was the one used by Building Use Studies [1]. During these visits a structured, recorded interview was conducted with a key architect and environmental engineer from the design team, and a detailed tour undertaken of each building and its facilities, photographing key features, and collecting relevant documentation.

The sixty or so questions of the standard two-page questionnaire used cover a range of issues. Fifteen of these elicit background information on matters such as the age and sex of the respondent, how long they normally spend in the building, and whether or not they see personal control of their environmental conditions as important.

However, the vast majority asked the respondent to score the following aspects of the building on a seven-point scale:

- Operational – Image to visitors, space in building, space at desk, furniture, cleaning, meeting rooms availability, storage arrangements and facilities.
- Temperature (overall and whether it is too hot or cold, stable or varies) and air (overall and whether it is still or draughty, dry or humid, fresh or stuffy, odorless or smelly) in both winter and summer.
- Lighting (overall, whether there is too much or too little natural light and artificial light, and whether there is glare from sun and sky or from the artificial lights).
- Noise (overall, and whether there is too much or too little from colleagues, other people, inside sources, and outside sources; and the frequency of unwanted interruptions).
- Personal control – of heating, cooling, ventilation, lighting and noise; and
- Satisfaction – design, needs, comfort overall, productivity and health.

Overall, there were some 2,035 respondents to the questionnaires.

Analysis of the responses yielded a mean value on a 7-point scale for each variable (other than productivity). In addition to calculating these mean values, the analysis also enabled the computation of a number of indices in an attempt to provide indicators of particular aspects of the performance of the building or of its 'overall' performance. These include a Comfort Index which was dependent on a set of seven 'environmental' factors; a Satisfaction Index, which was dependent on the scores for design, needs, productivity, and health; and a Summary Index which is the average of these two indices (see Appendix). These are intended to 'provide snapshots of how a building works for its occupants' [3].

In what follows, the features of the buildings with the highest Summary Indices will be described in more detail, in an attempt to discern those particular features that may have led to such high user perception scores and indices. The actual values of these indices, and the scores for the various factors from which they are calculated, are listed in Table 13.1.

TABLE 13.1 List of Buildings in Descending Order of Summary Index, Together with the Scores for the Eleven Factors on which that Index is Based.

Building	Summary Index (−3 to +3)	Comfort Factors (1 to 7)							Satisfaction Factors (1 to 7)			
		comf	light	noise	Tw	Ts	Aw	As	des	need	heal	Prod%
NRG	2.93^{1}	6.56^{1}	5.86$^{4=}$	4.67^{13}	6.11^{1}	5.00^{8}	6.32^{1}	5.65^{2}	6.77^{1}	6.68^{1}	5.47^{3}	19.51^{4}
TRC-AC	2.83^{2}	5.72^{5}	6.46^{1}	5.39$^{4=}$	5.54^{3}	5.86^{2}	5.62^{2}	5.73^{1}	6.10^{5}	5.79^{8}	5.53^{2}	20.88^{2}
NRDC	2.82^{3}	6.50^{2}	6.30^{2}	5.05^{8}	4.84^{8}	5.39^{4}	5.32^{4}	5.61^{3}	6.65^{2}	6.20^{3}	5.85^{1}	23.00^{1}
MFRC	2.45^{4}	5.92^{3}	5.69^{10}	5.67^{2}	5.27^{6}	5.92^{1}	5.00^{6}	4.90^{9}	6.23^{4}	6.38^{2}	5.17^{5}	20.00^{3}
Erskine	2.39^{5}	5.86^{4}	5.71^{8}	5.39$^{4=}$	5.25^{7}	5.14^{7}	5.07^{5}	5.23^{6}	5.61$^{9=}$	5.80^{7}	4.52^{11}	9.80^{10}
TRC-PDEC	1.95^{6}	5.16^{12}	5.86$^{4=}$	5.09^{7}	5.84^{2}	4.61^{13}	5.54^{3}	4.44^{14}	5.86^{7}	5.44$^{11=}$	4.74^{7}	13.66^{6}
St Mary's	1.73^{7}	5.67^{6}	4.92^{20}	5.85^{1}	5.33^{5}	4.83^{11}	4.92^{7}	5.00^{7}	5.62^{8}	5.50^{10}	4.67^{10}	10.83^{8}
40Alb	1.42^{8}	5.65^{7}	6.04^{3}	5.12^{6}	4.42^{16}	5.35^{5}	4.36^{16}	5.56^{4}	6.27^{3}	6.00^{4}	4.73^{8}	10.00^{9}
MEWC	1.33^{9}	5.20$^{10=}$	5.10^{18}	4.99^{9}	na	5.16^{6}	na	4.96^{8}	5.44^{12}	5.26$^{15=}$	4.77^{6}	16.00^{5}
60L	1.23^{10}	5.62^{8}	5.75^{7}	4.22^{20}	4.56^{12}	5.50^{3}	4.64^{11}	5.33^{5}	5.61$^{9=}$	5.87^{6}	5.24^{4}	11.39^{7}
AUT	1.18^{11}	5.20$^{10=}$	4.88^{21}	4.92^{10}	4.35^{17}	4.95^{9}	3.96$^{22=}$	4.27^{17}	5.46^{11}	5.26$^{15=}$	4.18^{13}	3.64^{15}

Notes: Superscripts indicate the relative ranking of each factor out of all 31 buildings, and the scores for all factors are on a 1 to 7 scale where 7 is best.
Abbreviations: NRG - NRG Systems Facility, Vermont, USA; TRC AC & PDEC - Torrent Research Centre, Ahmedabad, India; NRDC - Natural Resources Defence Council, California, USA; MFRC - Military Families Resource Centre, Toronto, Canada; Erskine - The Erskine Building, Canterbury University, New Zealand; St Mary's - St Mary's Credit Union, Navan, Ireland; 40Alb - 40 Albert Road, South Melbourne, Victoria, Australia; MEWC - Ministry of Energy, Water and Communications, Putrajaya, Malaysia; 60L - 60 Leicester Street, Melbourne, Australia; AUT - AUT Akoranga, Auckland, New Zealand.

13.4 DESIGN FEATURES OF BUILDINGS WITH HIGH SUMMARY INDICES

In this section, the reasons for high and low perception scores amongst those buildings with high Summary Indices will be explored, in an attempt to reveal the factors that influenced them, and the design features that could be involved.

The following buildings will be described in more detail – all had Summary Indices greater than +1 on a −3 to +3 scale:

- NRG Systems Facility, Vermont, USA
- Torrent Research Centre, Ahmedabad, India
- Natural Resources Defense Council, California, USA
- Military Families Resource Centre, Toronto, Canada
- The Erskine Building, Canterbury University, New Zealand
- St Mary's Credit Union, Navan, Ireland
- 40 Albert Road, South Melbourne, Victoria, Australia
- Ministry of Energy, Water and Communications, Putrajaya, Malaysia
- 60 Leicester Street, Melbourne, Australia
- AUT Akoranga, Auckland, New Zealand

13.4.1 NRG Systems Facility, Vermont, USA – Figures 13.1 and 13.2

Located at latitude 44.5°N in a cold-temperate climate (winter/summer design temperatures −21 °C/+29 °C), this 4320 m^2 floor area building houses a manufacturing facility with offices, workshops and a warehouse. With the highest

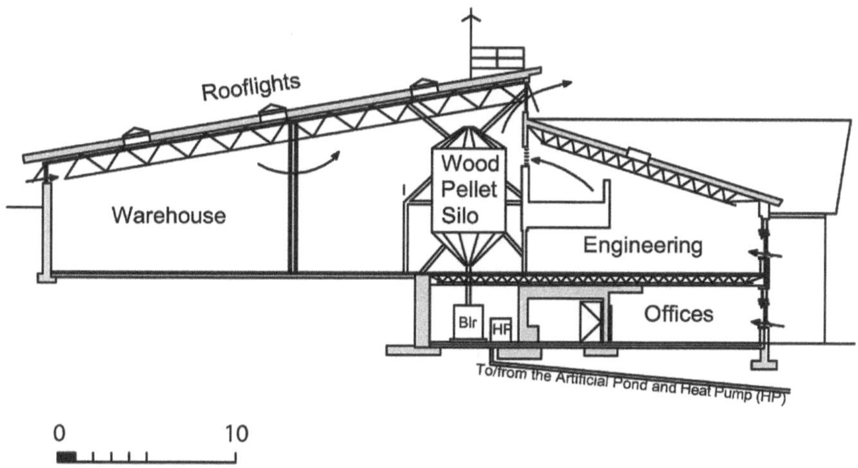

FIGURE 13.1 NRG Systems Facility – cross-section. *(adapted from William Maclay Architects and Planners)*

Summary Index of all the buildings in this set, it was to be expected that it would have some of the highest scores for each of the eleven factors that were taken into account. This was indeed the case, with nine of the factors ranked in the top five. What design features might have influenced these perception scores?

In this instance the design process was a fully integrated one, and the client was strongly committed to the use of renewable energy. The building itself was oriented to the sun, with its long axis lying E-W and its north side beamed into a hillside. It was well insulated and airtight, with carefully designed daylighting via high level strip windows and skylights integrated with dimmable artificial lights. Pipes embedded in the exposed floor slab enabled underfloor heating and cooling of the entire building. The offices and workshops had CO_2-controlled, mixed-mode ventilation via air handling units and opening windows, with red and green indicators letting the occupants know when opening or closing the windows was preferable (but not mandatory). The warehouse area was designed for natural ventilation mainly, but had extract fans at either end. Clearly, these systems were combining to provide an environment that was perceived to be near-perfect by the building's users.

Comfort overall, temperatures in both winter and in summer, design, and needs all scored well above 6, on a scale where a 7 would be the ideal; and are ranked first for this set of buildings. Not ranked first, but still within the top five, are lighting overall (4th equal), air in summer (2nd), health (3rd) and productivity (with a value of +19.51% ranked 4th).

The only 'low' scores were for temperature in summer and noise overall. With a score of 5, temperature in summer was ranked 8th – a check on the scores for the components of this factor revealed that the building was perceived to be too hot in summer. The mean score for noise overall, while still well on the satisfactory side of the scale at 4.67, ranked this building 13th of the set. A check

FIGURE 13.2 NRG Systems Facility – south façade.

of the individual components that contribute to the users' perception of noise revealed that the internal sources (colleagues, other people, other inside noises) were not the problem, but rather it was a perceived lack of noise from outside.

Almost the perfect building from the users point of view, the project team for the NRG Systems Facility had conceived and executed a building in which all of the basic 'rules' of good passive design had been applied (more of which later), coupled with transparent control systems and well documented procedures [4].

13.4.2 Torrent Research Centre, Ahmedabad, India – Figures 13.3 and 13.4

Located at latitude 23°N, Ahmedabad has three distinct climatic seasons – hot and dry from March to June with design temperatures of +41.0°C, warm and humid from July to September during the monsoon, and cool and dry from October to February with design temperatures around +13°C. The hot dry season in particular provided a challenging climatic context for the environmental ambitions of the project – to maximize the use of natural light and ventilation, use locally available natural materials, and control the ingress of dust.

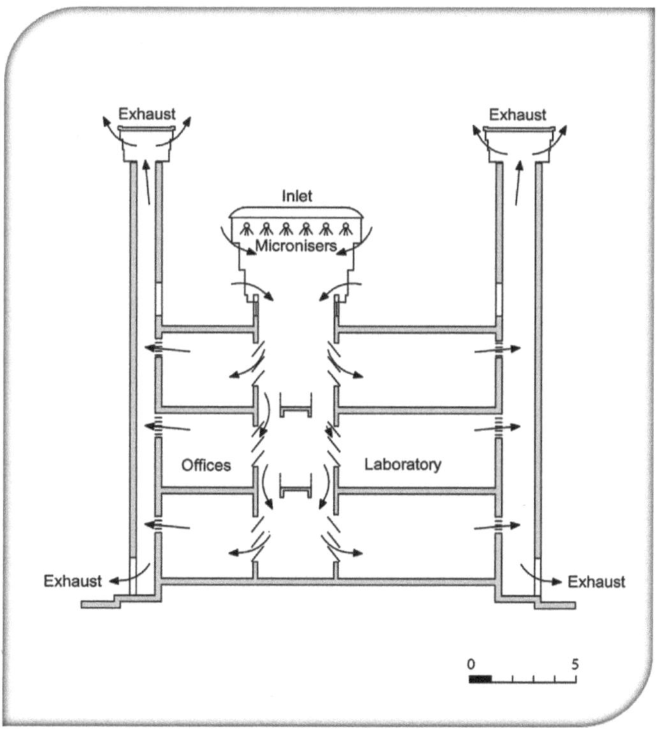

FIGURE 13.3 Torrent Research Centre – typical cross-section.

FIGURE 13.4 **Torrent Research Centre – typical laboratory block.**

Housing a pharmaceutical research organization, the facility comprised five three-storey laboratory buildings radiating from a central core building, and linked to separate administration and utilities blocks. Two of the laboratory buildings were air conditioned (AC) while the administration block and the other three laboratories were equipped with a passive downdraft evaporating cooling (PDEC) system. Their total floor area was some 12,000 m^2. Separate surveys were carried out of the AC and PDEC buildings and as can be seen from Table 13.1, their Summary Indices (2.83 and 1.95 respectively) placed these buildings 2nd and 6th. In the case of the AC buildings, all but one of the factors were in the top five, with lighting and air in summer ranked first out of the set; most factors for the PDEC building were ranked between 6th and 14th, apart from temperature and air in winter which were 3rd and 2nd respectively. As well as looking for design features that could have resulted in these two sets of buildings achieving relatively high Summary Indices, it was interesting to compare how they were perceived by their respective sets of occupants.

An integrated multi-disciplinary approach was evident here, with the design team committed to a building that could function during daylight hours with the minimum use of electricity. The concept for all of the laboratory buildings was of a central corridor flanked by working spaces. Centrally-located towers provide fresh air via the corridor; having passed through the working spaces this is exhausted via a set of towers on the perimeter. In the case of the PDEC laboratories and the administration building, a fine mist of water is released at the top of these central towers during the critical hot season. This cools the air, which then circulates through the building by natural convection. In the case of the two AC laboratories, the towers serve as routes for the distribution of a conventional system of air conditioning ductwork. The overall structure is thermally massive and the external walls and roof are white. Both high and mid-level exterior windows are utilized – these are well shaded from direct sun penetration by fixed horizontal overhangs and the vertical perimeter towers.

Given the effort that had gone into the design of the lighting, it was good to see high scores for this aspect, with the amount of natural light being perceived as about right and that of sun and sky glare relatively low. The prevalence of hard surfaces inside both kinds of building did not lead to poor scores for noise. Rather, the tendency was for there to be a perception of too little from outside.

The AC buildings appeared to outperform the PDEC buildings in most respects. A notable exception to this is the case of wintertime temperature, where the PDEC buildings score higher overall than the AC building (5.84 vs. 5.54). Further scrutiny of the data revealed that the AC buildings were perceived to be colder than the PDEC buildings. However, the largest differences were evident in the scores for summer temperature and air, where the AC buildings outperformed the PDECs. Nevertheless, with scores of 4.61 for temperature overall and 4.44 for air overall, the PDEC buildings were well over the mid-point of their respective scales. In this particular case, respondents were also asked about thermal conditions during monsoon conditions, for which the PDECs were equipped with ceiling fans. With scores of 5.75 and 5.47 respectively for temperature and air, clearly the ACs were perceived as much more comfortable than the PDECs with their corresponding scores of 4.97 and 3.38.

While the AC buildings are perceived to be better in most respects (the wintertime scores possibly hinting that the temperature set point may be on the low side) than the PDECs, it is the performance of the latter that is arguably the more notable. To be ranked 6th overall out of this set of buildings, given the severity and variability of the climate, is remarkable. The particular efforts to minimize solar heat gain through shading, exterior color, insulation and mass, together with careful attention to the disposition of the fenestration with regard to daylighting, have paid dividends. Of course these same efforts have benefitted the AC buildings too in terms of their overall performance, and have also resulted in low energy use in practice [5] as well as users' perception scores *par excellence*.

13.4.3 Natural Resources Defense Council, California, USA – Figures 13.5 and 13.6

Located in Santa Monica in the relatively benign climate of Southern California – latitude 34°N with design temperatures around 31°C and 6°C – this 1,400 m^2, three-storey building houses the offices of the NRDC.

As befits its high Summary Index, only one factor (winter temperature overall) scored less than 5. All the rest bettered that score and four factors scored well over 6. Given the NRDC's mission, there was a determination to demonstrate sustainable design. The project involved, inter alia, renovating an existing building in a downtown site, maximizing the use of daylight and natural ventilation, and achieving low energy consumption. The rectangular plan building is sandwiched between adjacent buildings on an approximately 37 m by 14 m site. Two main floors house the NRDC offices and associated facilities.

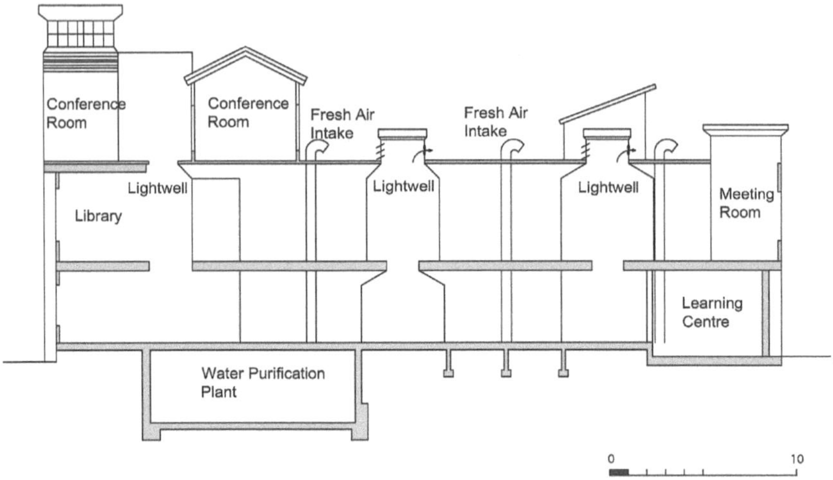

FIGURE 13.5 Natural Resources Defense Council – long cross-section. *(adapted from Moule and Polyzoides)*

A further upper level has two conference rooms, a staff work room and outdoor patio grouped around one end. The remainder of that level, the roof, is used for mounting photovoltaic panels, the solar water heater, condenser units, and other equipment.

Three lightwells punctuate the plan and these, together with a 2 m gap between the long façades and the adjacent buildings, enable daylight penetration to the building. The perimeter windows are all made to open to allow natural ventilation, and the upper glazed section of two of the lightwells is fitted with a louver opening and an extract fan.

The main cladding material, fiber reinforced cement, is used on all the façades and the lightwells. Windows have mostly low-emissivity double glazing, with interior roller blinds to allow control of any glare – opening a window automatically turns off the nearby convector heater.

While the upper level conference rooms rely totally on natural ventilation, the deeper plan of the main floors necessitated the installation of a supplementary mechanical ventilation system. Six small, fresh air supply units are integrated into the structure to deliver heated or cooled air at peak times. Thermostats in each office enable personal control of these systems, and care has been taken to ensure the staff are aware of conditions in the building and can respond appropriately. Not only are inside temperature and humidity readings on display at strategic points throughout the building, but CO_2 readings are also displayed, together with a warning light set to come on at around 750 ppm as a way of reminding the occupants of a space to open the windows. For the artificial lighting, both occupancy and daylight sensors have also been installed and the manual switches for the lights in each office also have dimming controls.

FIGURE 13.6 Natural Resources Defense Council – bird's-eye view.

Given the above attention to detail it is gratifying to see that the users' perception scores are correspondingly high overall. Looking a little deeper into the component scores, it turns out that NRDC has the distinction of having the freshest and most odorless air of the set of buildings. Despite good temperature overall scores, it would seem that the building is perceived to be a little too cold in winter and a little too hot in summer. Similarly, despite excellent lighting overall scores, there is perhaps a perception of too much daylight and a hint of some glare from sun and sky. Compared with the previous buildings (NRG and TRC), the fenestration in this case is fairly conventional and has no external shading. The artificial lighting on the other hand scores close to ideal. The underlying noise issues in this case appear to be from colleagues mainly, and to a lesser extent from other people inside the building and from outside. These issues appear to be a concomitant of natural ventilation strategies which tends to necessitate the use of large open passages across and even between floors, together with façade openings. The potential for airborne sound transmission within the building, and the entry of external noise, is evident.

13.4.4 Military Families Resource Centre, Toronto, Canada – Figures 13.7 and 13.8

Located at latitude 44°N in a cold-temperate climate (winter/summer design temperatures of −17°C/+29 °C) this 1840 m^2 floor area building caters for the needs of the spouses and children of military personnel in terms of child care, counseling services and educational programs. MFRC had the 4th highest Summary Index of all the buildings and a matching profile of high

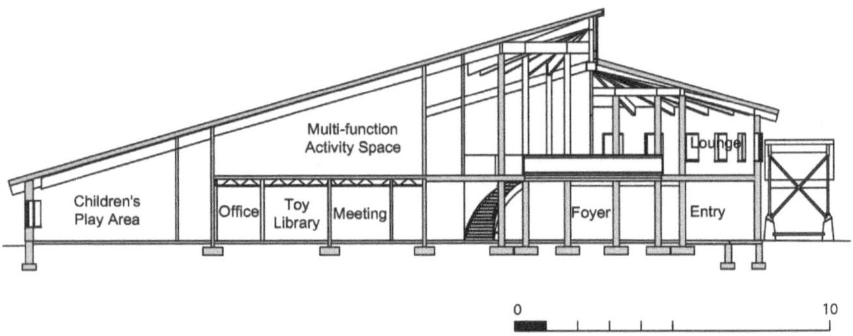

FIGURE 13.7 Military Families Resource Centre, Toronto – cross-section. *(adapted from Public Works Government Services Canada)*

FIGURE 13.8 View of Military Families Resource Centre, Toronto.

scores – only one had slipped slightly under 5, and two (design and needs) were well over 6.

The outcome from what was reported as an intensely integrated design process was a compact two-storey circular plan, with sloping timber roof structures, with mainly offices and child care facilities on the ground floor and a large multi-function space on the upper level. Roof, walls and floor slab were well insulated and clear double glazing was used throughout. The perimeter and roof level glazing enables daylight penetration to the majority of the spaces that are in frequent use, such as the child care areas and the offices. Windows at both high and low levels have openable sections to enable natural ventilation to take place when climatic conditions allow – typically spring through autumn. An underfloor system is the primary heating source for the building, while a small air handling unit distributes fresh air, heated or cooled as appropriate, to the occupied spaces.

Temperature overall in both winter and summer scored relatively highly at 5.47 and 5.92 respectively. However, the more detailed questions revealed that the users found it to be too hot in winter and too cold in summer, pointing to the necessity for (say) some adjustment of the control systems. While lighting overall scored a high 5.69, the scores for the more detailed components of that aspect hinted at too much natural light. The scores for glare from sun and sky and from artificial lighting were both on the high side indicating the need for care with these aspects. A score of 5.67 for noise overall placed the building 2nd for this aspect. Unsurprisingly, the main sources here were other people and noises inside, but despite that the overall score was excellent, given the type of activities housed. All the satisfaction factors achieved high ranking scores and the perceived productivity increase was 20 per cent.

13.4.5 The Erskine Building, Canterbury University, New Zealand – Figures 13.9 and 13.10

Located at latitude 44°S in a medium-temperate climate (winter/summer design temperatures of −1°C/+26 °C) this 11,551 m^2 building is split approximately equally between a seven-storey academic block, containing staff and postgraduate student offices, and a four-storey undergraduate teaching block. Known formerly as the Mathematics & Statistics and Computer Science Building, the two blocks are linked by a glass roofed atrium space and a basement area, which contains mainly teaching and service spaces. Lying 5th in the Summary Index rankings, all but one of its overall scores were in the 5 to 6 range.

The offices and the majority of the adjacent seminar rooms in the academic block are naturally ventilated and heated by a conventional radiator system. With their deliberately northerly orientation (the sunny side for this southern hemisphere building) and fixed overhangs, exposed thermally massive interior walls and ceilings, fixed and adjustable exterior and adjustable interior solar shading devices, and large number of window/natural ventilation opening options, the ninety or so office modules are equipped with a full range of passive thermal environmental control systems. The undergraduate teaching block and basement computing laboratories have separate air handling units, with the supply air distributed in such a way as to make good use of the thermal mass of the building's structural slabs.

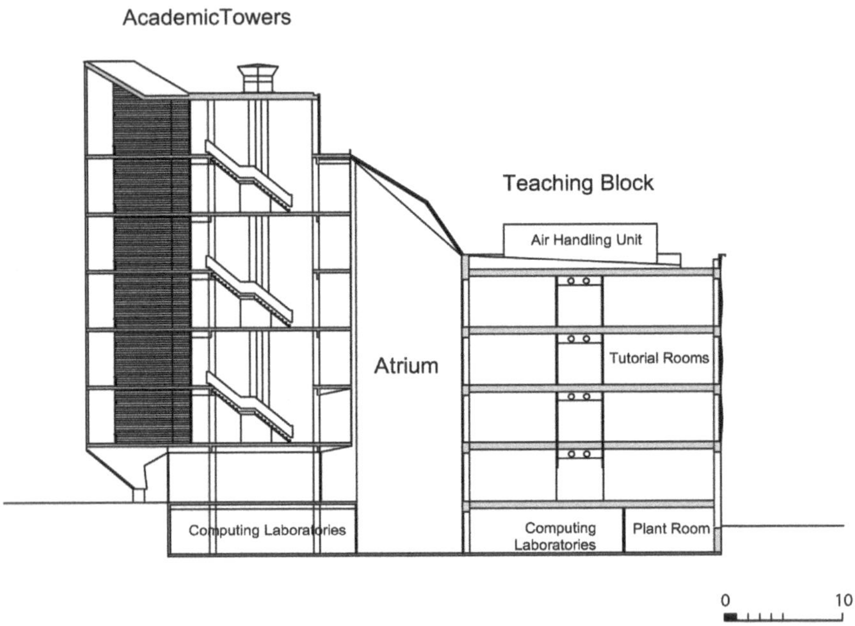

FIGURE 13.9 Erskine Building – cross-section. *(adapted from Architectus)*

FIGURE 13.10 Erskine Building – north-east Façade.

The building's score for comfort overall (5.86) was one of the highest, reflecting the good overall scores for all of the environmental factors. Exceptions to this occurred where, despite high overall scores for temperature and air in winter, staff perceived the air to be too still and dry – though this could be regarded as an inevitable consequence of using heating and natural ventilation in a cold climate. The users' responses also suggested that there was too much glare from sun and sky.

By the use, inter alia, of a well insulated external envelope, internally exposed thermally massive construction, predominantly manually operated window openings and shading devices, and deliberate orientation of staff studies to the north, this building has been able to satisfy the varying needs of a diverse staff and student population. However questions regarding control of sun and sky glare remain.

13.4.6 St Mary's Credit Union, Navan, Ireland – Figures 13.11 and 13.12

Located at latitude 54°N in a medium-temperate climate (winter/summer design temperatures around 0°C/+21 °C) this 1300 m^2 floor area, five-storey building houses the banking hall and offices of a financial services provider. Ranking 7th with a Summary Index of +1.73, all of its overall scores are in the 4.5 to 6 range.

The architect's aim was for this project to be an exemplar ecological office building. The basic plan is a roughly 15 m by 15 m square, oriented on the diagonal of the cardinal points of the compass, with its main structure and construction made predominantly of dense pine. A full height atrium space is designed to enable natural ventilation of the surrounding spaces and daylight

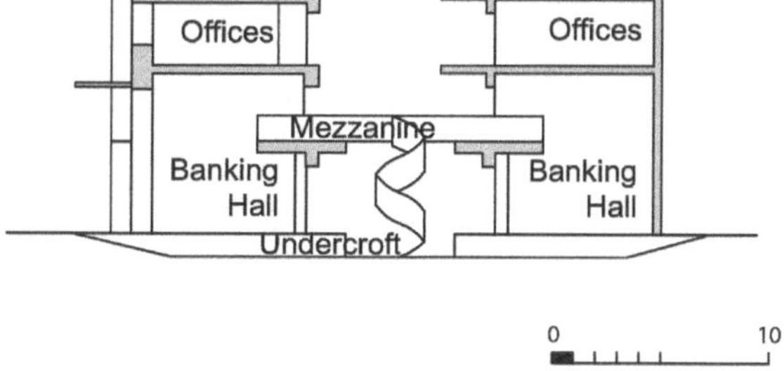

FIGURE 13.11 **St Mary's Credit Union – cross-section.** *(adapted from Gaia Ecotecture)*

penetration to the interior. The building orientation means that south-east and south-west façades are 'available' to pick up solar heat gains via the glazing on these façades.

Openings around the perimeter act as air intakes to the atrium, while extraction is via a rooftop venturi device or a solar chimney. Active heating and cooling systems serve terminal units distributed throughout the building. The control system is designed to maximize the utilization of ambient conditions and minimize the use of the active systems.

Noting the relatively small numbers of staff involved here (14 respondents) it can be seen that this building is perceived as performing very well on all counts. Noise overall scored particularly well at 5.85 – clearly a relatively quiet environment has been achieved, despite the open nature of the building and its banking hall function. The scores for noise from colleagues and noise from inside were virtually ideal. The apparently low ranking for lighting overall, despite a score of 4.92, is a reflection of the high scores generally for this

FIGURE 13.12 St Mary's Credit Union.

aspect of these sustainable buildings. However, the component scores indicated that sun and sky glare may be an issue. The extensive use of glazing on the north-east and north-west façades in order to optimize solar heat gains may be contributing to this perception. Thermal factors were mostly fine, the only issue worth commenting on being a perception of the summer temperature being a little too variable; which is perhaps an inherent characteristic of natural ventilation systems. Overall though, the building scores well and ranks very high in this set.

13.4.7 40 Albert Road, South Melbourne, Victoria, Australia – Figures 13.13 and 13.14

Located at latitude 38°S in a warm-temperate climate (winter/summer design temperatures around +3.3°C/+34.5 °C) this refurbished building has approximately 1400 m^2 of office space on four floors above enclosed ground and basement level parking areas. With a Summary Index of +1.42, its overall scores range from 4.36 (for air in winter) to 6.27 (for design). In fact three aspects scored 6 or higher, four were between 5 and 6, and three between 4 and 5.

The objectives of the refurbishment (the existing building was a 1970s office block) included minimizing the building's energy consumption and environmental impact, and optimizing its thermal environment and use of daylight. The final outcome retained the basic structural integrity of the existing building, but introduced a number of strategic interventions designed to maximize usable floor space and improve the daylight and access to external views in the narrow floor plate. Skylights were added and a stairwell adapted to serve as both a lightwell and a natural ventilation 'stack'. The existing east-facing concrete façade was replaced by full height, clear, low-E double glazing, with additional

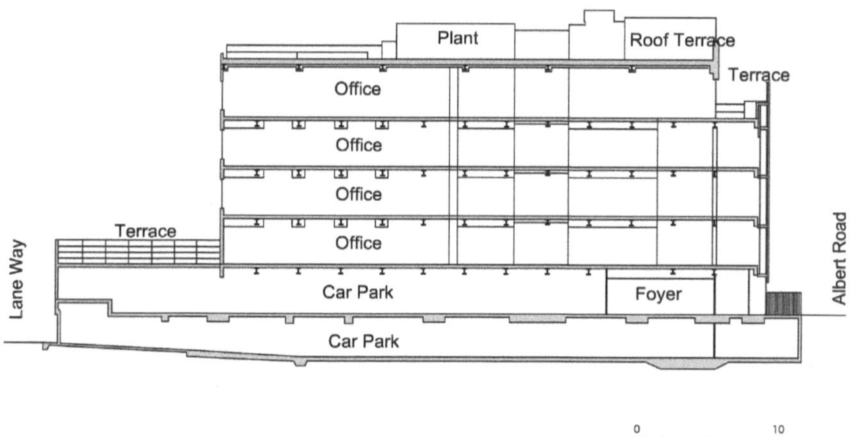

FIGURE 13.13 40 Albert Road, Melbourne – long cross-section. *(adapted from SJB Architects)*

shading provided by a steel perforated mesh. Windows on the west façade were also refitted with clear, low-E double glazing. Some of the existing ceiling tiles were removed to expose the thermal mass potential of the underside of the concrete slab.

The office spaces were designed to operate within a temperature range of 19–25°C during occupied hours. A mixed-mode of operation combines natural ventilation, when ambient conditions permit, with an air conditioning system for periods where heating or cooling is required. All aspects of operation are controlled by a building management system. Natural ventilation is achieved by admitting air through operable windows located on both façades of the building and using the stack effect in the open-tread stairwell to draw the air across the office and exhaust through louvers at roof level. The system is also employed in summer and in mid-season to provide night cooling.

Single, linear, T5 fluorescent fittings with specular low brightness louvers have been installed, with high frequency dimmable electronic ballasts designed to minimize glare and power density. Motion sensors, positioned around the building to detect occupancy, maximize energy savings by switching and dimming the lighting system, as well as switching off fan coil units when an area becomes vacant.

These efforts to optimize the lighting systems appear to have paid off, with the score for lighting overall averaging 6.04 and ranking 4th. The scores for the amounts of natural and of artificial lighting were close to the ideal, and those for glare on the low side. Temperature and air in summer scored well – ranked 5th and 4th at 5.35 and 5.56 respectively. However, winter temperature and air both ranked only 16th with overall scores of 4.42 and 4.36, a result, it would appear, of perceptions of winter temperatures being too cold and dry. On the other hand, the air was perceived to be particularly fresh and odorless in both winter and summer.

FIGURE 13.14 **40 Albert Road South Melbourne, Victoria, Australia.**

13.4.8 Ministry of Energy, Water and Communications, Putrajaya, Malaysia – Figures 13.15 and 13.16

This governmental office building is located at latitude 3°N in a hot-humid climate with year-round temperatures ranging from +22°C to +34 °C, accompanied by high relative humidity levels. There is approximately 19,200 m^2 of air conditioned space on four upper floors, surmounting two levels of parking. Ranking 9th with a Summary Index of +1.33, its overall scores range from 4.77 (for health) to 5.44 (for design); a comparatively narrow range.

The building was designed to be a showcase for energy efficiency and low environmental impact. It was to demonstrate, without compromising the comfort of the occupants, energy savings of 50% compared to more conventional office buildings. The layout was designed to give the majority of the façades a northerly or southerly orientation in order to minimize direct solar heat gain. The walls are well insulated and light in color while the top floor has 100 mm of insulation and an additional canopy roof.

Single glazed throughout and predominantly unable to be opened, the building's fenestration has been designed to minimize solar heat gain and optimize the penetration of daylight by the use of a combination of external shading louvers, light shelves, and recessing. There is no glazing on the western façade, while the windows on the easterly façade have been given a deeper light-shelf than those on

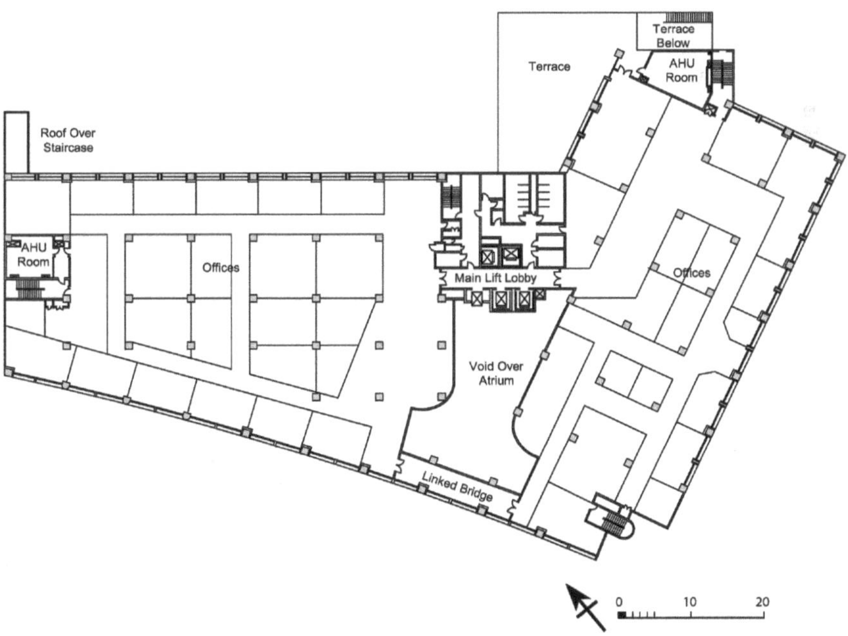

FIGURE 13.15 Ministry of Energy, Water and Communications – plan view of typical floor. *(adapted from SNO Architects)*

FIGURE 13.16 **Ministry of Energy, Water and Communications.**

the north and south to help cope with low angle morning sun. In keeping with the care taken over daylighting, the artificial lighting has been kept down to an installed load of around 11 W/m^2, while maintaining an average illuminance of 350 lux.

Apart from the atrium area and a few emergency openings, the building has a sealed envelope and is fully air conditioned, with a chilled water supply from a nearby gas-fired district cooling plant. Air Handling Unit Supply (AHUs) supply chilled air to their respective office floors via a zoned variable air volume system, designed to maintain an inside temperature of 24°C. The fresh air supply rate is controlled via CO_2 sensors, and electrostatic air filters are used to clean the mixture of fresh and return air before supplying it to the offices.

The high overall scores for temperature and air (5.16 and 4.96 respectively) indicate that the air conditioning system is doing a good job as far as the occupants are concerned. However, it seems that the temperatures are perceived to be too cold and variable. Lighting overall has achieved a good mean score of 5.10, and the amounts of natural and artificial light were fairly close to ideal. However both methods of lighting seem to be sources of glare. The building was perceived as having relatively little noise from any of the usual sources (colleagues, other people, inside and outside) and infrequent interruptions. Perhaps this contributed to its having the 5th ranked increase in perceived productivity of 16%.

13.4.9 60 Leicester Street, Melbourne, Australia – Figures 13.17 and 13.18

Located at latitude 38°S in a warm-temperate climate (winter/summer design temperatures around +3.3°C/+34.5 °C) this four-storey refurbished building houses around fifteen tenants on its approximately 3400 m^2 of floor area. Ranking 10th with a Summary Index of +1.23, its overall scores range from 4.22

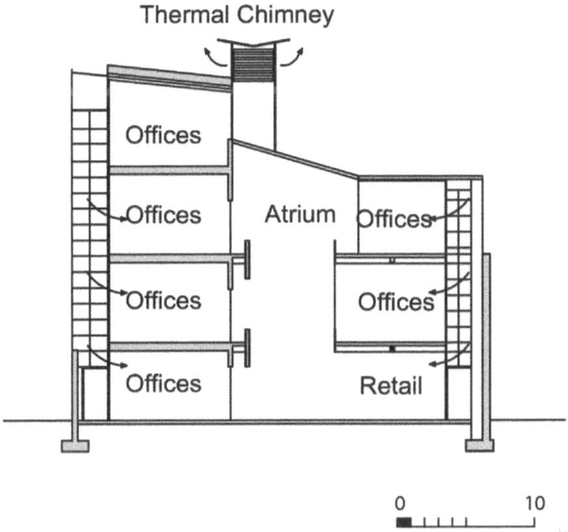

FIGURE 13.17 60 Leicester Street, Melbourne – cross-section.

FIGURE 13.18 60 Leicester Street, Melbourne – rooftop with thermal chimneys and atrium glazing.

(for noise overall) to 5.87 (for needs), with an increase in perceived productivity of just over 11%.

Set up to demonstrate the viability and practicality of environmentally sustainable office design, the overall design approach was to use passive systems, and only once these had been optimized to consider the integration of complementary active services. Rectangular in plan, on an approximately 30m by 110m site, the building is sandwiched between adjacent commercial and residential developments. Despite these constraints, the building has been planned to enable natural ventilation and natural lighting to be available whenever outside conditions permit. Lightwells have been cut into the northern and southern façades; these are glazed and have openable windows at each level to enable air and light entry to the occupied spaces. These spaces are grouped around a central glazed atrium which enables daylight to penetrate down into the Center of the building and air to exit via four thermal chimneys. Exposed brickwork and concrete provide usable thermal mass, while the roof and walls have R-values of 3.5 and 1.5 m^2 °C/W respectively and the windows are double glazed. The building is naturally ventilated when the outside air temperature is between 19°C and 27°C, and mechanically ventilated (heated or cooled as appropriate, using reverse-cycle heat pumps) when it is outside that range.

The efficacy of this ventilation strategy is given some credence by the summer scores for temperature and air which, at 5.50 and 5.33 respectively are ranked 3rd and 5th for this set of buildings. The corresponding winter scores are less comfortable at 4.56 and 4.64 (lying 12th and 11th respectively), and indications were that the temperature was too cold during that time of year. However, the quality of the air was good in both seasons.

The building scored satisfactorily for lighting too (mean of 5.75). Scores for the amounts of daylight and artificial light were close to ideal and glare was not a major issue. The scores for noise were less satisfactory, with noise overall ranked 20th and internal noises the main source of disturbance.

13.4.10 AUT Akoranga, Auckland, New Zealand – Figures 13.19 and 13.20

At just under 1,000 m^2, this building houses academic offices, a small registry and the main reception area for a satellite campus of the Auckland University of Technology. Auckland is situated at latitude 37°N and has winter and summer design temperatures of around +3 and +24°C. While it is the last of the case studies described in this chapter, it is by no means least. With a Summary Index of +1.18, on a −3 to +3 scale, its overall performance is still very good from the point of view of the users.

As the aim was for a low energy building, the design team took an integrated approach which also included extensive consultation with the future users and facilities managers. The outcome was a single-storey, naturally ventilated building, roughly U-shaped in plan and housing five separate administrative areas, all linked by a circulation corridor. The corridor is taken up higher than the surrounding single-storey accommodation, enabling it to perform both ventilating and daylighting functions.

Several passive elements contribute to the control of heat losses and gains via the building envelope – in particular, the use of 100 mm thickness insulation throughout, the installation of a ventilated fly-roof, plus the use of light colors externally and thermal mass internally. In addition, window areas on the sunny façades are fitted with either fixed external louvers or with motorized retractable

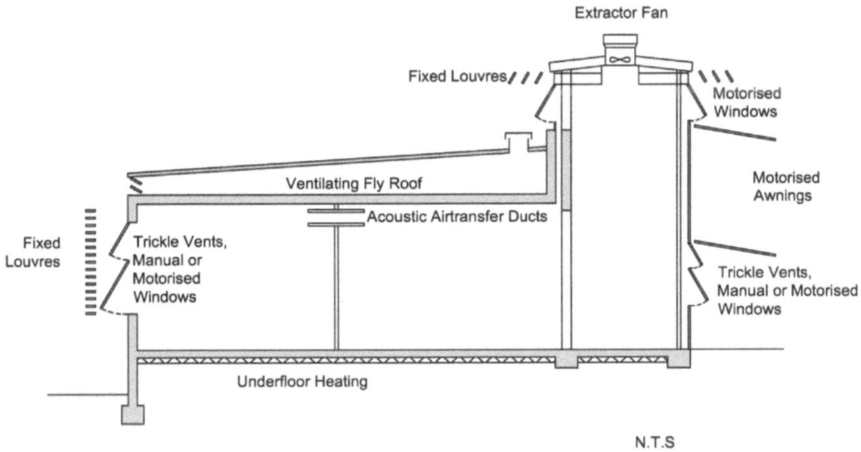

FIGURE 13.19 AUT Akoranga – cross-section illustrating the principles of the passive environmental control systems. *(adapted from JASMAX)*

FIGURE 13.20 AUT Akoranga – view from the west.

awnings. Control of the zoned underfloor heating, the ventilation (natural and mechanical), the boardroom air conditioning, and the motorized awnings is via a computer-based building management system.

Natural ventilation and daylighting is enabled through a range of perimeter window openings coupled with automatic windows at high and low level along the corridors. The perimeter and high level glazing is also designed to provide daylight to most of the spaces, with glare control supplied by means of the external louvers and awnings.

While most of the overall scores were between 4 and 5, they were all comfortably better than the mid-point of the 1–7 range and better than the corresponding Building Use Studies (BUS) benchmark. This included the overall scores for air and temperature in both summer and winter, although the latter was assessed as too cold and variable. Lighting overall scored well at 4.88, but component scores indicated there was too much glare from sun and sky. Likewise, noise overall at 4.92 scored well, but noises from colleagues, other people, and inside the building were all rated slightly worse, as was the disturbance due to unwanted interruptions.

Perceptions of comfort overall, design, and needs all scored above 5, and the score for health was higher than the mid-point of the scale – implying the staff felt healthier in the building.

13.5 KEY CHARACTERISTICS AND COMMON FEATURES OF THESE SUSTAINABLE BUILDINGS

Finally, it is worth summarizing the common features that characterize the more successful of these projects from the users' point of view.

Clients and Designers

- The clients were strongly committed to sustainability and energy efficiency.
- Design teams were committed to sustainable design – with local knowledge and a track record.
- Integrated design processes and adequate time to carry them out.

The Building Envelope

- Good insulation and double glazing.
- Airtight envelope.
- Appropriate solar orientation and shading.
- Judicious use of exposed thermal mass.

Systems Incorporated

- Change over mixed-mode HVAC systems with automated natural ventilation openings where appropriate.
- Carefully designed daylighting systems.
- Atria or similar systems to enable natural ventilation and daylight to building interiors.

In Operation

- A commitment to commissioning, post occupancy evaluation, and continued building management.

It is strongly recommended that designers of future buildings study the projects summarized in this chapter in more detail if they plan to adapt some of the solutions to their particular circumstances.

ACKNOWLEDGMENTS

My particular thanks must go to all the building users who responded to the questionnaire, the building owners who facilitated the surveys, and the designers and building managers who gave so freely of their time. Thanks too to all the collaborators who joined with me to survey particular buildings, to Adrian Leaman of Building Use Studies and the many research assistants who worked on the data analysis, the latter supported by grants from Victoria University of Wellington.

REFERENCES

[1] Building Use Studies, Useable Buildings – retrieved from , 2011 http://www.usablebuildings.co.uk/ (Accessed 21 August 2011).

[2] A. Leaman, B. Bordass, The Probe occupant surveys and their implications, Building Research and Information 29 (2) (2001) 129–143.

[3] NRG Systems, Tour our Facilities, 2008.

[4] http://www.nrgsystems.com/AboutNRG/TourOurFacilities.aspx accessed 23 August 2011.

[5] L. Thomas, Torrent Research Centre, Ch 31 of Sustainable Buildings in Practice (Baird, G.) Routledge, Oxon, UK, 2010.

Appendix

Calculation of Indices

First, it should be made clear that each of the factors has been assigned a benchmark (copyright BUS) on its 7-point scale. At any given time, these benchmarks are simply the mean of the scores for each individual factor, averaged over the last 50 buildings entered into the BUS database. As such, each benchmark score may be expected to change over time as newly surveyed buildings are added and older ones withdrawn. Nevertheless, none of these were observed to have changed significantly during the five years or so over which these buildings were surveyed.

The Comfort Index involves temperature, air quality, lighting and noise factors. It encapsulates, in a single figure, an overview of users' perceptions of that aspect of the building's performance. This index is formulated from the Z-scores for Comfort Overall (comf), together with the main environmental factors of Lighting Overall (light), Noise Overall (noise), Temperature Overall in both winter (Tw) and summer (Ts) and Air Overall in both winter (Aw) and summer (As). The Z-scores are derived from (actual score – benchmark) / (benchmark standard deviation). They are standardized scores with mean = 0 and standard deviation = 1, and are used here to give equal weights to the seven constituent values of the index.

The formula for calculating this index is simply the average of the Z-scores for these seven factors, i.e.,

$$\begin{aligned} CI = {} & (Zcomfort + Zlight + Znoise + Ztempwinter \\ & + Ztempsummer + Zairwinter + Zairsummer)/7 \end{aligned}$$

The Comfort Index is based on a scale of '–3' to '+3', where '+3' is considered 'best' (the mid-point lies at zero).

The Satisfaction index Involves design, needs, health, and productivity factors. It encapsulates, in a single figure, the users' overall satisfaction with the building. It is formulated from the Z-scores of the overall ratings for Design (des), Needs (need), Health (heal) and Productivity (Prod%). The formula for calculating this index is simply the average of the Z-scores for these factors, i.e.:

$$SI = (Z\ design + Z\ needs + Z\ health + Z\ productivity)/4$$

As before, the Satisfaction Index is based on a scale of '–3' to '+3' where '+3' is considered 'best'.

The Summary Index is simply the arithmetical average of the Comfort and Satisfaction Indices.

Abbreviations: NRG – NRG Systems Facility, Vermont, USA; TRC AC & PDEC – Torrent Research Centre, Ahmedabad, India; NRDC – Natural Resources Defence Council, California, USA; MFRC – Military Families Resource Centre, Toronto, Canada; Erskine – The Erskine Building, Canterbury University, New Zealand; St Mary's – St Mary's Credit Union, Navan, Ireland; 40Alb – 40 Albert Road, South Melbourne, Victoria, Australia; MEWC – Ministry of Energy, Water and Communications, Putrajaya, Malaysia; 60L – 60 Leicester Street, Melbourne, Australia; AUT – AUT Akoranga, Auckland, New Zealand. (see also in table 13.1)

Chapter 14

Sustainable Buildings and their Relationship with Humans and Nature

Lessons from the Past

Rahman Azari* and Maryam Singery†
*College of Built Environments, University of Washington, Seattle, WA, USA,
†Faculty of Architecture and Urban Studies, Islamic Azad University, Tabriz Branch, Iran

Chapter Outline

'We abuse land because we regard it as a commodity belonging to us. When we see land as a community to which we belong, we may begin to use it with love and respect. There is no other way for land to survive the impact of mechanized man, nor for us to reap from it the aesthetic harvest it is capable, under science, of contributing to culture.'

Aldo Leopold

14.1 BACKGROUND AND PRESENT SITUATION

We live in a century in which humankind's main concern has started to shift towards ways to maintain life and the environment at the present time and for future generations. The earth's population is growing at an exponential rate

Sustainability, Energy and Architecture. ISBN: 9780123972699

and our cities are expanding, despite the lack of necessary resources to meet people's basic needs and maintain a reasonable quality of life. Moreover, we are rapidly killing many species, consuming valuable resources and polluting the vital resources of water, air and soil.

The consequences are disastrous. The amounts of carbon dioxide emitted from the processes of material extraction, building our homes, driving our cars and growing our food crops has resulted in global warming. This has resulted in sea level rises and changes in day-to-day weather and climate. Hazardous materials produced by our activites affect our health and even cause fatal diseases.

The construction and operation of buildings is a big contributor to most of these environmental impacts and a major obstacle to sustainability. According to the US Energy Information Administration (EIA) [3], buildings are responsible for more than 40% of primary energy consumption, 40% of carbon dioxide emissions and 75% of electricity consumption in the United States. These numbers are mainly caused by the operation of Heating, Ventilation and Air Conditioning (HVAC) and lighting systems in these buildings. The EIA estimates that space heating accounts for about 30% of total energy consumption in residential buildings, followed by space cooling, water heating and lighting, which consume 12.3%, 12.2% and 11% of the total energy, respectively.

The intense reliance on fossil fuel for energy sources, either directly or through the use of electricity in buildings, occurs not only because of contemporary lifestyles, but also as a consequence of the way that we design and construct our buildings. As a result of the new types of buildings introduced in the last 200 years, as well as the complexity of their design and construction, various professions and design disciplines have been formed in the construction industry. Professionals working within these disciplines have tended to resolve the design problems within their own discipline, while showing little tendency to cooperate in cross-functional teams. As a result of this fragmentation, various design disciplines, rather than taking a holistic approach to the inception and development of a building based on the characteristics of the surrounding environment, have instead tried to optimize their subsystems regardless of the impact on other interdependent subsystems and on the building as a whole. As an example, buildings' mechanical systems are designed by mechanical engineers to meet the highest levels of heating and cooling loads without consulting with architects and the rest of the design team to seek opportunities for downsizing the systems and to decrease the energy demand. The buildings created in this way can offer seemingly higher levels of living standards to their occupants by including massive mechanical heating and cooling energy systems, which are expensive and highly polluting.

Contemporary buildings also impact the occupants' heath and quality of life. According to a study by the US Environmental Protection Agency [4], Americans spend 90% of their time indoors. This implies that building occupants are constantly exposed to the pollutants and emissions produced inside the buildings. Without sufficient ventilation or the elimination of the sources, these pollutants and emissions can cause serious diseases.

The negative impacts of contemporary architectural design procedures and their consequent buildings are not, however, limited to the environment and occupants' health. Indeed, under the influence of modernism, architects have started to move away from the previous site- and culture-specific identity of buildings towards representing architecture through a globally shared image in which cultural identity had no place.

In recent decades, various attempts within the framework of sustainable architecture have been made to lower the environmental impacts of built environments by implementing design, delivery and construction practices which support the environment and improve the quality of occupants' lives. The attempts provide design guidelines on how to create buildings that are less reliant on fossil fuel energy, with fewer emissions and releasing less waste to the environment. They also provide a framework to assess and measure the achievement of these goals. While these efforts should be continued, in order to reverse the destructive impacts of buildings on the environment, more serious steps need to be taken towards developing a sustaining healthy relationship between buildings and the culture and society that contain them.

14.2 TRADITIONAL ARCHITECTURE; THE OUTCOME OF A COMPLEX THINKING SYSTEM

Traditional architecture has evolved into an approach capable of response to climatic, cultural, economic and functional stimuli. It created buildings which have functioned respectfully as part of a larger healthy ecosystem that included humans and the surrounding natural environment. Also, it developed an identity founded on interactions with the geographical, cultural and environmental context that encompass it. The meaning of traditional architecture, therefore, should be sought not just in the physical features it created or in its historical values, but also in its interactions with people and the environment which created its identity.

Our ancestors first lived in caves and then constructed their early shelters to protect them from the hostile environment surrounding them. Their later shelters, however, were designed to meet their functional needs, carry esthetic content, provide relative comfort and stability, and connect them with the outside community. Indeed, the improvements in function and form of living places, from primitive shelters into masterly designed traditional buildings, portray how human concerns extended over time; from meeting their basic functional needs, to handling environmental and social interactions. They also show how humans extended their ability to use available resources to meet their needs.

Good examples of traditional buildings, especially houses, are, in fact, reflections of a complex thinking system that created physical spaces on a foundation of the concerns which coped with the individual, climate, religion, culture and society. For instance, in Middle Eastern houses, the traditional building design addressed not only how to moderate the usually harsh outside climate, and protect occupants, but also ways to organize building spaces according to Islamic teachings which emphasize women's privacy.

This intelligent architectural design in traditional buildings was the outcome of an 'integrated' view of design. Based on this approach, which dates back to ancient times, a single individual took on the responsibility of meeting owners' requirements in building design. The single individual, in the case of vernacular architecture, was often an individual who tried to build a home for his own family with the help of his community. In other types of traditional architecture, a professional master builder designed the building and supervised its construction. In either case, the integrated identity of the designer was deeply tied to social norms and cultural values of their society. He was also aware of the natural and climatic potentials of the site and its surrounding environment. Indeed in this approach, a single individual embraced knowledge passed through generations, gleaned through applied previous experiences facing similar design cases and created his own architecture. The outcome was buildings that were compatible with their socio-cultural and climatic context, self-sufficient and solely reliant on natural forces to meet the heating and cooling needs of their occupants.

Traditional architecture, therefore, can be considered as a phenomenon that is dynamic at both temporal and geographical levels, capable of respectfully responding to the cultural and environmental stimuli surrounding it, with responses that vary from time to time in history, from climate to climate, and from culture to culture.

14.3 TRADITIONAL ARCHITECTURE AND ADAPTIVE RESPONSE TO CLIMATE

Formation and growth of the traditional urban fabric of the world's cities has been influenced by many drivers, including natural and climatic conditions, political factors, economy and so forth. Among these, natural and climatic conditions were often the fundamental factors on which the other drivers depended. The conditions determined urban geometry, the fabric's density, the orientation of streets and their aspect ratio. They also affected the buildings' geometry, form, orientation and spatial organization. The adaptive response of traditional buildings and cities to climate implies that physical characteristics of buildings and urban forms are developed to take advantage of the pleasant aspects of a climate and to cope with the unfavorable ones.

Cold and hot-arid climatic regions in Iran are characterized with the cities traditionally built with dense urban fabrics and attached buildings. In these cities, the use of forms with low surface area to volume ratios, both at the individual building and at the urban scale, reduced the occupants' exposure to the harsh effects of the environment. Indeed, every attempt was made to restrict heat transfer between outside and inside, and to protect the people using urban open spaces from the climate. To do this, urban spaces such as bazaars, which should be normally open and uncovered, become covered to protect shoppers against summer sun and winter cold. Daylight and ventilation in these covered spaces are then provided through apertures at the top of the domes which cover the space. Moreover, the streets and alleys of traditional cities were designed with a high aspect ratio and were covered in some areas, to provide shade for pedestrians.

In the humid, temperate regions of the north, urban structures had to deal with uncomfortable levels of humidity. As a result, cities and rural areas of this region were designed at lower densities, mainly to facilitate air circulation within the urban fabric.

At the building scale, traditional cities embraced structures with optimum access to sun and wind. Courtyards, building apertures, materials, and wind towers are among the main elements in the buildings, and especially in the traditional houses of Iran. These elements facilitated desirable thermal comfort during various times of the day and year.

14.3.1 Courtyards

Among the key features of traditional buildings, are the courtyards. Their functions, apart from being site specific, are to create comfortable outside microclimates within the building. Courtyards were primarily used in cold or hot, arid climatic regions with the aim of balancing the outside unpleasant effects of the harsh climate, including the large diurnal temperature swings. Additional elements such as trees and water ponds or wells provided shade and fruit, as well as allowing the summer hot breezes to cool by evaporative cooling.

Courtyards functioned thermally by a combination of convective, evaporative, conductive and radiation mechanisms of heat transfer. Summer winds, as well as local convective air movements, are produced as a result of temperature differences between shaded and unshaded areas of the courtyards. Hot air then entered the open windows and made the building's high thermal mass absorb the heat. This convective effect usually had a humidifying, cooling component as well. When passing over water ponds and through greenery in the courtyard, dry outside air will pick up moisture, which will decrease its temperature (see Figure 14.1).

Eventually, as a result of the collective action of the thermal mass, windows and landscape elements in a courtyard, hot air movement is turned into cool pleasant breezes for occupants. At night, when the temperature drops below comfort levels, the thermal mass of the building which has been thermally loaded during the day through conduction, convection and radiation, will provide the needed warmth for the occupants in winter time by the thermal lag effect.

Sometimes, a single house is designed to have several courtyards of various sizes. This design not only facilitates spatial organization, but also causes other

FIGURE 14.1 The courtyard and its elements provided a comfortable microclimate inside the house in a desert climate; Abbasian House, Kashan, Iran.

spaces connected to the courtyards to have access to cooler breezes resulting from the evaporative cooling mechanism and thus, to benefit from increased levels of cross-ventilation in summer.

In addition to thermal advantages, courtyards provide cultural benefits. They provide privacy for women inside the house, a place for family gatherings and sleeping, and security from bandits.

14.3.2 Apertures

The windows in traditional Iranian architecture were primarily designed to face the courtyard to admit daylight, air into the internal spaces and provide pleasant views of the courtyard. These usually large windows were shaded on summer days by trees, porches or interior curtains. Stained glass was occasionally used in these windows to balance the thermal and glare effects of daylight and decorate the internal space. Windows to outside alleys were designed to be minimal in size and number in order to restrict the view to the indoors and to provide security.

FIGURE 14.2 Various types of apertures, in combination with wind towers guaranteed maximum air circulation in a hot and dry climate; Borujerdiha House, Kashan, Iran. *(Photos: Masih Mostajeran)*

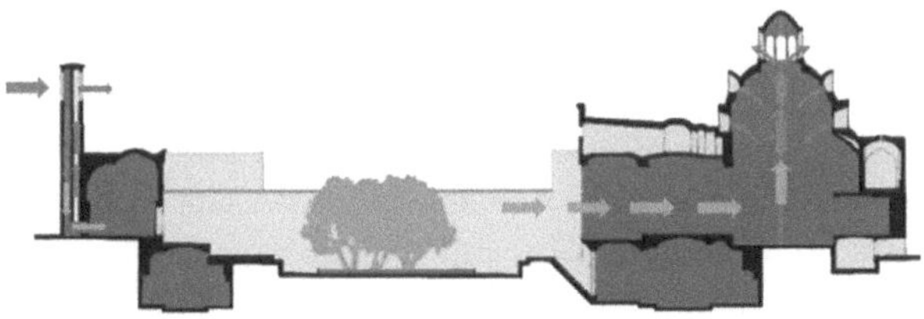

FIGURE 14.3 High ceilings with apertures at the top facilitated stack ventilation (right of the diagram). The wind tower was used to catch the high air streams and direct them into internal spaces (left of the diagram); Borujerdiha House, Kashan, Iran.

Especially in hot climates, ventilation apertures were provided in small domes covering some house spaces and some urban spaces such as bazaars. The warmer air inside the space, which tends to rise due to its lower density, can find a way to escape through these apertures. In the case of houses, this effect was aided by windows at lower levels and high ceilings, improving the stack ventilation effect (Figures 14.2 and 14.3). High ceilings also caused warmer layers of air to move above the zones occupied by inhabitants.

14.3.3 Material and Thermal Mass

Materials in traditional buildings were selected mainly based on their availability and ability to respond to climatic forces. Adobe has long been the traditional material of choice in cold and hot-arid regions in Iran. Even in hot and humid regions, where adobe is not an ideal material with respect to thermal performance, it is used in buildings due to its availability.

In summers, adobe structures provided high thermal mass which functioned through a time lag effect. In fact, the outside hot air penetrates the adobe surface through radiation, convection and conduction mechanisms and it is stored in the wall's thermal mass. The stored heat is then released into the space with a delay – time lag – at night. Using this time lag effect, adobe building materials balanced high diurnal temperature swings between interior and exterior spaces in hot and cold regions.

The time lag effect was not specific to summer. In winter, solar radiation enters the room through windows and is turned into heat within the space. The heat was trapped inside the internal spaces with closed windows, was stored in the thermal mass of adobe walls and then released at night to contribute to thermal comfort. However, it should be mentioned that thermal performance of adobe in winter is not always desirable. In fact, in winter, the adobe walls cool down at night, and remain so during the next day.

14.4 WIND CATCHER/TOWER

Another element of traditional architecture which was common in hot and arid regions in Iran was the wind catcher/tower system, or *Baudgeers* (in Farsi). This system was intended to supply internal spaces with fresh outside air and to remove hot air from inside the house, aided by pressure and temperature differences. In fact, the windward side of the tower will have a positive pressure, while the leeward side will experience a negative pressure.

To supply air, high outside air streams were directed down the wind tower's shaft deep into the internal spaces, as the inside hot air escaped the shaft. Adjustment of wind tower's physical characteristics, such as height, aperture sizes and configuration of internal partitions according to contextual parameters enhanced the performance of the system. In other words, the physical characteristics of the wind towers varied from city to city and from house to house in order to adapt to varying wind directions and velocities, and the occupants' cooling needs.

Most traditional wind towers usually had several faces to catch winds from all directions (Figure 14.4), and internal partitions to provide air exhausts. As the hot outside air traveled down the wind tower's shaft and heated the thermal mass of the wind tower, the hot inside air rose due to lower density and escaped through the exhaust partitions. In some buildings, wind coming in the tower would pass over a water pond, offering evaporative cooling. In general, wind towers applied convection and evaporation to decrease the temperature of internal spaces to below outdoor temperatures.

The function of wind towers was not limited only to houses, and they were not only used to provide thermal comfort for the occupants. They could also be used in water reservoirs, adjacent to the dome covering reservoirs space, helping to provide clean, cool, potable water through ventilation and natural cooling effects.

It is worth noting that sometimes in addition to their bioclimatic function, the wind towers' height, volume and decorations played a symbolic cultural role in the city's skyline to emphasize their owners' affluence [5].

FIGURE 14.4 Wind towers could have various numbers of faces, mainly depending upon the availability of wind and the function of the space beneath. *(Left photo: Hajar Maghbouli; Right photo: MasihMostajeran)*

14.5 SPATIAL ORGANIZATION: A MEANS TO ADAPT TO CULTURE AND CLIMATE

Spatial organization in traditional architecture was founded on socio-cultural and religious structures. It created a hierarchy and a spatial spectrum that ranged from public to private, from open to closed, and from out to in. It also adapted to climate to naturally provide thermal comfort in the internal spaces.

Amos Rapoport's study on house forms and the drivers behind them [7] is one of the major studies on the issue. According to him, socio-cultural factors, including religious beliefs, social organization, interactions between individuals, etc. are the primary forces that determine house forms. He further points out the following socio-cultural aspects as affecting the built forms in house buildings [7]:

1. Basic needs of the occupants (air, comfort, etc.)
2. Family structure
3. Position of women
4. Privacy as affected by the attitudes towards sex, shame, feelings of personal worth, territoriality, etc.
5. Social intercourse

He then states that climatic factors are '*secondary*' or '*modifying*' factors in the creation of built forms.

While it is arguable which factors, socio-cultural or climate, are the prime drivers of forms in traditional houses in different regions, it is clear that surrounding environment is a major parameter in developing the forms of these buildings. In Iranian traditional houses, the main concerns, under the influence of Islam, were to segregate the realm of women from unrelated men, to provide privacy and to welcome the guests. Also, buildings were designed to give women maximum freedom inside the house without compromising the similar needs of neighbors for privacy. In other words, the objective in the arrangement of spaces was to minimize the visual contact between urban spaces and internal spaces of the house, as well as between neighboring houses, so that the intimacy of the house and privacy of the women is not intruded upon. This often translated into a courtyard in the middle of the building, around which other spaces were organized. Edwards et al. [2] characterize traditional courtyards as having four major cultural advantages, including providing privacy for women as well as other occupants, providing a place for treatment of the guests, fulfilling the owner's responsibility to his neighbors and providing modesty in life.

Even within the boundaries of the house, varying degrees of privacy were provided for the occupants. This was done mainly through the establishment of a spatial hierarchy in relation to the courtyard, to the entrance and to the main axes of the house (Figure 14.5).

In this spatial hierarchy, the public spaces of the house, which were used for social interactions and family gatherings – and to admit guests – were usually located at the ends of the main north-south and east-west axes of the courtyard (Figure 14.5). It is important to note that the main axes of the courtyard were usually set according to solar geometry in order to provide the optimum solar access. The spaces in the courtyard located at the north end of the axis faced toward the south sun, and thus could benefit from low winter sun. These spaces were especially suitable for cold seasons. At the same time, summer sun with its high angle could not penetrate deep into the space. The unwanted summer sun in these spaces was avoided by the aspect ratio of the courtyard, using attached porches, or simply by using curtains or stained glass (Figure 14.6).

In summer, the building's occupants usually migrate to the spaces in the southerly end of the axis. These north-facing spaces have limited or no access to direct sunlight; a great benefit in intensely hot summers.

More private spaces in the public/private hierarchy, depending on the level of privacy needed, were distributed off the courtyard axes. Sometimes, courtyards smaller than the main courtyard of the house were added, to provide these private spaces with the needed sunlight, ventilation and view. However, the idea of several courtyards for a house serves another function as well. Indeed, courtyards in an Iranian traditional house were classified into extroverted and introverted, based on the level of privacy or openings they provided. Introverted courtyards surrounded by private spaces provided maximum freedom for women and were usually used for their outdoor activities. Public and semi-public spaces had no or limited view of these courtyards, and unrelated male guests were not directed there. On the other hand, extroverted courtyards, such as the main outdoor space of the house, were intended to be a place for social

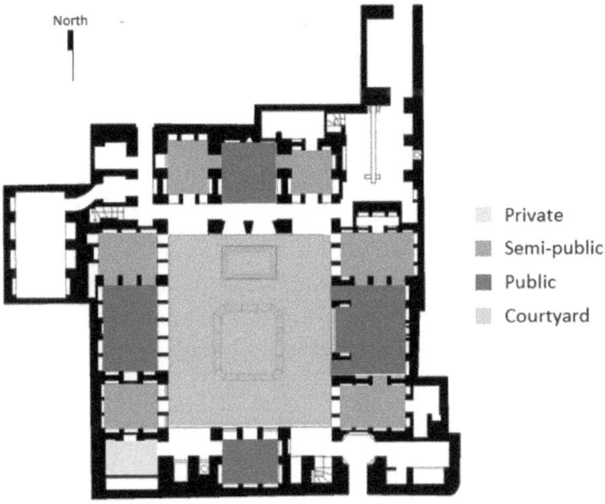

FIGURE 14.5 Distribution of spaces into a spectrum of public to private met cultural needs in a traditional Iranian house. This type of distribution considered optimum solar access as well.

FIGURE 14.6 Stained glass on windows of traditional houses could restrict the entry of daylight and view; Abbasian House, Kashan, Iran.

interaction. They were surrounded by public and semi-public spaces, and functioned as a space to welcome and treat guests, or as a transition space to direct them to public spaces.

This distribution of spaces based on thermal, functional and cultural considerations was not limited to the ground floor. The basement spaces of traditional houses in hot regions provided a naturally cool space due to effects of radiation and convection; this made them suitable for storing water and food. Rooms in the upper levels of the house were living areas with better access to higher convective air flows. These rooms were usually designed with porches to restrict the penetration of the sun into the space.

It should be mentioned that the rich experience of space provided for the occupants inside Iranian traditional houses was not merely due to the arrangement of spaces and the fact that they provided thermal comfort for the occupants. The architecture of these houses functioned to stimulate and enrich people's senses. For example, the pleasant odors of beautiful flowers blended with the birds' sounds in the courtyard of an old traditional house, hence providing an ideal place to spend time with the family. These spaces gave to their inhabitants a unique sense of belonging and connection, which is nostalgic to many old people who remember living inside these buildings.

14.6 CONCLUSION

The relationship between humans, buildings and nature in contemporary architecture needs to be redefined. While the current discourse on sustainable architecture focuses on creating built environments that lessen the impact on the surrounding environment and occupants' health, we should take bolder steps toward an architecture that establishes a healthy, respectful relationship between buildings, their occupants and the environment.

The focus on traditional architecture in this chapter was intended to provide an example of such a relationship. The collective effect of the adaptive strategies and techniques employed, both at the urban and the building scale in Iranian traditional architecture, provided the people with the capability to naturally achieve the desired level of thermal comfort in usually harsh climates, and the ability to follow their cultural and social identity.

The realities of contemporary culture and climate in many parts of the world, however, impose problems different from those of the past and, thus, different approaches should be followed in dealing with them.

Today, our buildings and cities should be able to respond to burgeoning of the global population and improved living standards, and, at the same time, cope with diminishing natural resources and destructive environmental impacts that restrict the location and design of our built environments. Therefore, this chapter does not intend to recommend that elements of traditional architecture should be imitated as a solution to current challenges, nor do the authors wish to say that traditional architecture has no deficiencies with respect to their thermal and functional performance. Instead, we would like to suggest that contemporary architecture should be created using the essence of traditional architecture, which emphasized respect for culture and environment. While fundamental logic and principles of traditional architecture should be followed to create buildings tied to the characteristics of their context and with low energy and resource demands, more importantly, the thinking system behind creation of the buildings of the past should be applied and adapted to contemporary context.

ACKNOWLEDGMENTS

The authors would like to sincerely thank Norbert Lechner, Professor Emeritus at Auburn University, and Robert Mugerauer, Professor at the University of Washington, for reviewing this chapter and providing the authors with their valuable comments which really improved the quality of this manuscript. Also, we would like to acknowledge the efforts of Caron Gussoff, who patiently read the draft of the manuscript and edited it. Masih Mostajeran and Hajar Maghbouli contributed their photographs to the authors. We would like to thank them also.

REFERENCES AND FURTHER READING

[1] E. Beazley, M. Harverson, Living with the Desert: Working Buildings of the Iranian Plateau, Orchid Press, Thailand, 2006.
[2] B. Edwards, M. Sibley, M. Hakmi, P. Land, Courtyard Housing: Past, Present and Future, Taylor & Francis, UK, 2006.
[3] EIA, Energy Outlook. Energy Information Administration (EIA) of US Department of Energy, Retrieved in September 2011 from, 2008, http://www.eia.gov/forecasts/aeo/data.cfm.
[4] EPA, The Inside Story: A Guide to Indoor Air Quality, US Environmental Protection Agency (EPA), 1995, Office of Air and Radiation, EPA 402-K-93–007.
[5] V. Ghobadian, Climatic Analysis of the traditional Iranian Buildings, first ed., Tehran University Publications, Iran, 1998 (In Farsi).
[6] N. Lechner, Heating, cooling, lighting: sustainable design methods for architects, John Wiley & Sons, US, 2009.
[7] A. Rapoport, House form and culture, Prentice-Hall, US, 1969.
[8] J. Reynolds, Courtyards: Aesthetic, Social, and Thermal Delight, John Wiley and Sons, US, 2002.
[9] S. Roaf, D. Crichton, F. Nicol, Adapting buildings and cities for climate change: a 21st century survival guide, Architectural Press, UK, 2009.

Chapter 15

Architectural Buildings in Romania

Ruxandra Crutescu
Faculty of Architecture, Spiru Haret University, Bucharest, Romania

Chapter Outline

15.1 ONE FAMILY HOUSE IN BURLUSI CIOFRINGENI, ARGES COUNTY, ROMANIA

In 2003, I started to design the first Romanian passive house in Burlusi, Ciofringeni, Arges County, to provide a sustainable future for one family. This was built in 2004, and is certified by the Passive House Institute, Darmstadt, Germany. It was a documentation intense activity and the certification shows that the decisions made in its design were correct. I understand that this is the future. Only a total change in our thinking will ensure sustainable development on the Earth. The energy efficiency of the Burlusi passive house is an example to follow, particularly for the local community. We are

Sustainability, Energy and Architecture. ISBN: 9780123972699

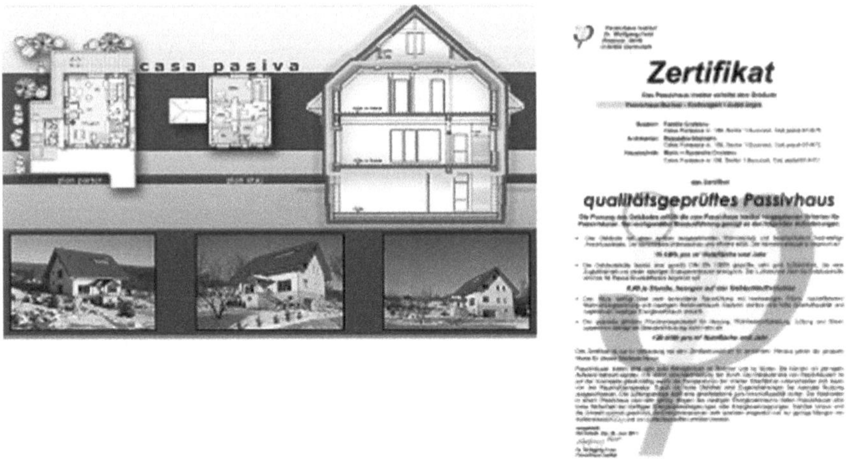

FIGURE 15.1 **Burlusi Ciofringeni family house Romania + passive house certificate.**

FIGURE 15.2 **2009 First participation of Romania at the 13th International Passive House Conference of the two Romanian passive buildings.**

now working to make this first Romanian passive house into a zero energy house (Figure 15.1).

The house has a steel reinforced concrete structure, and the exterior walls are constructed from efficient brick masonry with insulation on both sides. 30 cm of styropor with 24 kg/m^3 density is applied to the exterior and 10 cm mineral wool to the interior. The windows have low-emissivity triple glazing with Ug = 0.53 W/(m^2 K) and super insulated frames. Windows and external doors were produced by Corina Gealan SRL, a Romanian company who understands that this is the standard for the future. The roof is insulated with 45 cm thick basalt mineral wool. The house is equipped with an energy recovery (93%) controlled ventilation installation. Solar installation with vacuum tubes was provided to have warm water and a floor heating contribution. The house has a wood heating installation and floor heating in all spaces (Figure 15.2).

To build a passive house in Romania in 2004 was a real achievement, and set an excellent example. We used PHPP to plan and verify the results of the complete project.

The work needed to calculate all the energy losses and gains in a classical building system for such a new idea and implementation was too complicated

FIGURE 15.3 Neopor insulated concrete forms produced by Amvic – MARC Bragadiru Romania.

for myself, the planners and the Romanian building market. For this reason, for all the passive buildings that we have developed, we have chosen a building system that simplifies the work involved for both the planner and the builder – the system of neopor insulated concrete forms (Figure 15.3).

15.2 AMVIC PASSIVE OFFICE BUILDING – BRAGADIRU, ILFOV COUNTY, ROMANIA

In the context of global warming and the necessity for sustainable development, a valuable concept has been developed in response to the need to save energy. This concept is the 'Passive House' standard, developed by Dr. Wolfgang Feist in Germany. We have demonstrated that it is possible to create an innovative, sustainable office building at the Passive House standard, by applying very simple, energy efficient technology. The technology is neopor insulated concrete forms produced in Romania, for the thermal envelope of the walls. The first Romanian passive office building was the headquarters of Amvic Group, planned in 2007 and built in 2008 in Bragadiru , a town located 10 km from Bucharest.

The building system consists of reinforced concrete walls, laid directly onto insulated concrete forms (ICF – produced in the factory behind the offices) constructed from neopor (a type of grey polystyrene with graphite) with additional styropor (classic white polystyrene) insulation for the walls. All utilities are supplied by renewable energy: floor heating is supplied by water-geothermic heat pumps from two 120m deep drillings, and hot water is produced by 10 vacuum solar pipes. The building has a ventilation system with heat recovery, which is helped by an earth heat exchanger – Awaduct – which pre-heats or pre-cools the air, depending on the season. The result was economical to maintain and ensures excellent interior comfort (Figure 15.4).

15.2.1 Planning Concept

The criteria for the Passive House standard for houses in Central Europe according to [PHPP 2007]4 are:

Space heat demand ≤ 15 kWh/(m²a)
Heating load ≤ 10 W/m²
Excessive temperature frequency ≤ 10% (> 25°C)

FIGURE 15.4 Amvic MARC Bragadiru Head Office, Romania.

Primary energy demand ≤ 120 kWh/(m²a) for all energy applications including electricity (sustainability criterion).

This is achieved by substantially improving the construction details for each relevant component.

In particular, the strategies for achieving the Passive House standard are:

- very good insulation for the building envelope and careful execution plan
- high level of airtightness of the building (n50-value ≤ 0.6/h)
- windows with very low heat losses (including frames and thermal bridges) and at the same time, high heat gains (Passive House windows)
- ventilation system with highly efficient heat recovery
- efficient building services
- efficient electrical devices and lighting

Good planning means 50% success.

15.2.2 Building Construction

100% success is achieved by good quality construction. We tried to add value and adapt local conditions to our advantage, to make the most efficient choice concerning the energy savings for all the processes in this building (Figure 15.5).

The insulated concrete form factory offers partial protection for this building. A heat-earth recovery system is installed in the garden, which pre-heats or pre-cools the air for the entire building. The building is oriented towards south (its main façade). The building was designed to be a commercial and office building, so hot and cold water consumption is usually low. It is lit by energy efficient electric lights; there is a system that senses peoples' movements and activities. All equipment, such as heating pumps, regenerative heat exchangers (between the introduced and evacuated air) and solar vacuum pipes are chosen to have low energy consumption (Figure 15.6).

The heating of the air inside the building is made with hot air generally prepared with unconventional sources: soil heating taking over, preheating in a

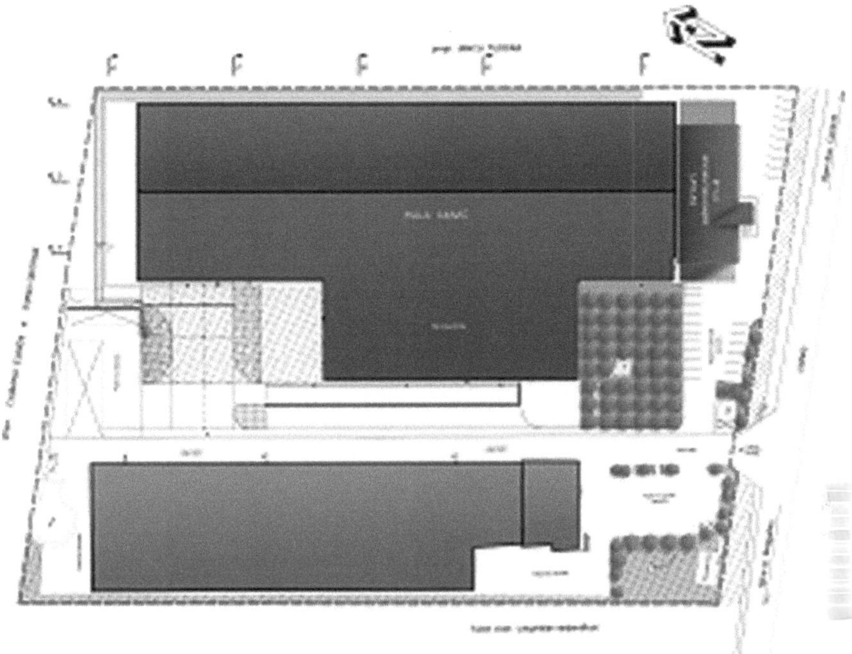

FIGURE 15.5 **Site plan for the passive office building in Bragadiru, Romania.**

FIGURE 15.6 **Amvic MARC Bragadiru Head Office Romania.**

battery with heat carrier from a surface of solar receptor and from an industrial process, heating recovery from the air exhausted, the final heating being accomplished using water – water heating pump.

During the summer, air conditioning is carried out by a reversible heating pump. The electric power needed to do this is approximately three times lower than the heating effects.

More than 50% of the energy used is recovered from the environment (soil, sun and the interior air). Because the energy demand of this building is so low, it can be supplied by renewable energy, which reduces pollution and means it is a sustainable development for the future.

15.2.3 Building Envelope

The envelope and structure of the building is made from neopor ICF, with additional polystyrene (24kg/m^3) thermal insulation added to the exterior, and fireproof cellulose to the interior.

We used fireproof cellulose and polystyrene for the thermal insulation of the roof; the U-value of the glazing is 0.5 W/ (m^2 K), of the whole window (including the aluminum frame) is 0.8 W/ (m^{2}K); adjustable external shutters for solar protection; for the preheating/ precooling of the fresh air we have an 'awaduct' pipe system. The pipe is covered inside with silver ions and buried 3.5 m deep in the garden nearby; the building has very good protection on the northern side, consisting of the production hall. Renewable energy is used. Heat pumps are in the floor for the radiant heating system and a controlled ventilation system is used with heat recovery (Figure 15.7).

The external walls have the following structure (from the exterior to the interior) (and also Figure 15.8):

External coating: $\delta = 0.015$ m; $\lambda = 0.93$ W/m.K;
Thermal insulation layer made of expanded polystyrene: $\delta = 0.20$ m; $\lambda = 0.04$ W/m.K;
Neopor layer: $\delta = 0.063$ m; $\lambda = 0.027$ W/m.K;
Concrete steel layer: $\delta = 0.20$ m; $\lambda = 1.74$ W/m.K;
Neopor layer: $\delta = 0.063$ m; $\lambda = 0.027$ W/m.K;
Thermal insulation layer made of mineral wool: $\delta = 0.05$ m; $\lambda = 0.04$ W/m.K;
Rigips layer: $\delta = 0.012$ m; $\lambda = 0.50$ W/m.K;

The windows of the building are insulated by triple thermopan windows with a heat resistance greater than 0.9 m^2 K/W (Figure 15.9).

The stairwell is made of structural walls of concrete steel cast in neopor heat insulated casing, with 20 cm insulation made of extruded polystyrene under the plate that is ±0.00 m and supplementary insulation of 20 cm expanded polystyrene whose density is 25 kg/mc on each external wall.

The roof of the house has a fireproofed, wooden, rigid frame, with cellulose insulation of 40 cm total thickness, and a Lindab profiled plate cover. On the upper part, the ceiling is made of Rigips plates, with incorporated lighting units.

FIGURE 15.7 Envelope of the Amvic Marc office building in Bragadiru, Romania.

FIGURE 15.8 **Amvic Marc Insulated Concrete Forms building system.**

FIGURE 15.9 **Amvic Marc office building, Bragadiru, Romania – passive windows and solar protection.**

Due to the great degree of thermal insulation, following the analysis accomplished in order to establish the energy performance, an average heat resistance resulted, as follows:

Rm = 6.94 m^2.K/W and a global coefficient of thermal insulation: G = 0.11 W/m^3.K;

The specific consumptions are as follows:

Specific consumption for heating: 9.56 kWh/m^2.year;
Specific consumption for hot water: 11.86 kWh/m^2.year;
Specific consumption for illumination: 9.40 kWh/m^2.year;
Specific consumption for air conditioning: 9.23 kWh/m^2.year.
Total specific consumption: 40.06 kWh/m^2.year.

15.2.4 The Heating System and Controlled Ventilation System

The building is heated by air heating using unconventional energy: soil heating taking over, preheating in a battery with heat carrier from a surface of solar receptor from an industrial process, heating recovery from the air exhausted, and the final heating being accomplished with water heating pump water – water.

During summer, the interior air conditioning is accomplished by means of a reversible heating pump.

The electric power necessary to generate the cooling task is approximately three times lower than the normal heating effects.

To determine what heating, ventilation and hot water systems would be necessary for the functioning of the building, an assessment of the administrative house was undertaken under different climatic conditions, summarized by equivalent temperatures. The resulting values are presented in Table 15.1, and provide the background for evaluating the heating, cooling and domestic hot water equipment and installations.

For sanitary purposes, the fresh air level for all occupants must range between 30 and 35 mc/h. This value, combined with the volume of the building, meant that a minimum supply of 3000 m3/h fresh air was needed, which would be provided by normal ventilation levels of passive houses; that is between 0.4 and 0.7 air exchanges per hour.

Domestic hot water is mainly produced by solar panels located on the southerly part of the roof of the building. Hot water is stored in 2000 liter tanks.

The heating pump is a key piece of equipment, which functions either in cooling or heating mode, depending on the season. It uses waste energy; which can be renewable or with low heat potential and they are freely disposed in nature (Figure 15.10).

To reduce power consumption in the warm period of the year, and so increase the energy efficiency of the building, its interior can be cooled using external air (free cooling). The air flow can be increased by up to 150% of the nominal flow, to improve the efficiency of the cooling process.

The ventilation system and domestic hot water production were designed to produce continuous indoor thermal comfort, irrespective of external conditions. Heating and cooling of the interior are accomplished by means of two independent systems. The first uses a flow of fresh air from the exterior, without recirculation, and the second uses the radiant floors situated on each level.

TABLE 15.1 The Heating and Cooling Necessary for the Administrative House

Summer				
Te [°C]	25	30	35	40
$Q_{Total\ Cool}$ [W]	– 415	1284	3040	4766
Winter				
Te [°C]	–15	–10	–5	0
$Q_{Total\ Heat}$ [W]	12818	11065	9311	7557

FIGURE 15.10 Earth heat exchanger, fresh air introduction pipe, controlled ventilation system with energy recovery + heat pump.

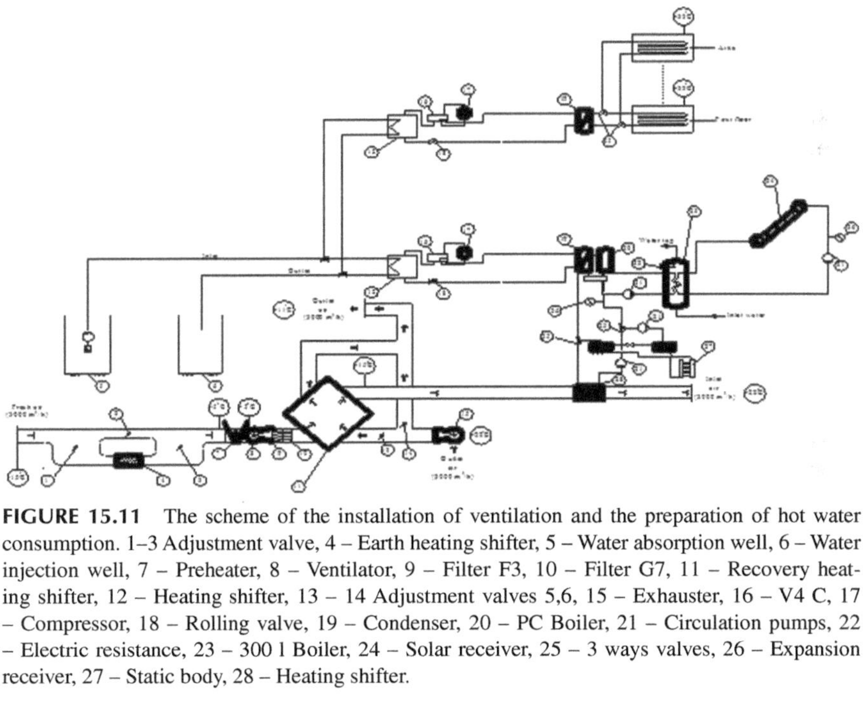

FIGURE 15.11 The scheme of the installation of ventilation and the preparation of hot water consumption. 1–3 Adjustment valve, 4 – Earth heating shifter, 5 – Water absorption well, 6 – Water injection well, 7 – Preheater, 8 – Ventilator, 9 – Filter F3, 10 – Filter G7, 11 – Recovery heating shifter, 12 – Heating shifter, 13 – 14 Adjustment valves 5,6, 15 – Exhauster, 16 – V4 C, 17 – Compressor, 18 – Rolling valve, 19 – Condenser, 20 – PC Boiler, 21 – Circulation pumps, 22 – Electric resistance, 23 – 300 l Boiler, 24 – Solar receiver, 25 – 3 ways valves, 26 – Expansion receiver, 27 – Static body, 28 – Heating shifter.

The two systems function simultaneously or separately to produce the heat load necessary for indoor thermal comfort.

The introduction track for fresh air is provided with adjustment and guiding valves to modulate the flow of air according to the season or external temperature. There are mainly three distinctive types of functioning for the installation afferent to the warm season (summer), cold (winter) and to the transition seasons (during spring and autumn).

A schematic of the ventilation and hot water systems is presented in Figure 15.11.

The fresh air is introduced in the system by a heating shifter (HS) that is buried in the soil by means of which a first recovery loop is made for both winter (heating of the flow of fresh air taken over from the exterior) and for the summer (cooling of the fresh air debit taken over from the exterior). If the temperature of the external air is close to that of the ground, the flow of fresh air is guided directly through the entrance piping without covering the whole circuit buried in the soil. This is done by changing the position of the guiding valves.

During the winter, with an external air temperature of −15°C and relative humidity of 80%, the energy recovered by this shifter has a maximum value of 9.75 kW.

By stimulating its functioning, the values of the parameters could be obtained in the exhaust section from the respective shifter for the flow of fresh air of 3000 mc/h and for the exterior parameters presented above the temperature of −5°C and a relative humidity of 20% resulted.

Next, the fresh air passes through the heating shifter named PREHEATER (PH), which increases the temperature of the air in the system to +5°C, before being introduced in the filtration battery. This is done to avoid the formation of ice on the surface of the filters. This heating can be accomplished using waste energy derived from local industrial activity (hot water). Energy recovered from this loop attains a maximal value of 10.84 kW. After the filtration stage, the air is sent to a recovery heating shifter (RHS) where an important quantity of heating is recovered in the flow of internal air that is evacuated from the building. As a consequence of this process of heating of the flow of fresh air, the value of the temperature increases to a +15°C value and the relative humidity increases by 5%.

The primary agent that is the air extracted from the building is cooled down to a 22°C temperature value and the humidity of 40% to a value of the temperature of +11°C and a relative humidity of approx 80%, thus avoiding the problems which may occur as a consequence of the condensation of water gases before their evacuation to the exterior.

The amount of energy recovered by this equipment is approximately 10.2 kW. Thus, as a consequence of these successive recoveries, the fresh air taken from the exterior is heated from −15°C to +15°C. In this situation, the energy recovered is equivalent to 30.79 kW.

Following these recoveries and before being introduced into the interior, the air is again heated in a final stage, for which the primary heat carrier is produced by a heat pump battery. The fresh air intake to this heater has a temperature of +22°C and a relative humidity of around 35%. The heating power developed by the heating battery pumps is approximately 20.5 kW.

In addition to the air heating system maintaining the interior temperature, there is also a parallel heating system in the floor. The heat carrier from this circuit is prepared by another heat pump (HP) functioning of this system; this can take place alongside or separately to the air heating. The heating power of the heat pump within this heating circuit is 11 kW.

Cooling of the space can be accomplished by two means: either by using reversible heat pumps, or by using a passive cooling system. The water from the wells which represent the heating source are used for the functioning of the heat pumps.

For passive cooling of the well water with an approximate heat potential of +10°C is pumped both for cooling of the internal air after its passing through the shift with ground or for cooling the floors which are to be used as cool radiation surfaces.

Passive cooling has a much lower energy consumption than some reversible heat pumps. During summer (cooling of the interior air) the preheating is not supplied with heat carrier and the internal air exhausted from the building will bypass the recovery heating shifter, by means of some guiding valves. The maximum heat need resulted from the calculation for this building is 62.3 kW out of which 30.79 kW represents the quantity of energy recovered and only 31.5 kW heat produced and used directly. The heat need calculated for a usual building of the same weight and under the same climatic conditions but accomplished in a classic manner is 205.9 kW. As a consequence, taking into account the energy produced for the heating of the passive building, comparatively with the energy need for heating the classic building, the energetic consumption is only 15.3% out of the total energy needed.

15.2.5 Vacuum Solar Collectors

Domestic hot water is either heated by solar collectors during periods when there are significant levels of solar radiation (generally during summer), or by using a heat pump (see Figure 15.12).

This type of hot water preparation is mainly accomplished alongside space heating. The heat pump linked to the heating track by means of the floor is mostly used for domestic water heating. There is an accumulator receiver between the heat pump and the user; its capacity is approximately 300 liters; the temperature of the water produced is around +50°C.

Due to the reduced value of the consumption, water temperature that is for a 7–10 days interval, the volume of water from the interior of the accumulation receiver will be subjected to the process of raising the temperature value until reaches of +70 °C for the prevention of existence and development of some bacteria.

FIGURE 15.12 **Solar vacuum panels on the passive office building in Bragadiru, Romania.**

15.2.6 Analysis and Monitoring Data

Figure 15.13 presents the energy saving achieved by each piece of equipment, as well as the global saving that is accomplished by using the energy loops developed in the functional scheme. Approximately 50% of this is energy recovered from the environment (soil, sun) and from the energy of the interior air. The isolation degree is highlighted once again by the energy need for the heat pump 3 that only covers the power losses to the exterior.

Figures 15.14–18 below show some characteristic results.

This building offers a real example to show the remarkable advantages of the passive house standard, as applied in Romanian climatic conditions. We can see also that the requirements of a passive house are fulfilled in the extract from PHPP (Figure 15.19).

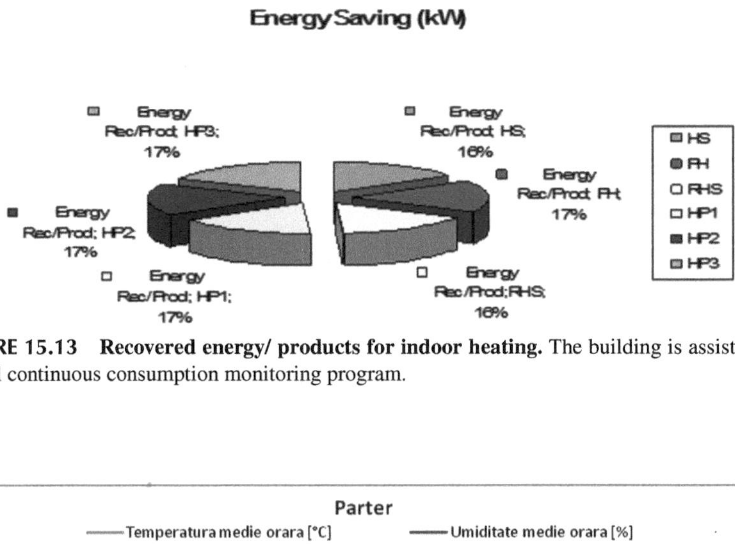

FIGURE 15.13 Recovered energy/ products for indoor heating. The building is assisted by a special continuous consumption monitoring program.

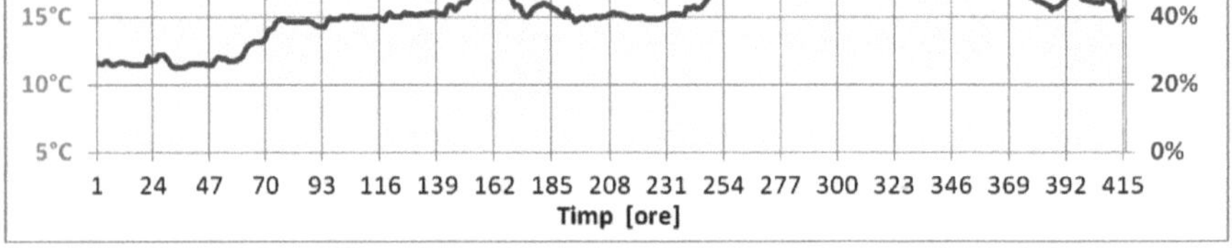

FIGURE 15.14 Monitoring graphs for the ground floor. Red – hourly average temperature; blue – hourly average moisture; yellow – time (hours).

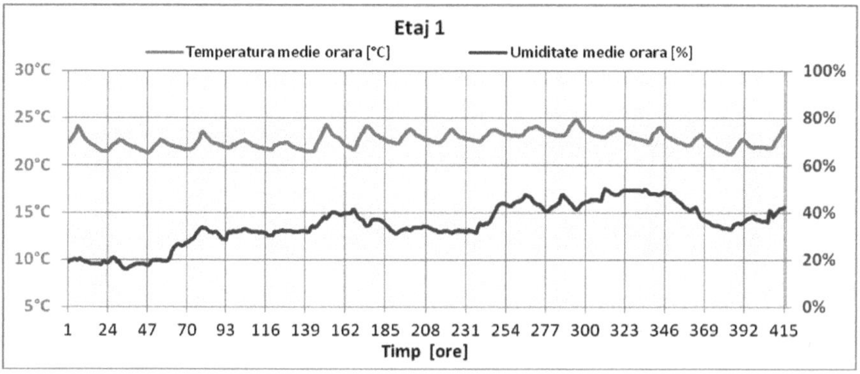

FIGURE 15.15 Monitoring graphs for the first floor. Red – hourly average temperature; blue – hourly average moisture; yellow – time (hours).

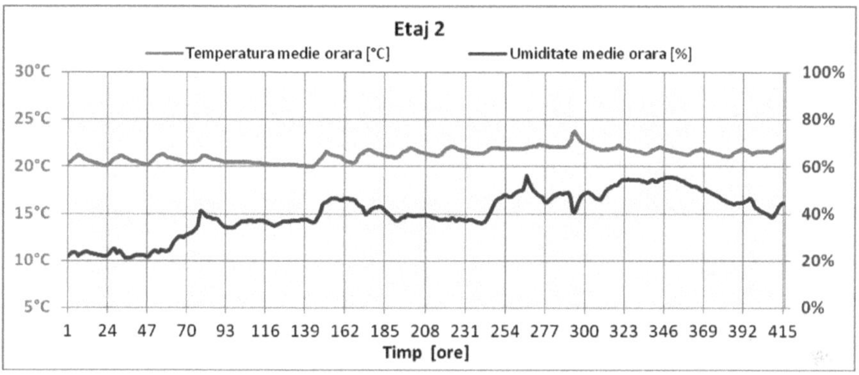

FIGURE 15.16 Monitoring graphs for the second floor. Red – hourly average temperature; blue – hourly average moisture; yellow – time (hours).

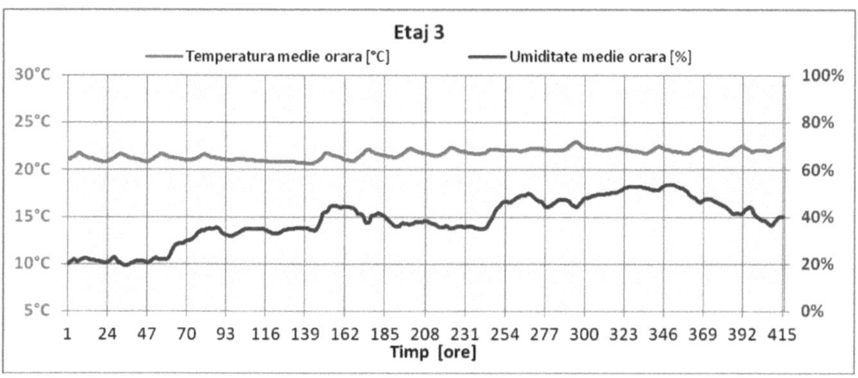

FIGURE 15.17 Monitoring graphs for thr third floor. Red – hourly average temperature; blue – hourly average moisture; yellow – time (hours).

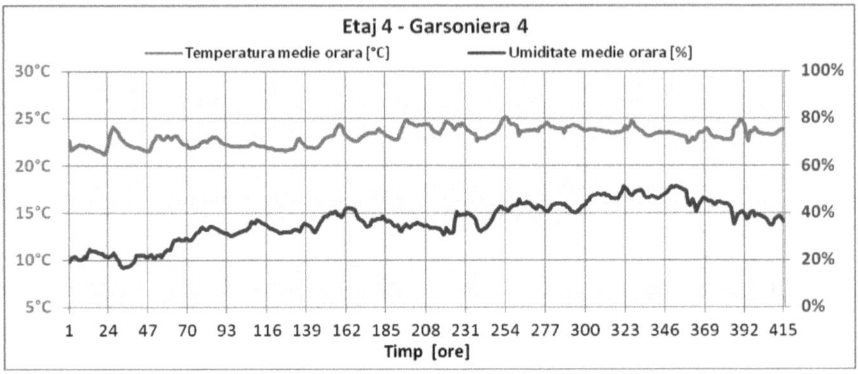

FIGURE 15.18 Monitoring Graphics for Fourth floor. Red – hourly average temperature; blue – hourly average moisture; yellow – time (hours).

Specific Demands with Reference to the Treated Floor Area

	Applied:	Monthly Method		PH Certificate:	Fulfilled?
Treated Floor Area:	2085.9	m²			
Specific Space Heat Demand:	7	kWh/(m²a)		15 kWh/(m²a)	Yes
Pressurization Test Result:		h⁻¹		0.6 h⁻¹	
Specific Primary Energy Demand (DHW, Heating, Cooling, Auxiliary and Household Electricity):	66	kWh/(m²a)		120 kWh/(m²a)	Yes
Specific Primary Energy Demand (DHW, Heating and Auxiliary Electricity):	14	kWh/(m²a)			
Specific Primary Energy Demand Energy Conservation by Solar Electricity:	0	kWh/(m²a)			
Heating Load:	9	W/m²			
Frequency of Overheating:	0	%	over	25 °C	
Specific Useful Cooling Energy Demand:		kWh/(m²a)		15 kWh/(m²a)	
Cooling Load:	1	W/m²			

FIGURE 15.19 Extract from PHPP for Amvic Passive Office Building in Bragadiru.

15.3 RESIDENTIAL LIVING UNITS IN CLUJ NAPOCA, CLUJ COUNTY, ROMANIA

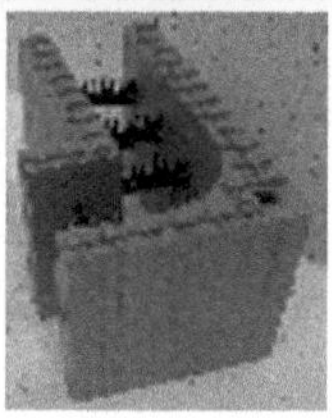

Low-energy cascade houses 2010
Architects: Dannon Construct, Cluj
Building system: Structural steel reinforced concrete walls poured in Amvic MARC Neopor insulated concrete forms

15.4 TWO PASSIVE HOUSES IN CARANSEBES, CARAS-SEVERIN COUNTY, ROMANIA

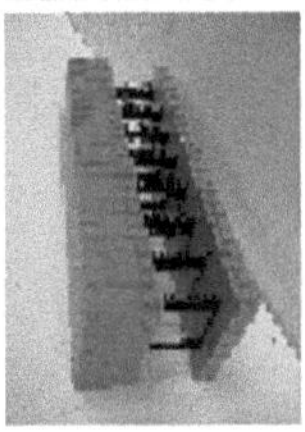

Architects: Ruxandra Crutescu, Bucharest
Heat energy demand: 15 $kWh/m^2/a$
Blower door test: n50 = 0.4 1/h
Building system: Structural steel reinforced concrete walls poured in Amvic MARC neopor insulated concrete forms for passive buildings

15.5 CHURCH IN BISTRA, NEAMT COUNTY, ROMANIA – LOW-ENERGY BUILDING

Architects: Bojescu Constantin – Piatra Neamt, Neamt County
Building system: Structural steel reinforced concrete walls poured in Amvic MARC Neopor insulated concrete forms

15.6 CONCLUSIONS

These examples show that the passive house standard is an ongoing success story if you know how to apply it in an intelligent and creative way. Starting with the planning of an energy efficient and sustainable solution, then continuing with a relatively simple building process, we have obtained environmentally friendly results by using the neopor ICF energy efficient building system. More than 50% of the energy used is recovered from the environment (soil, sun and the interior air). The low energy demand of this kind of building means that it can be met completely by renewable energy sources. This reduces pollution, and contributes to sustainable development for the future. This building can be easily transformed from a certified passive building into a zero energy building.

The basic principle is to think globally and act locally.

Our experience in this research activity has highlighted the importance of combining the capability of a multi-disciplinary team with local knowledge and international best practice, to achieve and build better sustainable buildings in Romania.

In Romania, the architects and planning teams who care about what is happening to our Earth are promoting intelligent design and environmental architecture based on practical principles, in order to provide a better quality of life with the lowest possible carbon footprint.

Energy efficiency is an important concept and must be applied in sustainable development, in Romania and elsewhere.

Romanian people would like to have their own passive houses as soon as possible, which is why we have developed a special program for standard passive houses, including 32 types of one family houses that can easily be upgraded to 'zero energy buildings'.

One suggestion is to move away from classical building systems towards new construction technologies, using easy and economical ways to build passive houses quickly with guaranteed quality. Fortunately, a suitable building system exists in Romania – this uses neopor insulated concrete forms produced by Amvic-Marc in Bragadiru.

The environment is what we cherish, and we must preserve it for succeeding generations.

REFERENCES

[1] Dr. W. Feist, Protokollband Nr. 18 – Qualitaetssicherung beim Bau von Passivhausern, 2007.

[2] Dr. V. Badescu, Use of soil thermal energy for passive house heating, 2nd International Symposium, 'Renewable Energies and Sustainable Development', 24–26 Sept Tulcea, Romania, 2005.

[3] Dr. W. Feist, Protokollband Nr. 31 – Energieeffiziente Raumkuhlung, 2005.

[4] Dr. W. Feist, Protokollband Nr. 27 –Waermeverluste durch das Erdreich, 2004.

[5] Dr. W. Feist, Protokollband Nr. 22– Lueftungsstrategien fuer den Sommer, 2003.

[6] S. Pepper, Dr. W. Feist, CEPHEUS – Luftdichte Projektierung von Passivhausern, 2002.

[7] Dr. W. Feist, Passivhaus Sommerklima Studie, 1998.
[8] Dr. Hera, Usage of the heating pumps for dwelling heating, an ecological solution and power efficient, rev. The Plumber, no 9/2006, ISSN 1223-7418, pp. 26–37.
[9] V. Badescu, Modeling the earth-to-air heat exchanger of a passive house, Conference of the Installation Fac, Technical Building University, Bucharest, 2005, November 2005.
[10] Dr. Hera, I. Mateescu, G. Ivan, Optimization of power consumption within buildings by using heating pumps systems (Conf. a XI-a a Installation Fac., Efficiency, comfort, power conservation and environmental protection', Buc. 25–26 Nov. 2004), publ. CD. conf. p. 6.
[11] Directive 2002/91/EC of the European Parliament and of the Council of 16 December 2002 on the energy performance of buildings.
[12] EN, Eurocode 2: Design of concrete structures – Part 1: General rules and rules for buildings, 1992-1–1.
[13] EN, Eurocode 8: Design of structures for earthquake resistance Part 1: General rules, seismic actions and rules for buildings, 1998-1.
[14] Directive 2010/31/EU of the European Parliament and of the Council of 19 May 2010 on the energy performance of buildings. (recast)

Chapter 16

Sustainable Architecture in Africa

Manuel Correia Guedes
Department of Civil Engineering and Architecture, Instituto Superior Técnico, Av Rovisco Pais, Lisbon, Portugal

Chapter Outline

16.1 INTRODUCTION

The problems of environmental sustainability and energy saving are universal and common to all countries and regions worldwide. The interdependence of climatic and environmental factors is a reality that makes all countries and all citizens equally responsible for the present ecological problems, which will inevitably worsen if we are not all aware and supportive of their mitigation, if not prevention.

The first and possibly most effective measure for reversing the general trend is certainly the informing and mobilization of the public, and especially of the professionals, for a deeper understanding and accountability, leading to systematic interventions for the resolution of these problems.

Sustainability, Energy and Architecture. ISBN: 9780123972699

The awareness of any problems affecting human society is surely proportional to the mobilization capacity of the media, the degree of literacy of the people, the cultural level of the information professionals, and, especially, of the education system's ability to frame and focus the education of the child, adolescent and adult, in this case, on the problems of sustainability and environmental balance.

All these capabilities are still in their infancy in Africa. Therefore any threat to current life that is not objective, immediate and tangible tends to be perceived by most people as not requiring a short or even long term change of attitude. The issues of sustainability and climate change are frequently considered as belonging to rich countries. The African continent, despite its low levels of industrialization and consumerism, is in a more vulnerable position than developed and highly industrialized countries. Hyper-consumerism should not be an aspiration for developing countries, which sometimes mistakenly adopt Western trends. There is a latent need not to follow the bad examples of the industrialized world, and to preserve a quality that can be considered as intrinsic to the lack of financial wealth, which is the ability to recycle and take advantage of existing resources.

The richer countries have exploited the natural resources of the poor, and the (few) rich people of the poorest countries collaborate with this system, allowing the exportation of natural resources at minimal costs. The debate against hunger, poverty and endemic diseases features prominently in Africa. This is the real situation in Africa. The consequences are multiple and harmful.

It is essential to think of strategies of ecologic planning and sustainable development in a holistic and integrated way, avoiding short-term and low-reach solutions. The energy sustainability and the responsible use of natural resources must be an integral part of the sustainable development of the ecosystem.

Building and urban renewal have an urgency that requires a different approach from that in Europe. This is due to the scarcity of resources, energy shortages, the pressing demand for social housing and new or refurbished public buildings such as schools and hospitals, and the difficulties of implementing building and town planning regulations (often deficient or even non-existent).

An indispensable measure is self-sufficiency. The high import costs may be the motivation to produce, and naturally lead to more viable solutions in ecological and environmental terms – involving the use of local resources. There must be a popular awareness of this. What can and must come from outside are the new techniques and concepts of construction, allowing a more rational use of raw materials.

Although the few measures implemented by the construction sector in recent years have improved some aspects, this sector can only truly be fostered through the implementation of a new model of economic growth, based on ecologically sustainable development. In political terms, measures to promote the use of low-cost local materials must be implemented, simultaneously developing local typologies and construction techniques, which prove to be decisive and efficient. Cooperativism and associativism should be encouraged, leading to a network of solidarity and cooperation among citizens and between the eco-technosphere and the biosphere.

More than one billion people living in developing countries do not have adequate shelter, and an estimated one hundred million are homeless. The participatory and self-construction processes should be integrated in this synergetic web of solidarity and collective union, with the objective of overcoming the problems of shortages of financial resources. The architect, in his professional practice, beyond the use of local materials and the introduction of renewable energy systems, must consider priority areas in the project, and contemplate the building as an organism that can grow in a process of spatial evolution that accompanies the growth of families. The evolving shelter that includes spaces with potential for expansion to accommodate growing families is a cultural element in Africa. Simultaneously, the definition of priority areas of construction is essential for the management of financial resources.

There are many various definitions for sustainable architecture, but the essence of sustainability is intrinsically linked to the essence of architecture. A good building is naturally sustainable. We can also find practices of sustainability in the vernacular architecture of many communities. It incorporates building technologies that result from the empirical knowledge of many generations that, through the centuries, developed strategies to adapt to the surrounding environment, using local resources.

Today there is very little information on the issue of sustainable building that is adapted to the climatic, social-economic and cultural context of African countries. On the one hand, customers, either private or institutional, are not sufficiently informed or motivated to commission sustainable buildings. On the other hand, technicians, engineers and architects do not assume an attitude conducive to the adoption of sustainable design strategies in the project process. Finally, and as a consequence of this situation, there is not enough motivation or clarity to promote adequate legislation, with necessary incentives or sanctions to those who implement, or disregard, the correct application of environmental, sustainable building design strategies, whatever their level of intervention in the process. This is reflected in all aspects relevant to the solution of the problem, including the commercial aspects, leading to a near-total absence in the local market of materials and equipment to ensure a better environmental performance of buildings.

There is however a vast body of academic knowledge and analysis tools that allow for the identification of the main design strategies to use in building projects in Africa – efficient and low-cost solutions, providing a good indoor comfort performance. The information presented in this chapter results from a three year EU project, SURE-Africa, which aimed at producing and strengthening knowledge about low-energy architecture for hot regions in Africa, contributing to sustainable development through the vital area of energy efficiency in buildings and cities. The situation found in the participant countries – Angola, Cabo Verde,

Guinea-Bissau and Mozambique – is representative of many other countries in Africa, with developing economies often scarred by long-term armed conflicts.

It is important to consider energy conservation through passive building design as a proven equivalent to renewable energy power generation. Well-established knowledge in this area can be adapted to the African economic and climatic context. In non-domestic buildings, a high priority is the avoidance of air conditioning, or the reduction of air conditioning loads by fabric design and controls, to low values. In the case of housing, it is important that basic comfort performance criteria are met, as failure in this respect will prompt the occupants to purchase packaged air conditioners if and when reduced costs and improved finances allow. The purpose of this chapter is to suggest basic measures for a comfortable house that respects nature, with reduced costs of construction and maintenance. Taking into account the climate, natural resources and socio-economic context, best-practice strategies are drawn for the architectural project. This chapter does not pretend to be more than a simplified, and therefore easily accessible, introduction to the problem of environmental sustainability in the context of architectural design, focusing on bioclimatic design strategies.

Vernacular architecture in Santo Antão (Cabo Verde): the use of local resources for construction, and the adaptation to the climatic context are centuries-old practices.

Degraded slum in Luanda (Angola): the fight against poverty is a priority.

Sustainable, bioclimatic house, in Mindelo: a contemporary example of adaptation to the local context.

Eco-tourism resorts in Bijagós (Guinea-Bissau): contemporary architecture inspired by the local vernacular, with a good bioclimatic performance and use of local building materials.

Examples of imported architectural typologies, inadequate to the local climatic and social-economic context (Bissau, left; Luanda, center, right).

16.2 BIOCLIMATIC PROJECT: GENERAL GUIDELINES

In the variety of climatic contexts existing in Africa, it is possible to achieve a balance between building and climate by applying a series of project strategies – referred to as bioclimatic or passive design strategies. Passive design strategies aim to provide comfortable environments inside the buildings, and simultaneously reduce their energy consumption. These techniques allow the buildings to adapt to the external environment through architectural design and the intelligent use of building elements and materials, avoiding the use of mechanical systems that use fossil fuel energy.

In buildings, the use of electricity generated from fossil fuels greatly contributes to the intensification of global warming. Passive measures reduce the energy consumption of buildings throughout their existence. Two examples of passive strategies are the optimization of the use of natural lighting to reduce the need for artificial lighting systems, and the promotion of natural ventilation to avoid the use of air conditioning for cooling.

One can find good examples of African architecture that are suitable to the local environmental context. However, today, the practice of a passive or bioclimatic architecture, with environmental and energy concerns, seems to be increasingly forgotten. For example, in many new buildings, climatization issues are left to air conditioning engineers, who tend to adopt the 'safe' use of air conditioning. Despite the existence of many examples that prove the efficacy, improved levels of comfort, and economic advantages of using passive techniques, there is still a great need for implementation of this knowledge and to increase the number of passive, bioclimatic buildings, in terms of new construction and rehabilitation.

Considering that heat is the predominant feature of African climates, particular attention should be given to the issue of cooling of buildings. The aim of the passive cooling techniques is to avoid the accumulation of heat gains and provide natural cooling, avoiding the occurrence of overheating. The principles of passive cooling techniques have been successfully used for centuries, before the appearance of air conditioning. These traditional techniques were simply reinforced with contemporary technological knowledge, and optimized so that they could be successfully incorporated in the design and operation of buildings.

This chapter starts with a brief description of the various climatic regions existing in distinct locations in Africa, as examples of the starting point of the methodological process for the practice of bioclimatic architecture, of passive design. This is followed by the presentation of the main strategies of bioclimatic design (Figure 16.1, 16.2).

FIGURE 16.1 Above: Vernacular house in S. Antão, Cape Verde, adapted to the climatic context. Below: buildings in a recent tourism complex, with building typologies inspired in vernacular architecture, in Angola (center) and Mozambique (bottom).

16.3 CLIMATIC CONTEXT

A summary description is made of the climatic characteristics of four countries – Angola, Cape Verde, Guinea-Bissau and Mozambique – showing a diversity of situations representative of the main types of climate found in sub-Saharan Africa.

In **Angola**, the climate is classified as sub-tropical, hot and humid in most of the territory, and semi-arid and sub-humid dry in the south and on the coastal strip up to the Province of Luanda. Figure 16.3 shows the division by zones of

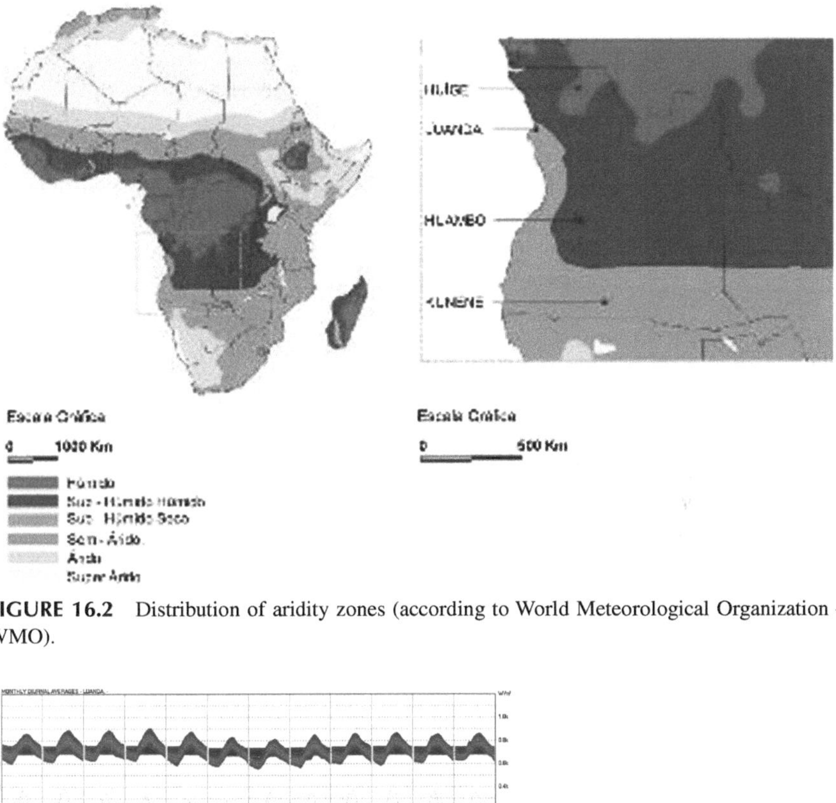

FIGURE 16.2 Distribution of aridity zones (according to World Meteorological Organization – WMO).

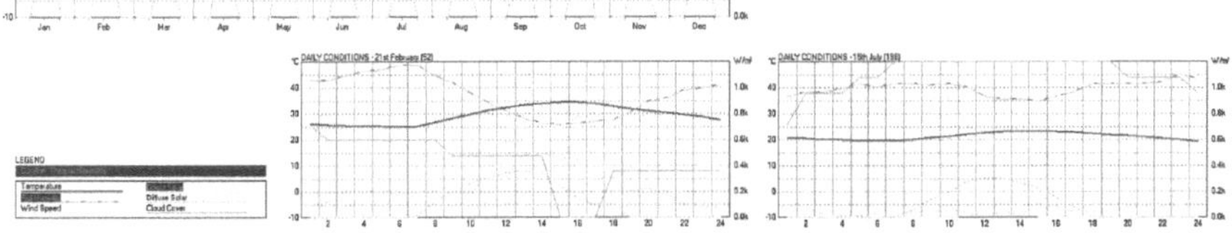

FIGURE 16.3 Above: chart showing the annual average temperature values for Luanda. Bottom: values of the air temperature (blue), relative humidity (dashed green), wind speed (dashed blue), direct solar radiation (yellow) and diffuse radiation (dashed yellow) for a hot day (21 February) and a cold day (July 15, right), in Luanda. Estimated values obtained using the software METEONORM.

aridity (climate division) for Africa. The graphs show the typical profile of the mean annual temperature and humidity values for the city of Luanda.

In **Cape Verde**, the climate is classified as tropical dry with maritime influence, with high temperatures throughout the year, subject to the effect of dry air masses from the Sahara, and long periods of drought. The islands are about 500km off the African coast, but most of them still present as a prolongation of the Sahara. There are two seasons during the year: the dry season and the rainy season, the latter occurring when the intertropical front rises from the equator. This front, which brings rainfall, does not cover all the islands, for several reasons: it does not move parallel to the equator; it is subject to the influence of the Azores anticyclone; local disturbances related to air circulation and heat fluxes; and due to reduced plant cover. This causes an anomaly in the distribution of rainfall in the islands. The slopes exposed to the moist air masses from the northeast are subject to greater rainfall. This orientation of the relief means that we can find different climate zones in the same island, such as in the case of St. Antão (Figure 16.4), one of the most illustrative examples, in which we find one area exposed to moist air masses with a humid climate, lush vegetation and a good amount of water, and another area exposed to warm, dry air masses, where the vegetation is difficult to implement, with characteristics of the arid climate zones. The graph shown in Figure 16.5 presents a typical profile of annual mean values of temperature and humidity of the island of Sal.

The sun is the dominant element, throughout the year, with little or no scattered cloud. Consequently, surfaces constantly receive radiation during the day, which is converted into heat. This is absorbed and lost overnight. Thus, the diurnal temperature range is higher than the annual range. This temperature variation between day and night is more pronounced in mountainous areas, where there is a greater drop in temperature, causing a thermal shock in geological material and resultant dilatation of the rocks, a phenomenon that can be

FIGURE 16.4 View of the island of Sao Vicente. The landscape reflects the effects of the dry tropical climate, and of maritime influence, with long periods of drought.

observed in the interior of the municipality of Porto Novo. The absorption of solar radiation varies with the color and texture of surfaces.

The municipality of Porto Novo, on the island of Santo Antão, and the county of Tarrafal on the island of S. Nicolau, harbor a vast extension of land where one can observe the highest rates of solar radiation and absorption in the country (Figure 16.6). The temperature increases rapidly during the first hours of the day. The rising hot air dissipates moisture, and the few clouds outline the area and are channeled towards the mountains, where trees create small local depressions.

Guiné-Bissau is located in a hot and humid climate zone, characterized by strong solar radiation, constantly high temperatures, heavy precipitation (1,000 to 1,600 mm per year), high relative humidity, frequently over 80%, and moderate winds. During the year, temperature fluctuates between 20°C average minimum and 35°C average maximum, with daily amplitudes of around 6–10°C (Figure 16.7). The hottest period of the year occurs between March and May, when the maximum daily values reach between 32°C and 42°C. During the wet season (late May to October) humidity levels are higher, heavy rains occur especially in the month of August, and the prevailing winds are from the northeast. In the dry season (November to April), the days are relatively cool,

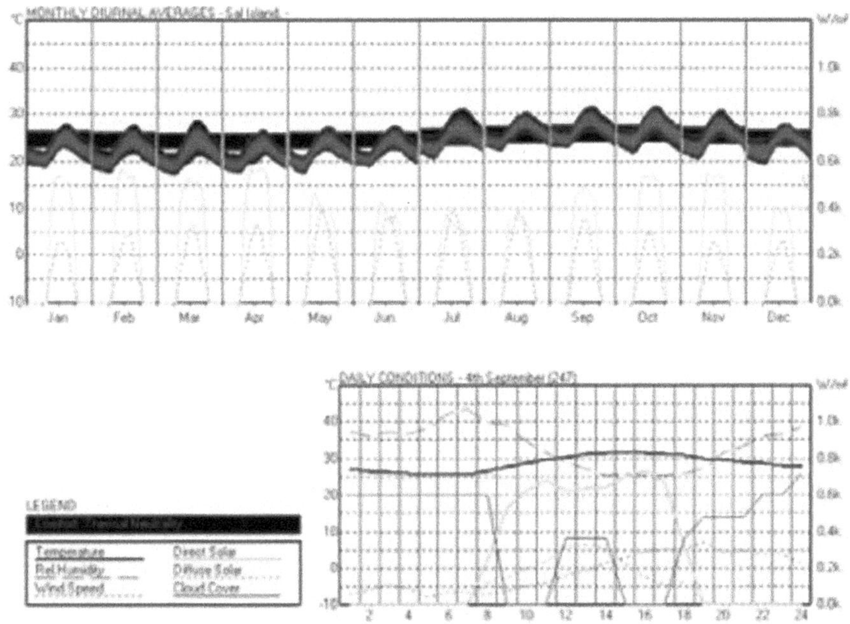

FIGURE 16.5 Chart showing the profile of mean annual temperature for the island of Sal (top). Values of air temperature (blue), relative humidity (dashed green), wind speed (dashed blue), direct radiation (yellow) and diffuse radiation (dashed) for a hot day on the island of Sal (Sept. 4, below). Estimated values, obtained using the software METEONORM.

FIGURE 16.6 Mass of moist air in a mountainous area of the island of Santo Antao, during the rainy season.

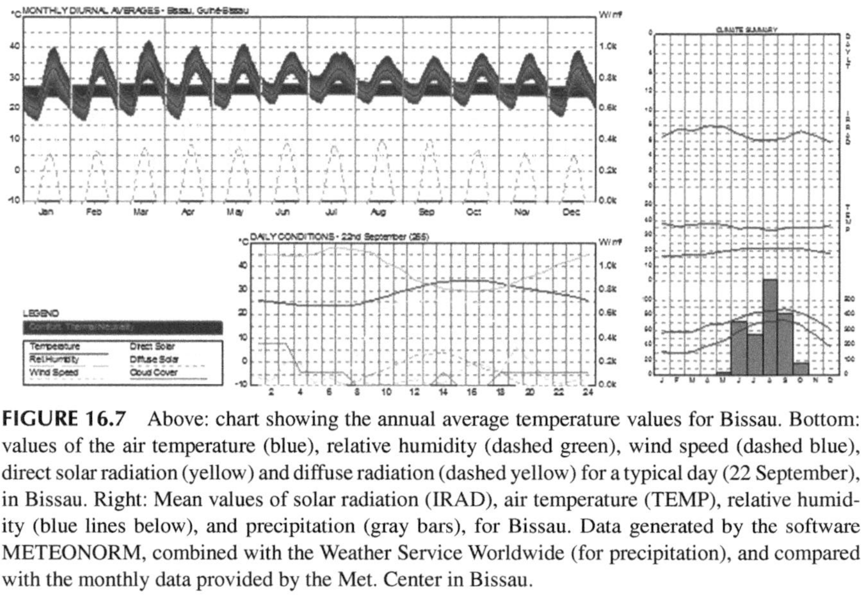

FIGURE 16.7 Above: chart showing the annual average temperature values for Bissau. Bottom: values of the air temperature (blue), relative humidity (dashed green), wind speed (dashed blue), direct solar radiation (yellow) and diffuse radiation (dashed yellow) for a typical day (22 September), in Bissau. Right: Mean values of solar radiation (IRAD), air temperature (TEMP), relative humidity (blue lines below), and precipitation (gray bars), for Bissau. Data generated by the software METEONORM, combined with the Weather Service Worldwide (for precipitation), and compared with the monthly data provided by the Met. Center in Bissau.

especially between December and February, with minimum values sometimes below 20°C.

In **Mozambique,** the climate is influenced by the monsoons of the Indian Ocean and the warm current of the Mozambique Channel. Generally, it can be considered a tropical and warm climate, varying, according to the region, between sub-humid dry and semi-arid. Average annual temperatures vary between 20°C in the south and 26°C in the north (22.5°C in Maputo, 24.1°C in Beira, 18.4°C in Vila Cabral), with higher values during the rainy season. The annual average humidity values (% RH) are usually relatively high, ranging from 65% (dry season) to 75% (hot and wet season). Figure 16.8 shows the division by zones of aridity (climatic division) for Africa and Mozambique.

There are two distinct seasons: the dry and cold season ranging from April to October, and the hot and rainy season, from October to March. From October the rains begin to intensify and continue until March or April. In the south, the rainy season often takes longer to begin, due to the influence of the high-pressure center of the Indian Ocean, and the intertropical convergence zone in the Transvaal.

Despite the warm tropical weather overall, Mozambique also presents a series of regional variations due to local factors like altitude, latitude and proximity to the coast. The northern region is under the influence of the equatorial low pressure, while the south is affected by tropical anticyclones and the existence of warm currents of the Mozambique Channel. One can distinguish three climatic zones:

I. North and Center: monsoon climate, with a dry season of four to six months.
II. South: drier climate, with a dry season of six to nine months.
III. Mountain areas: tropical climate of altitude.

Except in the inner regions of the southern coast and part of the depression of the lower Zambezi, with between 400 and 600 mm of annual rainfall, the rest of

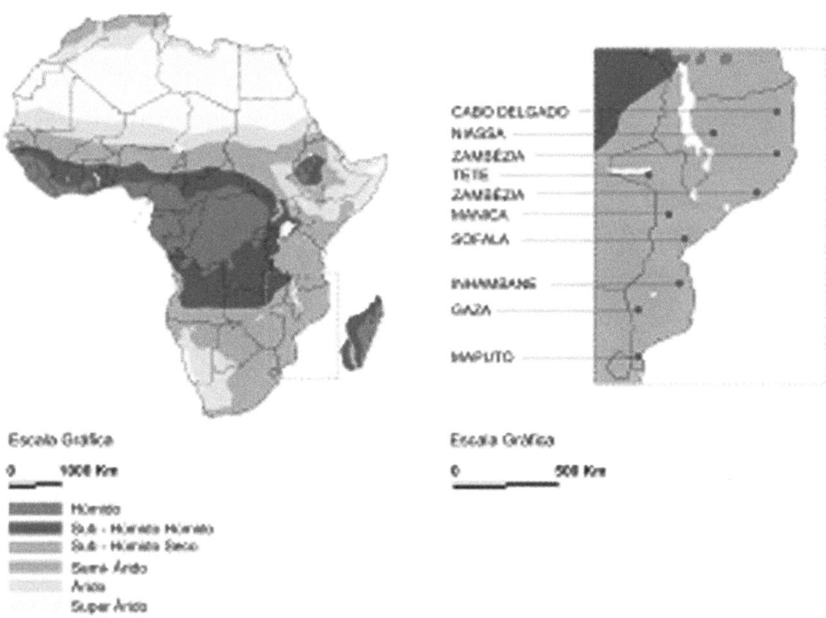

FIGURE 16.8 Distribution of aridity zones (according to the World Meteorological Organization – WMO).

the country receives more than 800 mm. The most abundant rainfall is recorded on the slopes of Manica and Sofala and in upland areas of high Namuli, which receive up to 2,000 mm. During the rainy season, the dominant winds are from the northeast in the northern half of the country, and from the south in the southern part of the country. The graph shown in Figure 16.9 shows a typical profile of mean annual temperature and humidity for Maputo. Average temperatures in Maputo vary between 13°C and 24°C in July, and 22°C and 31°C in January.

These climatic characteristics determine the level of comfort inside the buildings, as well as the appropriateness of different building design strategies. Several issues associated with the sun should be analyzed during the design process, such as: the orientation of the house, solar protection needs in different areas – coastal or mountain, the required spacing between buildings, the road and promenades' surface, the implementation of trees and green areas to lessen the impact of radiation and preserve fresh air, and types of materials to be used. These principles are next presented, beginning with the first items to consider – the location, shape and orientation of buildings.

16.4 BUILDING LOCATION, FORM AND ORIENTATION

The selection of the place, shape and orientation of the building are the first options to consider for optimal exposure to the sun path and prevailing winds. In hot climates, it is essential that the design of the houses takes into account the wind regime for efficient ventilation and consequent improvement of indoor comfort. In mountainous areas, the houses must be located in the lower zones of the mountain and above the riverside, where there is more air circulation. Priority should be given to the slope side with more hours off. On the coast, the façades facing the sea should be protected by generously proportioned porches, to lessen the impact of the sun's reflection on the sea inside the house. The exterior

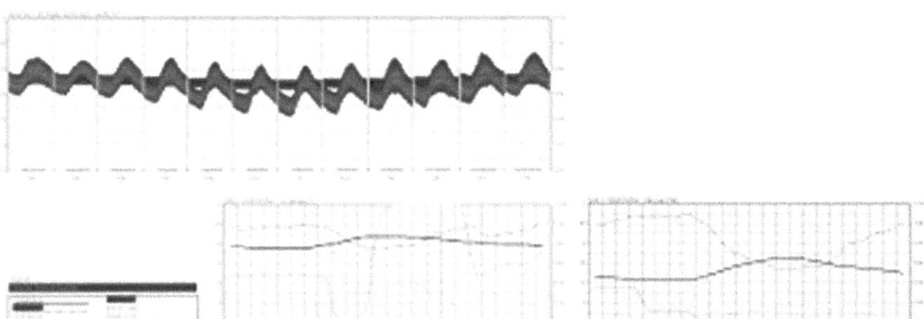

FIGURE 16.9 Above: chart showing the annual average temperature values for Maputo. Bottom: values of the air temperature (blue), relative humidity (dashed green), wind speed (dashed blue), direct solar radiation (yellow) and diffuse radiation (dashed yellow) for a hot day (1 January) and a cold day (July 15, right), in Maputo. The average annual relative humidity values (RH%) range between 65% (dry season) and 71% (hot and rainy season). Estimated values obtained using the software METEONORM.

arrangements are essential to protect the interior from excessive solar gain (Figures 16.10, 16.11).

New residential areas should also be designed at a convenient distance from the road with the largest circulation, to avoid noise and other nuisances. The streets should be narrow and oriented so that there is always a shaded side (Figures 16.12–16.14).

As the outdoor environment is hot, ventilation and indoor comfort are critical. In urban areas the impact of solar radiation on the roofs and the façades of buildings and the circulation of the cool breezes around buildings should be studied. Otherwise there is a risk of creating a very uncomfortable environment inside the houses (Figures 16.15–16.17).

In terms of the shape of the building, the configuration and arrangement of internal spaces influences exposure to solar radiation as well as the availability of natural lighting and ventilation. In general, a compact building will have a relatively small area of exposure, i.e., a low surface to volume ratio. For small and medium-sized buildings, this offers advantages for the control of heat exchange through the building envelope; however the sizing of the openings should be adequate to enhance natural ventilation. In hot humid climates such as in Guinea, the window size should be maximized, whereas in more arid regions

FIGURE 16.10 Left: Mountainside houses on the island of Santo Antão, Cape Verde. Right: recent residential condominium located on a slope, in Maputo, Mozambique.

FIGURE 16.11 Houses with porches, in the island of Sal, Cape Verde.

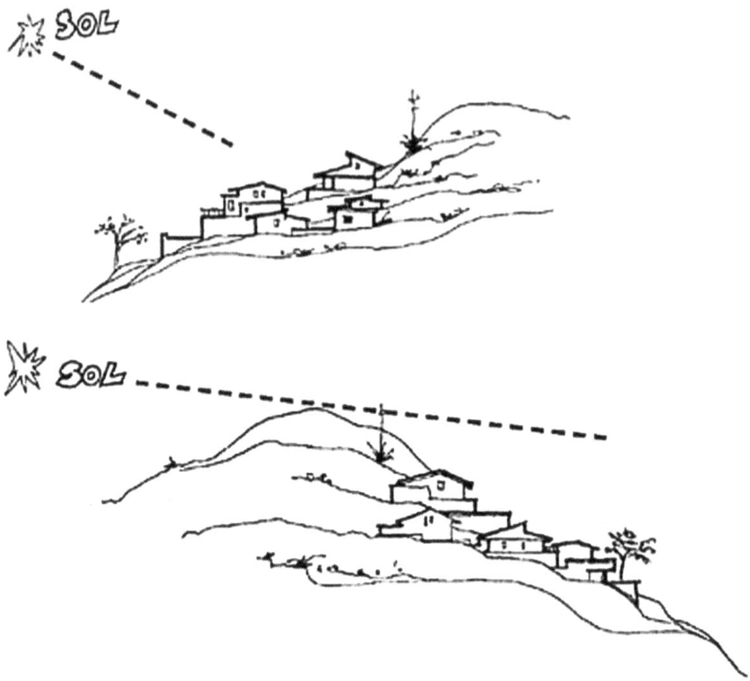

FIGURE 16.12 **Location on a hillside.** In the first scheme, the houses are too exposed to the sun at times of peak incidence. The second diagram, below, shows a more favorable location. In times of higher incidence of the sun, the houses benefit from the shade of the hillside.

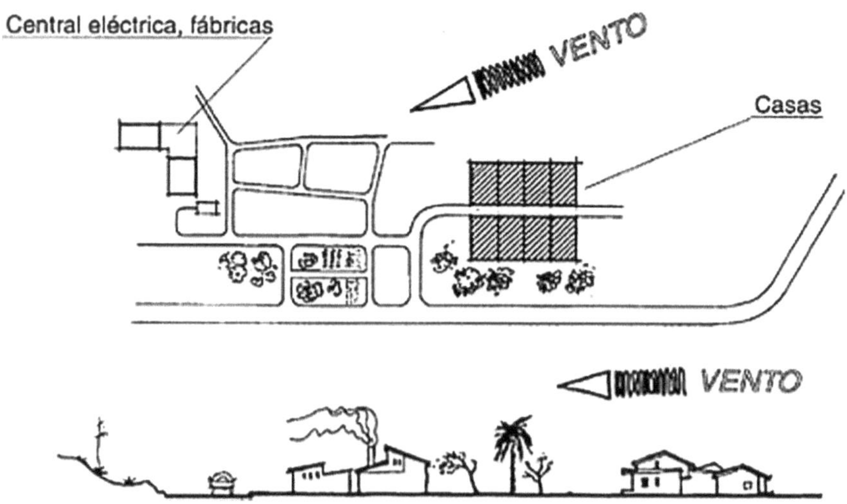

FIGURE 16.13 Correct orientation, considering the wind regime.

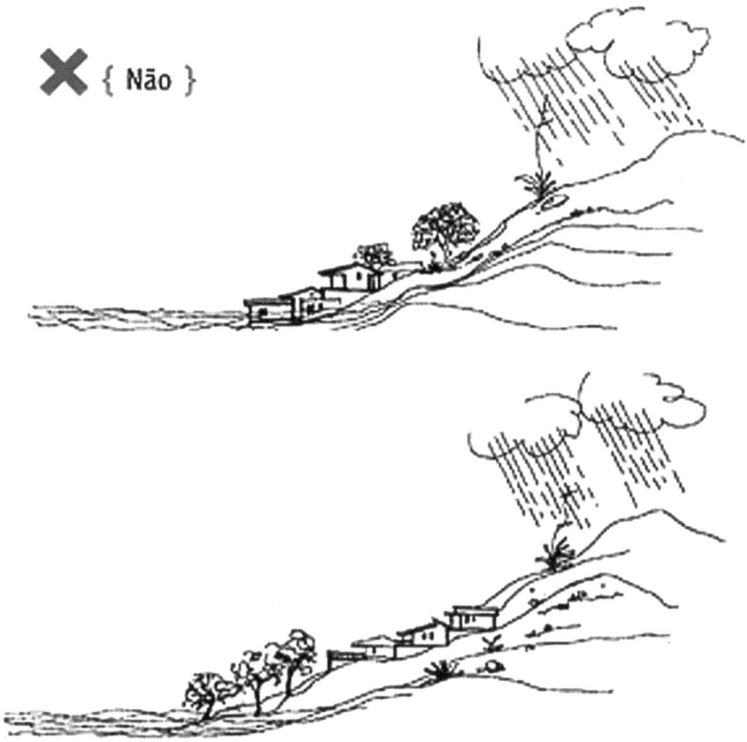

FIGURE 16.14 It is necessary to prevent building the dwellings too close to streams, riverbeds, flood-prone areas, and slopes subject to landslides. Choose safe areas, protected from flooding. The fact that it does not rain regularly in Cape Verde is misleading, because it causes people to build anywhere. In times of heavy rains, the water meets his old way. Correcting draining works are always more expensive and usually only done when the rains have caused much damage. The second scheme presents a convenient location for a housing cluster.

such as in Cape Verde, with higher daily temperature amplitudes (allowing for night ventilation), window size should be reduced. The twinning of the buildings in the band also has advantages, by reducing the area of exposure to the sun, hence reducing risks of overheating.

The areas of the building that can potentially be lit by natural light and naturally ventilated, the so-called passive areas, can be considered as having a depth of twice that of the floor-to-ceiling height (i.e. usually about 6 meters). This depth can be reduced when there are obstacles to natural light and ventilation, arising because of inadequate internal divisions, neighboring buildings, or in the case of spaces adjacent to atria. The ratio of the passive area of a building in relation to its total area provides an indication of the potential of the building for the use of bioclimatic strategies (Figures 16.18, 16.19).

The objective is always to maximize the passive area. In buildings with non-passive (active) areas of significant size, solutions using energívorous mechanical

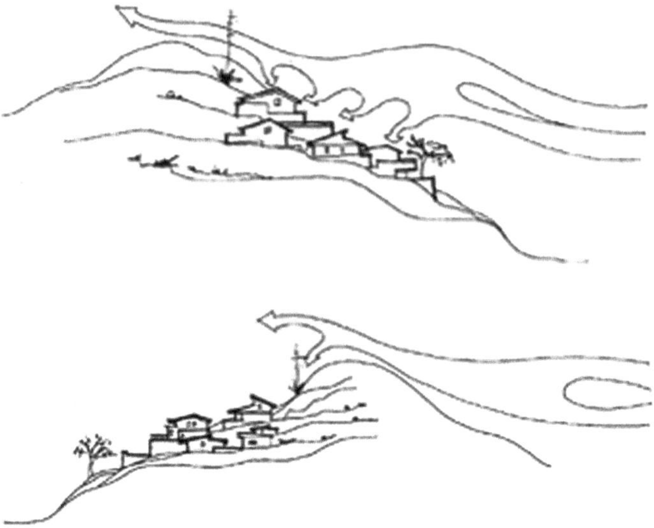

FIGURE 16.15 On a hillside location, we must study the prevailing winds, in order to maximize the potential for indoor ventilation.

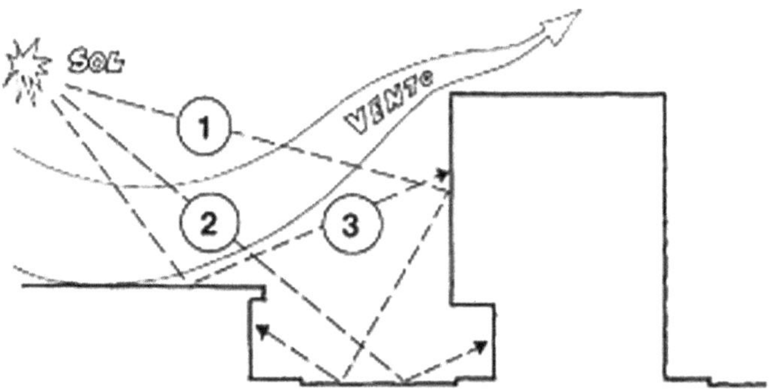

FIGURE 16.16 In this scheme, the sun beams (1) fall on building façade, which reflects them onto the pavement and then into the building interior. The beams (2) hit the pavement and are reflected in the people circulation. The beams (3) fall on the flat roof of the lower building, and are reflected by the façade of the taller building. The wind slips over the flat roof, and as there is no entrance in the front façade, it passes over the building. The environment gets too hot in and around these buildings.

systems tend to prevail (Figure 16.20). When rehabilitating buildings, any active areas should be converted to unoccupied spaces, for example storage. When the active area becomes large, it is advisable to incorporate lightwells or atriums.

The concept of the passive zone should be considered from the first stages of the project, when one defines the shape and orientation of the building. The passive design strategies to use vary, depending on the orientation of the different

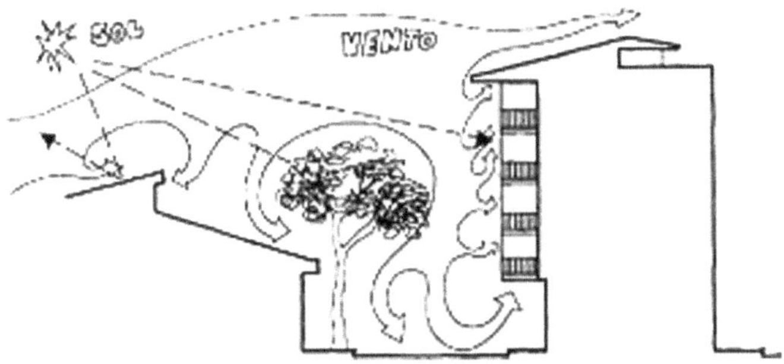

FIGURE 16.17 The configuration of the façade of the tall building and of the roof of the lower building were changed to improve the external environment in the area. The tree dampens the effect of sun radiation and promotes air circulation. The effect of wind in the area, aided by the sloping roof of the lower building and the balconies of taller building, becomes more turbulent, and thus can penetrate inside the houses.

FIGURE 16.18 The seaside village of Paúl.

areas of the building. They include, for example, changing the glazing area, or the use of different shading devices, which are described in the following sections.

The best orientation of the building to reduce solar heat gains will be parallel to the east-west axis, as this restricts the exposed area of the façades that receive low angle sun (east and west), and facilitates the shading of the façade that receives higher angle sun (Figure 16.22). It also improves natural lighting. In refurbishment, and in many urban situations where orientation is outside the control of the architects, an unfavorable orientation can be compensated for by using other design strategies to control solar gain, such as shading or window sizing.

FIGURE 16.19 Ventilation has a fundamental role: one should focus on solutions to optimize air circulation. The use of the courtyard-house typology is an effective measure in dryer climates such as in regions of Angola, Cape Verde and Mozambique.

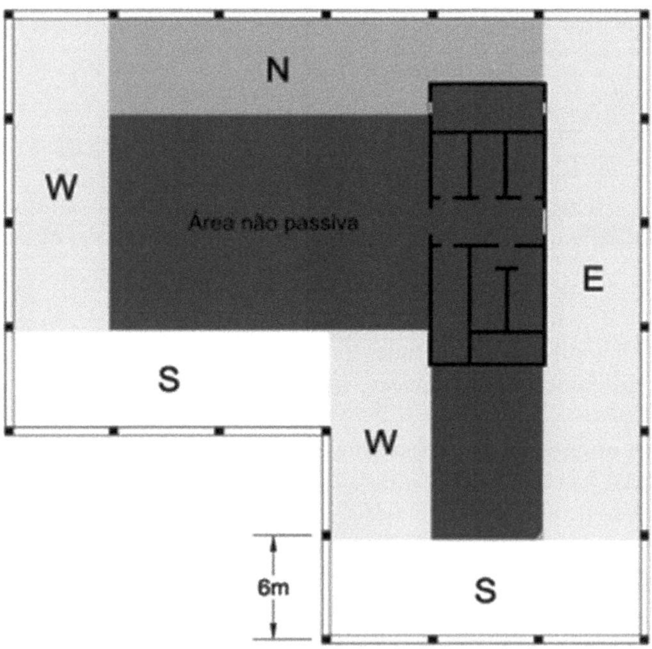

FIGURE 16.20 Definition of passive areas (light color) and non-passive areas (or active, darker color) in the plan of a building. *(Adapted from Baker, 2000)*

FIGURE 16.21 Top figure (1): residential areas in Maputo, Mozambique: the twinning of houses reduces the area of sun exposure, reducing the risk of overheating. Below (2): left: geminated houses in Sal, Cape Verde; right: a dominant façade an area of contributes to situations of discomfort from overheating.

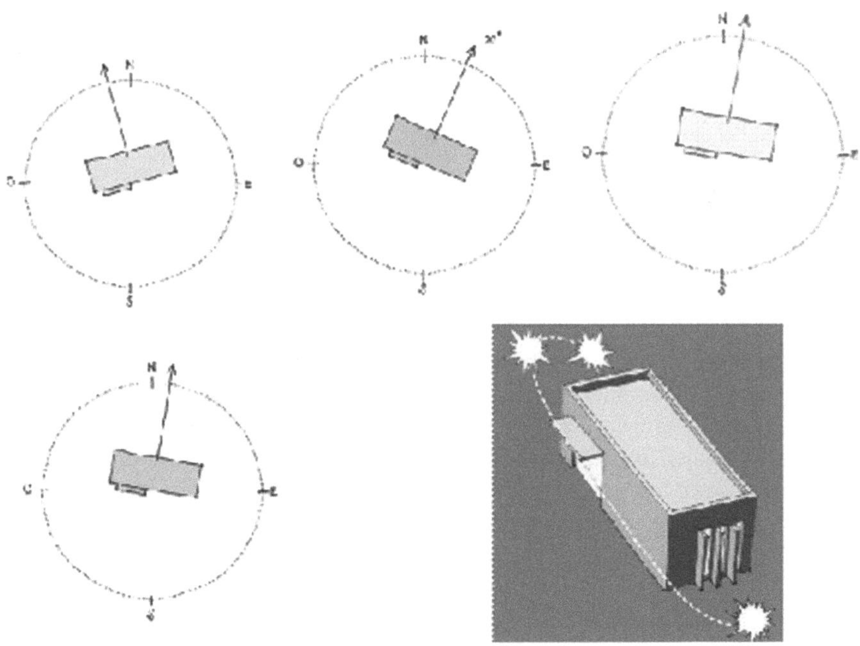

FIGURE 16.22 Optimization of solar orientation for different latitudes. Above, left: Luanda, Angola (8°50'S), the best orientation for the main façade is 352°05'N; center: island of Sal, Cape Verde (16° 35' N), the best orientation is 20°N; right: for Maputo, Mozambique (25°58'S) the best orientation is 5°N. Below, left: the best orientation for the main façade in Bissau (11°5'N) is 2°5'N. The acceptable orientation should not exceed a variation of 45° (east or west) from the optimum orientation.

The correct orientation of the occupied housing spaces, depending on the sun's path and wind regime, is the starting point to take advantage of renewable energies. The insolation of the façades is defined in the process of the choice of building location and orientation, and is decisive for the indoor comfort. Orientation to the south is generally recommended for the northern hemisphere, because it optimizes solar gains for heating during the cold season. However, in regions where the issue of overheating is a priority, such as in Cape Verde and Guinea, a southerly orientation should be avoided due to the high incidence of solar radiation; and orientation to the north is recommended, as it is for regions located below the equator (such as Angola and Mozambique). The use of porches should also be considered in order to prevent direct solar radiation falling directly on the external walls. These obstruction elements provide shade to the façades and eliminate excessive solar radiation and overheating.

The bedrooms, when oriented to the east, catch less heat, and are cooler spaces during the afternoon. The façades oriented to the west should be protected from excessive solar radiation. The design of the glazed areas must be compatible with the orientation of the façade. The kitchen space should be the coolest one in the house, so it cannot be oriented south (or north in the southern hemisphere), or west. The direction of prevailing winds should be taken into account, so that heat and odors are not dragged to the rest of the house.

Thus, for the more permanently occupied spaces, the best orientation is to the north in hot African countries located both in the northern and southern hemispheres, as shown in Figure 16.22 – yet a variation of 45° (between northeast and northwest) may be acceptable.

The optimization of the orientation and the passive area help to avoid overheating, and are thus the first step towards promoting strategies for heat protection and dissipation. The heat protection techniques such as shading, the sizing of windows, the reflective coating of the building envelope and thermal insulation provide protection against the penetration of unwanted heat gains into the building, and minimize the internal gains. In Cape Verde, building elements that provide obstruction and shade are essential to achieve thermal comfort inside the house. These elements can be tectonic: flaps or porches, vegetation or porches with vegetation (*cariço, sisal* or vines) (Figure 16.23). Planting near the façades, or even cladding façades with vegetation elements increases internal comfort, and acts as a filter for solar rays. In semi-arid regions, the walls must be thick or double to retard the penetration of heat by day and cold at night (allowing for night ventilation).

Heat dissipation techniques maximize the losses of heat accumulated inside the building, dissipating it through natural ventilation, thermal mass, evaporation, radiation, or to a 'heat sink' such as the soil. The use of these techniques avoids overheating, leading to indoor temperature values which are close to the outside air temperature, or even below it (Figure 16.24).

Direct solar radiation is by far the main heat source. The use of solar control techniques in architectural design is a high priority strategy to minimize the impact of solar gains in the building. The best solutions combine different passive

FIGURE 16.23 Use of porches and vegetation canopies for protection against solar radiation in residential buildings (Cape Verde).

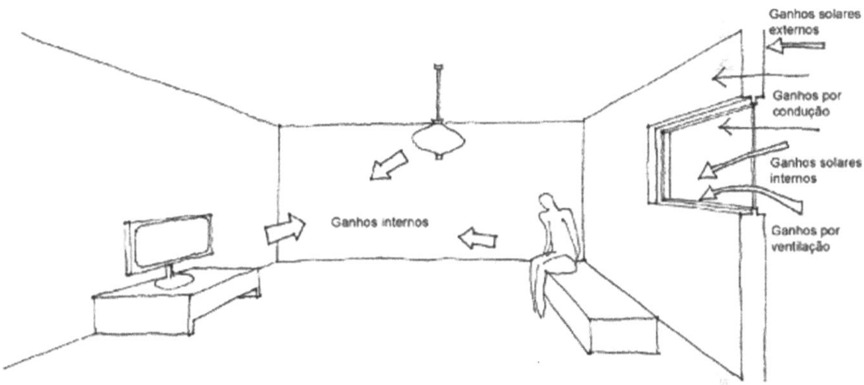

FIGURE 16.24 The different types of heat gains: I) Solar gains – caused by the incidence of solar radiation on the external surfaces, which is driven into the building (external solar gains), and the passage of radiation through the windows (internal solar gains); II) Internal gains – from the occupants, lighting and equipment; III) Conduction gains – from the conduction of heat from the warmer outside air into the building through the external surfaces of the building (façade and roof); IV) Ventilation gains – from the infiltration of warm outside air into the building.

cooling strategies to achieve greater efficiency – such as cooling by night ventilation together with thermal mass and external insulation. The effectiveness of passive cooling techniques can often be improved through the use of mechanical systems that use renewable energy, such as solar panels or photovoltaic systems, or low-energy systems such as fans. These systems are discussed further below.

16.5 SHADING

Shading is a very effective strategy to reduce the penetration of solar radiation into the building, providing protection to areas of glazing (windows), and also to the opaque envelope. The heat gain through windows can be very significant, because windows have very little resistance to the transfer of radiant heat. In hot climates, a shaded building can be between 4°C to 12°C cooler than one without shade.

Shading, of windows and of the opaque building envelope, can be achieved by fixed devices such as vegetation, or through adjustable devices. Balconies, patios or courtyards can be useful typologies for solar protection (Figures 16.25–16.30).

In terms of shading of the glazed areas, the building should be especially protected from solar gains on windows oriented to the east and west, due to the low angle of the sun in the early morning and late afternoon. This is especially true for poorly insulated and low inertia buildings. A wide variety of shading devices are available: fixed or adjustable, internal or external, admitting more or less light. Table 16.1 shows the characteristics of the different types of shading that can be used in homes or service buildings (see also Figures 16.31–16.48).

FIGURE 16.25 Shading is a centuries-old heat protection technique. Vernacular buildings in Guiné-Bissau (left) and Angola (right).

FIGURE 16.26 Left: a house with covered courtyard in Porto Novo. Right: Adjustable external shading in a house in Mindelo, Cape Verde.

FIGURE 16.27 Fixed Shading: market on the island of S. Vicente.

FIGURE 16.28 Patio and arcades of the Old School in S. Vicente (left). Veranda and porch in a residential building in Bafatá (Guiné-Bissau).

FIGURE 16.29 Shading using vegetation in Cidade Velha (left). Adjustable wooden venetian blinds in modern building in Bissau (Guiné).

FIGURE 16.30 Use of fixed shading devices ('Brise-soleil') in modernist buildings in Maputo (Mozambique, above), in Luanda, Angola (center), and Praia, Cape Verde (below).

16.6 ENVELOPE COATINGS

The light colors of some coating materials reflect a considerable amount of solar radiation. The use of whitewash (limestone based white paint) to paint buildings is an example. Light-colored coatings help to reduce the temperature of the building envelope and avoid the conduction of heat into the building. Light colors reflect more energy, and dark materials absorb more energy

TABLE 16.1 Characteristics of the Various Shading Strategies

Shading Type	Description	Performance
Fixed devices	Usually external elements like horizontal overhangs, vertical fins, egg crates, or fixed external louvers.	Horizontal overhangs, generally used above a southern window, can provide shading during summer and allow for solar heating during winter. In the east and west façades a horizontal fixed device is better than vertical, but the façade is never completely shaded. Vertical fins can protect the north façade from the rising sun and sunset. The use of grid systems (from wooden gelosias to pre-fabricated ceramic or cement elements) can also be very effective, and offers advantages in terms of privacy. However, they may reduce view to the exterior. Daylighting and natural ventilation must be considered in their design. Light-colored shading is preferable to dark colored, as it performs better in reflecting solar radiation, reducing its penetration in the building's envelope. It can also have a better performance in terms of daylight.
Intermediate spaces	Balconies, courtyards, atria or arcades.	These features can be very useful as a form of fixed shading, if their design is adequate. As in all shading strategies, design should also consider daylight and ventilation requirements. Shading performance depends on the building configuration, or on the balconies' design.
Neighboring buildings	Buildings across the street can provide shading of the façades, particularly for lower floors.	Neighboring buildings can provide efficient shading, although in some situations, such as in narrow streets, they may decrease daylight availability. The impact of this type of shading should be considered in the design process, in terms of the choice of shading devices and window sizing, e.g. increasing window size in permanently shaded areas, to improve daylight.
Vegetation	When possible, vegetation can be used to shade the lower building floors.	Trees generally have the effect of reducing wind speed. However a row of trees with bare trunks for the lower 3m of their height may, if the foliage is dense above, deflect and enhance the breeze at ground level. In situations where some passive heating may be necessary during the dry season, deciduous trees are used preferentially, in order to provide shading in the hot season, and let solar radiation and daylight in during the colder periods.

Continued

TABLE 16.1 Characteristics of the Various Shading Strategies—cont'd

Shading Type	Description	Performance
Adjustable devices	These devices can be external – such as shutters (hinged, sliding), rotatable fins, horizontal plates, retractable venetian blinds, canvas awnings, tents, blinds or pergolas – made of wood, metals, plastics, fabrics, etc. They can also be internal – like curtains, roller blinds or venetian blinds, or positioned between the glass panes of the window.	Adjustable devices are more effective than fixed, as they can admit all the solar radiation when it is desirable, like in winter, and offer more protection in summer. Their flexibility also allows a better use of daylight, when compared with fixed shading. They also allow for occupant control. External shading devices are more efficient than internal ones, as they prevent solar radiation from falling upon the glass, while internal shading devices aim only to reflect back the already entered radiation. Louvers between panes of double gazing can also have a good performance, similar to external shading. However, roller blinds, found in various cases of domestic-type offices in the present survey, can be a poor choice in terms of view, daylight and ventilation. Light-colored external opaque shading devices can reflect up to 80% of the radiation impinging on the building, if properly controlled. Translucent white external devices (such as adjustable canvas devices) can reflect up to 60% of this radiation.

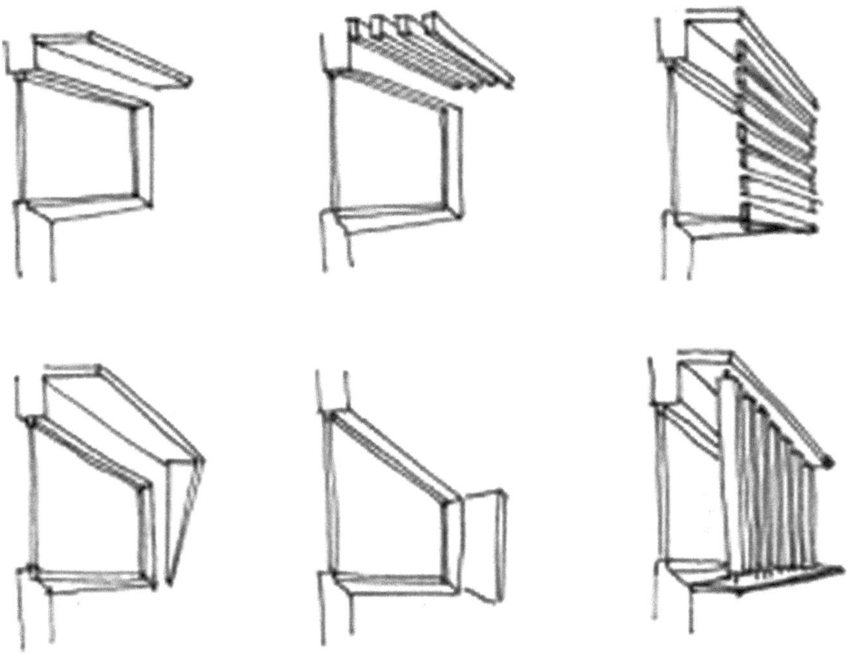

FIGURE 16.31 Some typical examples of external shading devices for windows.

(more heat). Examples are lava fields, black sand and paved roads. Table 16.2 describes the characteristics of light-colored reflective coatings (see also Figures 16.49–16.55).

16.7 INSULATION

Correctly located insulation protects the building against heat gains during warm periods, and improves thermal comfort throughout the year. The sealing of the walls is also improved (preventing the infiltration of hot air), and reduces condensation problems on surfaces in areas with humid climates (Table 16.3, Figures 16.56–16.62).

16.8 WINDOW SIZE AND GLAZING TYPE

Much of the heat gains of a building are obtained via the glazed areas of the façades, as windows offer very little resistance to the transfer of radiant heat. The orientation and sizing of the glazed areas, as well as the choice of glass, to a large extent determine the penetration of solar radiation into the building.

In a warm climate with high incidence of solar radiation, it is important to avoid large glazing areas on the façades, which lead to overheating and the need for air conditioning. In general, the area of glazing should not exceed 30%

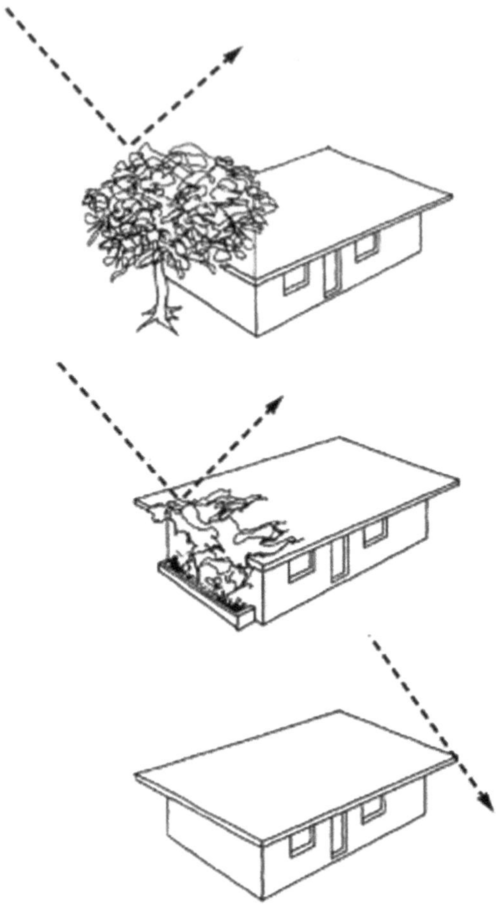

FIGURE 16.32 The trees and plants, and the projecting eaves, reduce solar incidence.

of the northerly and southerly façades' areas, considering that windows have adequate shading. This value can rise up to 40% in the case of tropical climates with high humidity values and low temperature amplitudes between day and night, such as in Guinea-Bissau. On the easterly façade, this value should be reduced to a maximum of 20% in any situation. Openings on the westerly façade should be avoided if possible. The design of windows is a complex task, but a number of software simulation programs are available to assist designers in sizing the openings; for example, Energy Plus, DOE, or, for architects, Ecotect or Design Builder.

The use of double-glazing can also reduce heat gains and losses. One can also use a type of glass that selectively transmits only the parts of the visible solar spectrum required for natural light, reflecting unwanted radiation – the so-called low-emissivity glass.

FIGURE 16.33 Fixed shading in the façades of modernist buildings in Luanda.

FIGURE 16.34 Fixed shading in Luanda: Above: porch (left); shaded roof (center): 'Brise-soleil' (right). Below: arcades (left) and a fixed shading device at the Faculty of Architecture's building.

Table 16.4 describes the strategies for solar radiation protection, through window sizing and choice of glass (see also Figures 16.63–16.67).

The sizing of the openings and the insulation of the opaque envelope, and protection against solar radiation, also prevent the entry of heat gain by conduction, caused by the flow of heat from the warmer outside air, through walls and

FIGURE 16.35 Examples of fixed shading in Cape Verde: 1. Projection of the roof to shade the façades, in the Amilcar Cabral house-museum; 2. Projection of the roof in institutional building; 3. Shading of esplanade in St. Antao, 4. Grid system for shading and ventilation in a building in Mindelo, 5. Vertical shading in a services' building in Praia.

FIGURE 16.36 Fixed shading in Cape Verde: double façade shading system, in an institutional building in the city of Praia. The exterior façade shades the glass and opaque areas through the use of an ingenious grid system using overlapping tubes, allowing for natural light and ventilation.

glass areas, when the outside temperature is greater than the internal temperature. They are a cause for concern, especially in those regions where summer temperatures reach 40°C, as occurs in many regions of Africa. However, the conduction gains generally tend to have a relatively minor impact on cooling requirements compared to solar and internal gains. They are particularly important in an air-conditioned building where the internal temperature is maintained at a lower temperature than outside for prolonged periods.

16.9 NATURAL VENTILATION

Natural ventilation is the flow of air between the outside and the inside of the building. Natural ventilation originates from two natural forces: pressure differences created by the wind around the building – wind-driven ventilation;

FIGURE 16.37 Adjustable external shading: wood shutters (venetian blinds) provide shade and simultaneously allow for natural lighting and ventilation. Right: Adjustable shading: leisure area in the fort museum, at the Cidade Velha. The shading strategy allows natural lighting and ventilation.

FIGURE 16.38 Removable and adjustable shading: 1 and 2: the terrace of the Club Nautico in Mindelo (shading also using ships' sails), 3: shading of outdoor space of a house, using a single sheet.

and temperature differences – 'stack effect' ventilation. Table 16.5 shows the various objectives and requirements of natural ventilation (see also Figures 16.68–16.73).

Wind-pressure ventilation is influenced by the wind intensity and direction, and also by obstructions caused by nearby buildings or vegetation. The knowledge of wind conditions around the building and its pattern of speed and direction (information can be obtained from meteorological institutes) are necessary data for the design of the openings. The wind direction varies throughout the day. In addition to the prevailing winds, the earth winds (night time) and the sea breeze (day time) are also important.

The distribution, size and shape of the openings are fundamental elements for achieving efficient ventilation. The openings should be widely distributed in different façades, according to wind patterns, ensuring they will have different pressures, to improve the distribution of airflow in the building. The entrance

FIGURE 16.39 Shading by vegetation: 1–4 buildings in Old Town, 5 and 6 street in the city of Luanda, 7 terrace in Maputo.

FIGURE 16.40 Examples of shaded porches in colonial buildings in Guiné-Bissau.

and exit openings (windows, doors, other openings) should be located to create an effective system of ventilation in which air flows through the occupied space. One should also consider the elements that can act as obstacles (internal partitions). The openings located in high positions allow high rates of ventilation for heat dissipation. Openings located on a lower level can provide air circulation throughout the occupied zone. The markedly vertical windows facilitate high-level ventilation, and achieve a better performance in terms of natural lighting and arrangement of interior space.

When designing windows for natural ventilation, a compromise must be reached with other environmental needs, such as: natural lighting, sealing, solar gains, spatial performance, maintenance, noise, safety, cost and control of air circulation. The problem of noise, typical in urban environments, can be

FIGURE 16.41 Fixed Shading: left – shading projection of roof in a contemporary residential building in Guiné-Bissau. Right: shaded streets in Bissau.

FIGURE 16.42 'Brise-Soleil' in modernist buildings in Maputo, Mozambique.

FIGURE 16.43 Fixed shading in modern buildings, Maputo, Mozambique.

FIGURE 16.44 **Fixed shading: Arcades.** Above: island of Mozambique (XVIc.), below: Maputo, Mozambique.

FIGURE 16.45 Shaded porches, Mozambique.

FIGURE 16.46 Shading in vernacular architecture (above) and in contemporary buildings (below), Mozambique.

FIGURE 16.47 Shading from vegetation: above: city of Maputo; below: rural area, Mozambique.

FIGURE 16.48 Adjustable shading: above: colonial buildings in Maputo; below: canvas, Mozambique.

minimized through the use of acoustic shelves on the outside of windows or absorbent acoustic panels on the inner surfaces. Pollution problems can also be avoided by the use of buffer space, and bringing air into the building from a less polluted area. Security problems can be solved by sizing the openings, or placement of external shutters (see Figures 16.74–16.85).

The situation in the first scheme is ideal for occupant comfort (cooling) – the entry of cooler air is made at lower level. The situation of the second scheme is for the cooling of the building – the indoor heated air rises and is accumulated near the ceiling, its exhausting ventilation is made at a higher level. The use of vertical, tall windows, are ideal to allow control of these two levels of ventilation.

'Stack effect' ventilation is appropriate for higher-rise buildings, especially in situations where the wind cannot provide an adequate air movement: if wind speeds are low or the wind has an unpredictable pattern. This method can also be used in conjunction with ventilation by wind pressure to enhance the performance of the ventilation system, especially in deeper-plan buildings where it is difficult to achieve cross ventilation. The 'stack effect' is generated by a vertical pressure difference, depending on the average temperature difference between the column of air and the outdoor temperature,

TABLE 16.2 Characteristics of the Use of Light Color Envelope Coatings

Reflective Coatings	Description	Performance
Light color paint or tiles	Light color paint of the opaque external envelope (e.g. white). Light color tiles can also be used in façades. The roof, when possible, should also have a light color.	White paint is a very cost-effective way of reducing the building's heat load in summer. The color that reflects most radiation is white. Painting the internal walls in a light color can also improve the internal levels of daylight, hence reducing the need for artificial light. Near the houses, one should avoid the use of dark colored pavements, in order to reduce solar radiation absorption. In some urban situations, the reflectance of solar radiation to other buildings may sometimes be undesirable, but it may constitute an advantage in terms of daylight. Undesirable reflection from neighboring buildings can be avoided through the use of shading devices. Metallic foils are only effective when they are facing a void between building elements. This is because heat is transferred across air gaps partly by radiation, which the foil reduces due to its low emissivity. They have no effect at all when in contact with materials on both sides. When heat transfer is downwards across avoid e.g. an attic space, the largest proportion of heat is transferred by radiation. This is when reflective foils are most effective.

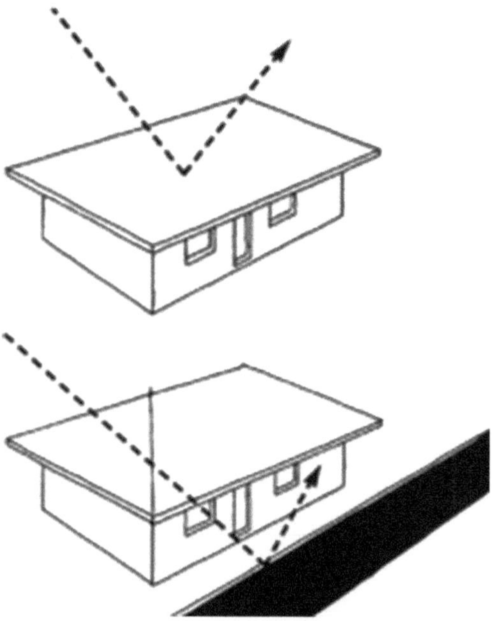

FIGURE 16.49 Whitewashed surfaces reduce solar incidence (top). The proximity of the house to dark colored pavements should be avoided, to avoid heat absorption and radiation into the housing (below).

FIGURE 16.50 View of the city of Mindelo, Cape Verde. Light-colored paint reflects radiation, avoiding overheating.

the opening sizes and location, and height of the air column. Hot air rises and exits through openings located off the stack; cooler air will penetrate into the building at ground levels. The problem with 'stack effect' ventilation is that the system reaches its maximum efficacy when outdoor temperatures are

FIGURE 16.51 Heat protection: (1) Whitewashed vernacular house, on Santo Antão, (2) contemporary building painted white, in Mindelo, Cape Verde.

FIGURE 16.52 Suburban area in the island of Santiago, Cape Verde: the use of light-colored paint would be an inexpensive strategy to reduce discomfort due to overheating.

FIGURE 16.53 Buildings painted white, in Luanda, Angola: the painting of buildings with light colors (e.g. using whitewash), would be an economical way to significantly reduce the discomfort from overheating.

lower and when there are greater differences in temperature inside the building. In warmer climates, such as Cape Verde, a solar chimney can be used to increase temperatures in unoccupied areas, thus increasing the temperature differences. The performance of this approach is not as good as wind-pressure

FIGURE 16.54 Buildings painted white, in Guinea-Bissau, to reduce the discomfort from overheating.

ventilation, as it requires larger temperature differences and larger areas of openings (e.g., cross ventilation achieved from a 2.7m/s wind is better than a chimney with a height of 3m and with 43°C at the top).

Tables 16.6 and 16.7 show the characteristics of wind pressure and 'stack effect' ventilation. Table 16.8 illustrates particular day and night ventilation techniques, including wind pressure and 'stack effect' ventilation. Table 16.9 refers to the use of assisted ventilation (see also Figures 16.86–16.90).

When the outside temperature is too hot, it is necessary to prevent heat gains for ventilation – caused by the infiltration of external hot air to the interior of the building. This type of gain can be minimized by reducing the rate of ventilation when the outside temperature is higher than the indoor temperature. The ventilation rate should be increased substantially in periods when the outside temperature is lower than the indoor temperature – for example, during the night (night ventilation) (see Figures 16.91–16.103).

Some possible measures to lower the temperature of the roof slabs are: insulate the coverage with thin lime, whitewash and pozzolana; provide openings for hot air on the highest part of the walls; improve air inlet openings in the lower

FIGURE 16.55 Buildings painted white, in Mozambique: vernacular buildings in the region of Pemba (top); in the Ilha de Moçambique (center); and in the city of Maputo (bottom).

walls – oriented in the direction of the prevailing winds in order to provide cross ventilation; use cavity wall insulation; introduce vegetation to the perimeter of the house. Lightened, hollow concrete slabs supported by pre-stressed beams are the most appropriate solution in more arid climates, such as in Cape Verde and parts of Angola and Mozambique. As well as being lighter, they are cheaper and allow good ventilation.

Vaulted construction is another energy efficient solution. The vaulted, curved surface of the roof increases the air movement over it. To exploit this effect, the vaults must be constructed in the opposite direction to the prevailing winds.

In regions with very hot periods, natural ventilation can be enhanced by using low-energy mechanical cooling systems, such as fans. These energy efficient devices can be very useful in cases of existing buildings, especially those where the potential of natural ventilation is limited (Figures 16.104–16.107).

FIGURE 16.56 Vernacular houses in Praia, Cabo Verde (above) and Pemba, Mozambique (below), with thatched roofs. Thatch is an insulating material that protects the building against heat gains.

FIGURE 16.57 Contemporary use of thatch: Hotel in Sal, Cabo Verde (left) and Barra do Kuanza (Angola). Using this typology of local tradition also brings the benefits of solar protection.

FIGURE 16.58 **Roof with mixed construction system.** The thatch is superimposed on the corrugated metal sheet (sub-layer): the benefits of sealing and durability conferred by the use of the metallic layer and the insulating ability of the thatch (tourism building, Guinea-Bissau).

In very specific situations, where the potential of natural ventilation is reduced and the use of ventilation low power systems, such as fans, is not sufficient to meet the ventilation and cooling needs of the building, it is preferable to use so-called 'mixed mode' systems – that is, to use the HVAC systems only when and where necessary. The use of 'mixed mode' strategies can avoid over-sizing centralized systems, reduce operating costs of the building and save energy.

FIGURE 16.59 Guiné-Bissau: The use without protection (insulation) of metal roofing should be avoided, as it leads to situations of internal overheating. Furthermore, with the oxidation, the metal cover darkens, transmitting more heat to the internal spaces.

FIGURE 16.60 Internal roof insulation in a recent building, using local materials (thatch), in Guiné-Bissau (left), and Mozambique (right).

16.10 THERMAL MASS

In most of the buildings existing in Cape Verde and Angola, and in parts of Mozambique and Guinea-Bissau, the opaque surroundings of the building, structures and internal divisions are built with massive materials such as stone, earth (taipa or adobe), bricks or concrete. Thermal mass acts as heat and cold storage, regulating and smoothing fluctuations in temperature. The high thermal inertia of the massive building components reduces the maximum temperature values in the summer, providing more comfortable conditions. The heat stored during the day can dissipate at night, by night ventilation. Inertia slows the heat exchanges by conduction to the outside, which is particularly beneficial during heat waves.

Unlike other heat sinks, like the atmosphere, the sky or the ground, which provide an almost unlimited resource for this purpose, the use of thermal mass is a temporary, transitional solution. After a certain point, heat starts to accumulate in the mass of the building and decreases its efficiency. Therefore, the use of thermal mass should be combined with ventilation strategies to remove the accumulated heat, especially night ventilation. Night ventilation strategies coupled with a good thermal mass can reduce the average inside temperature during the day to below the outside daytime average temperatures. However, in

FIGURE 16.61 Use of thatch in vernacular (left) and contemporary (right) buildings, in Guinea-Bissau.

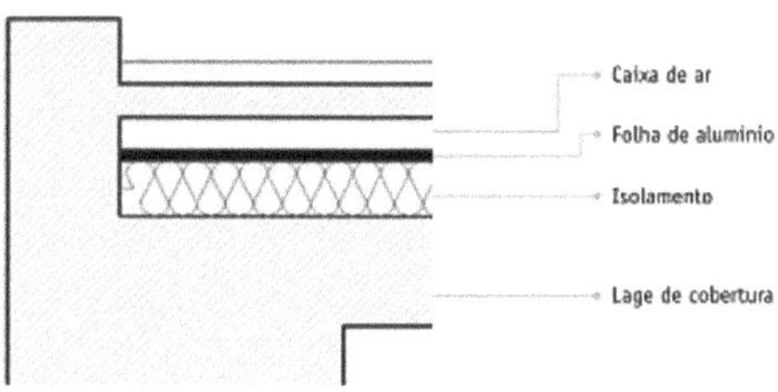

FIGURE 16.62 Left: Schematic representation of a radiant barrier on a roof, with ventilated air box. Right: the use of perforated brick may improve the building's insulation, though it should be further reinforced with external or cavity wall insulation materials.

buildings with high internal gains, such as office buildings with a lot of occupants and equipment, this is more difficult to achieve. But even in these particular cases, the average daytime temperatures inside can still be reduced to values close to the average outside, or a little above this, with a still reasonable performance in terms of passive cooling.

When auxiliary cooling systems are necessary, as in the case of 'mixed mode' buildings, the use of thermal mass can delay the need for and reduce the periods of time during which cooling becomes necessary.

TABLE 16.3 Characteristics of Insulation and Radiant Barriers

	Description	Performance
INSULATION	Insulation material can be added either to the inside or the outside surface, or by filling the cavities within the wall structure. In terms of heat avoidance, insulation materials prevent heat conduction due to the existence of trapped gas in many layers (fiberglass bat) or in cells (polystyrene), increasing the material's thermal resistance to conduction in proportion to their thickness, but do not necessarily restrict radiant heat. External insulation can be added using pre-fabricated insulating panels. Should have a light color.	The insulation of the external opaque elements, or the use of additional insulation to the façades, is one of the simplest and most effective measures for cooling and heating load reduction. The air existing in the brick's cavities or in the space between walls (double wall façade) also provides insulation to the building, but this can be significantly improved with additional material (external or cavity wall insulation). External insulation is preferable to internal insulation, making maximum use of the internal thermal storage mass, and has a better performance in preventing heat gains in summer. It plays an optimum role in passive cooling, when associated with thermal mass. It also minimizes or even eliminates thermal bridges and condensations. Internal insulation should be avoided, as it reduces the amount of exposed thermal mass, reducing the benefits of thermal inertia. Roof insulation is also very important, as it decreases the risk of high temperatures on the top floor of buildings in hot conditions.
RADIANT BARRERS	Radiant barriers, made from aluminum foil-coated products, can be installed in the ventilated air gap of cavity walls and roof. The aluminum foil reflects long wave radiation, and the airspace prevents heat movement downwards by conduction.	The effectiveness of this method depends on the ventilation required to transport the heat from the foil by convection. When cooling is the main concern it is preferable to use a foil radiant barrier, as an alternative to high insulation levels that could create moisture and increased roof temperatures. This system can however be more expensive than simple insulation.

TABLE 16.4 Description of Strategies Involving Window Sizing and Glazing Type.

	Description	Performance
WINDOW SIZING	Windows, glazing ratio, façade orientation.	Windows also influence performance of daylight and natural ventilation, acoustics, and the visual contact with the external environment. They must be designed to allow this integration. Windows must also be sized according to orientation. There is appropriate software for this purpose, such as DOE, Design Builder (Energy Plus), or Ecotect. These can be used both in new design and in refurbishment. Glazing area should be reduced to the indispensable. It is recommended that it is not greater that 30% in north and south façades, considering adequate shading (or up to 40% in more tropical climates). In the east façade this value should be reduced to a maximum of 20%. In the West façade openings should be, if possible, avoided. Horizontal glazing areas should only be used with adequate shading, and in zones with high floor-to-ceiling height (at least 6 to 8 meters), as they can easily cause overheating problems. Large areas of horizontal glazing should be avoided.
GLAZING TYPE	Double-glazing, low-emissivity glazing, HOE.	Double-glazing increases the insulation value of the glazing area, and also has the advantage of reducing condensation at the window back, draught risk and infiltration rates. Compared to single glazing, its use can significantly reduce heat gains. A greater reduction in heat gains is achieved if low-emissivity glazing is used. Amortization of double glazed windows can be achieved in between 5 to 25 years, according to the quality of the materials and size of the windows. Low-Emissivity glazing can be almost opaque to infrared radiation, reducing the solar transmission by more than 50%. This glazing type and HOE do not reduce daylight levels, although they are efficient in reducing solar radiation. However, they can be expensive. The use of tinted and reflective glass for shading and glare prevention should be avoided, as these materials also substantially reduce daylight levels, increasing the use of heat producing artificial light. It is preferable to use clear glazing (double, low-e.), shading, and a reduced glazing area.

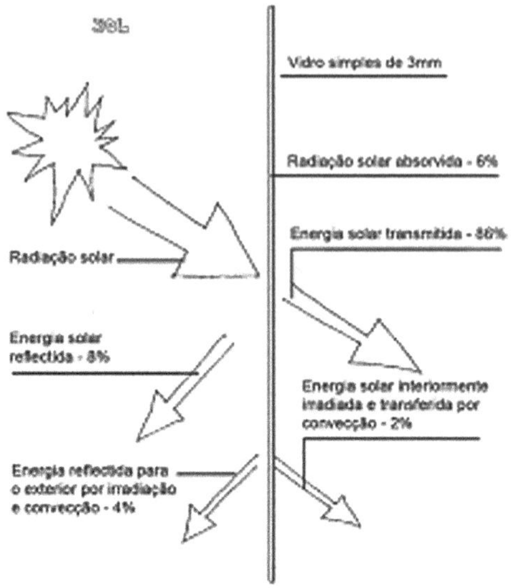

FIGURE 16.63 Energy exchange in a 3mm single glass window.

FIGURE 16.64 Large glazing areas should be avoided in the façades' typologies, as they are largely responsible for the overheating of the building, and subsequent use of energívorous air conditioning systems. The sealed façades with large areas of glazing are an imported typology, inadequate in the hot climates (Luanda, Angola; Praia, Cape Verde).

The performance of a thermal mass depends on the ability of the construction elements of the building to transfer heat into space: that is, it depends on the heat transfer coefficients of the materials used. The performance also depends on the physical capacity of these materials to store heat, i.e., their specific heat. The amount of thermal mass used in the process typically corresponds to a thickness of 50–150mm from the surface. The solid material must be as exposed as possible. Resultant acoustic problems, sometimes caused by increased exposure of massive elements (walls, slabs), can be reduced by the use of sound absorbing perforated ceilings (see Table 16.10 and Figures 16.108–16.113).

FIGURE 16.65 Most residential buildings we find in more established urban areas in Cape Verde (above), Angola (center), Guiné (below) and Mozambique (bottom) have very reasonable glazing areas. They are a good reference for the design of new buildings. The area of glass should not exceed 30% of the total area of the main façade, and should be fully shaded.

16.11 EVAPORATIVE COOLING

Evaporative cooling is achieved by an adiabatic process, in which the sensible air temperature is reduced and compensated by latent heat gain. The use of fountains and vegetation in the courtyards, or the act of pouring water on the floor, or the use of large porous clay pots filled with water in the rooms, are good examples of direct evaporative cooling techniques, many of which are used in some of the warmer countries of Africa, and hence which can also be applied successfully in Cape Verde, and also in Angola and Mozambique during the dry season, when the relative humidity level does not exceed 60%.

Indirect evaporative cooling techniques also exist, in which the air is cooled without increasing its water vapor content. In this type of system the air temperature can be lowered to match the wet bulb temperature. The water consumption is much lower than in direct systems. However, indirect systems involve the use of mechanical devices, which can be expensive and require complex maintenance (see Figures 16.114–16.118).

16.12 CONTROL OF INTERNAL GAINS

The main sources of heat inside the building are electric lighting, the number of occupants, and the mechanical equipment they use. Internal heat gains can

FIGURE 16.66 Rehabilitated vernacular house in Santo Antão. The window area is sufficient and adequate to meet the needs of natural lighting and ventilation.

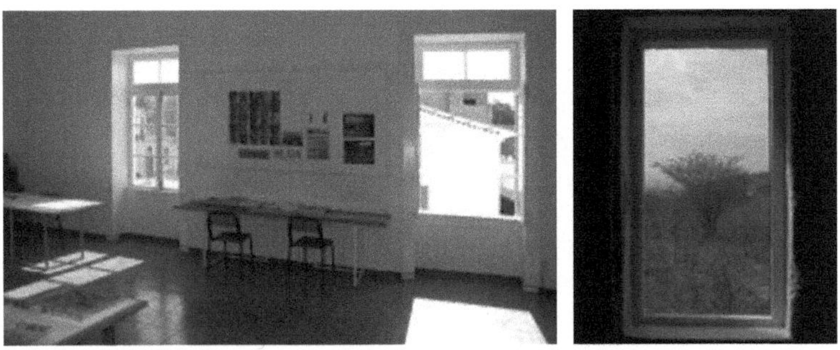

FIGURE 16.67 School building in Mindelo. The vertical arrangement of the horizontal window is preferred because it avoids problems of daylight discomfort (glare and contrast) and has advantages in terms of the arrangement of furniture inside. It also allows ventilation at a higher level for the renewal of air and cooling of the building mass at night.

FIGURE 16.68 External venetian blinds in Ribeira Grande and in Mindelo. As well as shading, they allow and direct the flow of natural ventilation.

FIGURE 16.69 External fixed shading systems in Luanda, Angola. As well as shading, they allow and direct the flow of natural ventilation.

contribute significantly to overheating, especially in office buildings with larger dimensions. The main strategies to reduce internal heat gains are:

a) Avoid excessive use of artificial lighting;
b) Optimize the use of natural light;
c) Avoid excessive heat gains from occupants and equipment.

See Table 16.11 for further details.

FIGURE 16.70 **External fixed shading systems in Guiné-Bissau.** As well as shading, they allow and direct the flow of natural ventilation.

16.13 THE USE OF ENVIRONMENTAL CONTROLS

Some passive cooling techniques, such as the use of thermal insulation or reflective coatings to reduce heat penetration into the building, do not involve the use of operational controls, i.e., the systems are fixed, inherent to the building and do not require occupant control or automatic interaction.

However, in many other passive strategies, such as opening windows for natural ventilation, shading adjustment or the use of fans, system performance is governed by operational controls. In these cases, the efficiency of the energy consumption reducing systems, and the creation of comfortable environments are determined not only by the efficiency of the controls, but also by the way the occupants use them. The use of environmental controls allows users to change the environment, adapting it to their thermal comfort needs. Consequently, there may be a significant improvement in thermal satisfaction, allowing the occupants to meet their specific comfort needs and reduce their discomfort from overheating.

It is important that the occupants are aware that the use of controls not only leads to an improved efficiency of the system itself, but also has a major impact on energy savings. To do this, the design must be simple in order to facilitate an intuitive understanding of its use.

FIGURE 16.71 Ventilated roofs in Guiné-Bissau.

FIGURE 16.72 'Brise-Soleil' grills in Mozambique, for shading and natural ventilation.

FIGURE 16.73 Maputo, Mozambique: adjustable devices for natural ventilation.

16.14 PASSIVE DESIGN AND THERMAL COMFORT CRITERIA

Passive design techniques can be applied with a good degree of efficacy. It is true that the use of these techniques does not promote the kind of uniform, low temperature, environments found in air-conditioned buildings. A question must then be asked: is that kind of uniform indoor environment really necessary and desirable?

In surveys conducted throughout the world in naturally ventilated buildings, where environmental conditions vary outside the conventional thermal comfort standard, a majority of people report feeling, in fact, comfortable with their thermal environment. Other studies, conducted in buildings with central air conditioning, showed a significant dissatisfaction with the thermal environment by the occupants. This dissatisfaction could be attributed to various causes, such as a lack of 'naturalness' and the health problems inherent in the system, or yet

TABLE 16.5 The Various Objectives of Ventilation, and Respective Requirements

Objectives	Description	Requirements
Provision of fresh air	Ventilation is necessary to provide fresh air to the office occupants, replacing stale air and controlling odors, moisture, CO_2 and pollutant concentration.	Typically 0.5–3 air changes per hour per person, depending on the intensity of occupation. The CIBSE (1997) building regulations consider a minimum standard of 5l/s per person (which is achieved by average infiltration rates), this standard being raised to 16l/s in smoking areas.
Heat removal from the building	This type of ventilation is used to remove excessive heat from the building interior, providing more comfortable temperatures and reducing the cooling loads.	Requires higher ventilation rates than the previous, at high level to remove the accumulated heat. When the external air temperature is lower than the internal air temperature, typical ventilation rates for space cooling are 5–25 ach/h, depending on the temperature difference. The greater the heat gains, the more ventilation is required.
Convective and evaporative cooling of the human body.	A higher air speed increases the sweat evaporation from skin, broadening the comfort temperature upper limit. A thermal sensation corresponding to an effective temperature of 27°C can be achieved if air movement of 1m/s is applied to a room with an air temperature of 30°C.	This process requires air speeds between 0.5 and 3m/s. It is accepted that each increment of 0.275m/s increases 1°C to the upper comfort limit. The upper velocity recommended in offices is 1.5m/s. In houses, this value can increase to 2.5–3m/s.

FIGURE 16.74 External shutters are a typology traditionally used in Cape Verde and Bissau. They provide window shading, and allow the control of the natural ventilation flow to the interior of the building.

FIGURE 16.75 The use of fixed grids, although not allowing adjustments to control shading and the ventilation flow, can also be an effective solution in some cases.

another very important factor: the lack of environmental controls in existing buildings with a centralized system, which inhibit the natural process of human adaptation.

There is now a major controversy regading which thermal comfort criteria should be adopted. Conventional standards accept only a limited temperature range as theoretically 'ideal', i.e., within which the vast majority of the occupants of a building would feel comfortable. These conventional standards of comfort, which form the current ASHRAE and ISO standards, are still regarded as applicable anywhere in the world, with only a small seasonal variation for summer and winter situations, despite the existing wide variety of climates. They consider summer temperatures around 22°C as ideal, with maximum temperatures of around 26°C. In warmer countries, this would require the extensive, in many cases permanent, use of air conditioning systems.

FIGURE 16.76 Modernist building in Praia, with façade ventilation openings at various levels: high, for air exchange and cooling of the building mass, and lower (windows) for the comfort of occupants. The need for natural ventilation and shading were factors that were considered in the design of the building, which is visible in the façade typology.

FIGURE 16.77 **The interior of a school in Mindelo.** (1): The inner openings have a generous height, allowing the flow of ventilation at a higher level, and contributing to a good performance in terms of natural lighting. (2): open windows for natural ventilation when the outside temperature is comfortable during the day.

FIGURE 16.78 Windows protected by a mosquito net, in Praia.

FIGURE 16.79 Vents with mosquito nets in a house in Mindelo.

FIGURE 16.80 Openings at the top level for the removal of hot air.

FIGURE 16.81 Ventilated roof in Maputo, Mozambique.

FIGURE 16.82 Ventilated atria in two buildings in Maputo, Mozambique.

FIGURE 16.83 Use of mosquito net in an old building in Maputo, Mozambique.

FIGURE 16.84 Naturally ventilated market, Ilha de Moçambique.

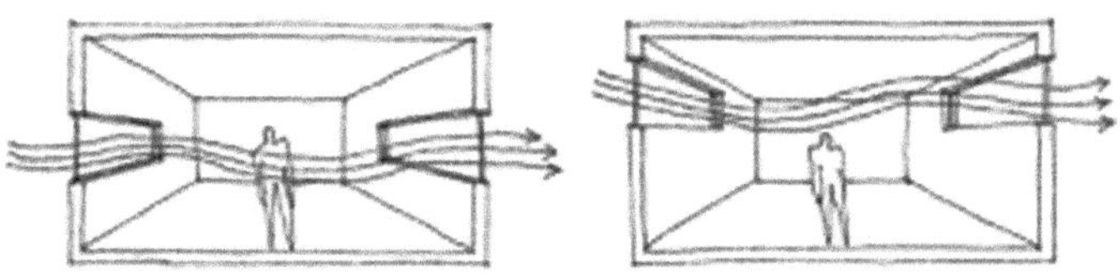

FIGURE 16.85 Opening positions for two types of cooling.

TABLE 16.6 Wind-Driven Natural Ventilation Strategies for Building and Occupant Cooling

Wind-Driven	Description	Performance
Single-sided ventilation	Ventilation provided by openings in only one side of the room or façade.	Single-sided ventilation has a more shallow penetration than cross ventilation – typically 3 to 6m or about twice the floor-to-ceiling height. It is created as air enters the room at one time and a few seconds later some air exits because of the fluctuating static pressure of the wind.
Cross ventilation	Openings on both sides of the building and an airflow route within the building.	Steady cross ventilation is usually the strongest mechanism of natural ventilation, especially in larger buildings. It has a useful depth of 9m or three times the floor-to-ceiling height – 18m zones can be ventilated if they are back-to-back. Courtyards can be used instead of deep floor plans, to promote cross ventilation. If the building is facing the prevailing wind direction, and the wind has a good intensity, the use of ducts and underfloor space for building cross ventilation can also be efficient alternatives to openings in the room façade. Circulation areas such as corridors and stairwells can also be used to supply rooms that have no access to the windward side. Patios can be used, instead of deep plans, to promote cross ventilation. If the building is facing the dominant wind, and the wind has a good intensity, the use of special conducts and slab cavities for cross ventilation can be efficient.
Wind towers	If the building is not in a favorable position towards wind and prevailing breezes' directions for natural ventilation, wind-channeling devices could be used, such as wind towers.	Wind towers, such as the ones used in some hot countries (2 to 20m tall) can also be useful to create air movement, when wind for cross ventilation is not available at the building's level. The stack supply and extract is wind driven, reverting to stack in the absence of wind. In certain hot and dry regions, pools or ceramic pots with water are placed in the base of the wind tower to provide additional evaporative cooling.

TABLE 16.7 Stack Effect Natural Ventilation Strategies

Stack Effect	Description	Performance
Single-sided double openings	Openings in a low and high position, in a window or wall, i.e. near the floor and ceiling.	It could be effective up to 9m or three times the floor-to-ceiling height. It can increase the depth of natural ventilation in open plan rooms. It depends on the difference in height between inlet (lower) and outlet (higher).
Atria	The introduction of atria offers good potential for stack ventilation.	Atria can be used specially in larger buildings. They must be carefully designed, as they can sometimes cause overheating in hot countries.
Solar and thermal chimneys	In solar chimneys, solar radiation is used to increase the stack effect; as the sun warms the surfaces of the chimney, the ventilation rate increases.	The stack must terminate above the roof peak, so that the stack top is always under suction compared with the lower inlet level. Otherwise a wind coming in the wrong way can introduce the hot stack air into the room. The wind pressures on the top of the chimney also influence it.
Vent-skin walls	Ventilated cavities within the walls (see also 'thermal mass').	Vent-skin walls improve the dissipation of heat stored in the building. This technique is exclusively for heat removal of the building (not comfort cooling).

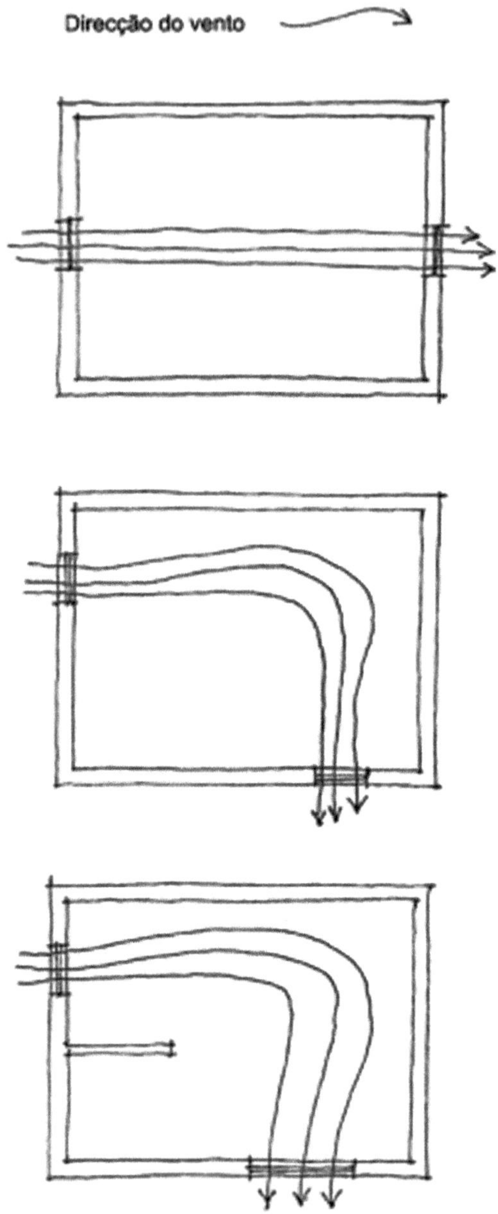

FIGURE 16.86 Some ventilation schemes for different window sizes and positions.

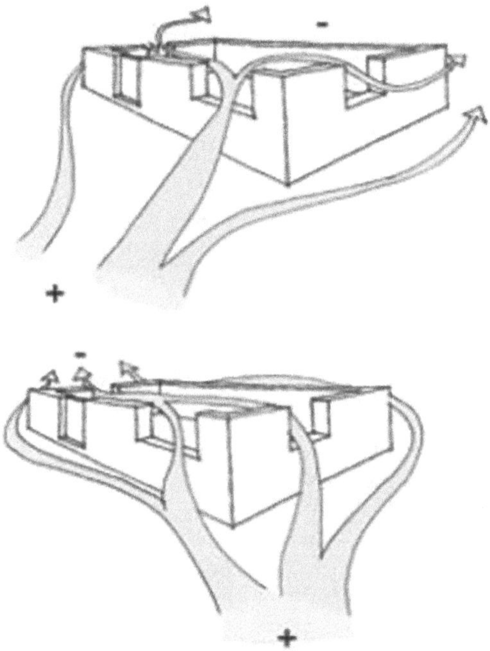

FIGURE 16.87 The positive and negative pressures caused by different wind directions and positions of the openings.

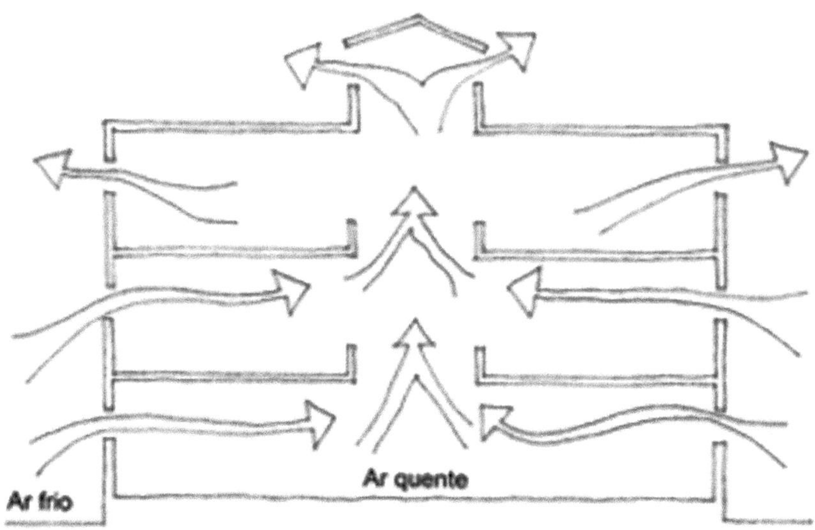

FIGURE 16.88 Stack effect ventilation in an atrium building.

FIGURE 16.89 Building for cultural activities in Lagedos. Openings at top and opposing fronts, for cross ventilation, and opening on the roof for stack effect ventilation.

FIGURE 16.90 (1.4) Atrium inside a building in Praia promotes natural lighting and stack effect ventilation. In hot climates, the top of the atrium must be at least 5m above the occupied space, and it must have outlet air vents in order to prevent the accumulation of hot air.

On the other hand, there is now a large body of evidence that shows that people living in countries with warmer climates are satisfied by temperatures that are higher than those which feel comfortable to people living in countries with colder climates, and these temperatures are significantly different (upper and lower, respectively) to the temperatures considered 'ideal' by conventional standards.

Buildings that use passive cooling techniques can be efficient and economic, energy efficient and environmentally friendly alternatives to air conditioned buildings. These bioclimatic buildings also offer more satisfactory thermal environments – not in their ability to meet strict standards, but in improving the physiological and psychological comfort of the occupants.

For a better understanding of what the internal comfort of a building might mean, Figures 16.119 to 16.122 show psychometric graphs for various cities, representative of the diversity of climates existing in Angola, Cabo Verde, Guiné-Bissau and Moçambique. The dark blue areas on the charts represent the

TABLE 16.8 The Use of Natural Ventilation Strategies According to the Difference Between External and Internal Temperatures; Daytime and Night Time (Wind-Driven or Stack Effect) Ventilation

Day / Night	Description	Performance
Daytime Ventilation	The simplest strategy to improve comfort when the indoor temperature is higher than outdoor temperature, providing comfort through higher indoor air speed. It can use wind-driven single or cross ventilation, or stack effect.	Suitable when indoor comfort can be experienced at outdoor air temperature, and with diurnal temperature swings of less than 10°C.
Night Ventilation	Used to cool the building mass during the night. At the end of the day the temperature of storage will be increased without degrading comfort, raising the cold storage capacity and heat dissipation capacity of the system. Heat is then flushed through by ventilation during the night, and the building is cool the following morning (see also thermal mass).	It is especially suitable to situations when daytime outside temperatures are too hot and ventilation is impossible. Its performance can be improved by fan-driven mechanical ventilation. Night time ventilation works when night temperatures are lower than daytime temperatures, by more than 8–10°C. This technique is used for heat removal of the building, not comfort ventilation.

climatic characteristics (wet and dry-bulb temperature, relative humidity and vapor pressure), and outlined in yellow is the ASHRAE conventional comfort zone, as determined by the Ecotect – Weather Tools software.

In these graphs one can observe the overlapping zones of influence of the various passive cooling techniques, based on research by Givoni (1969). The diagram shows how the conventional comfort zone can be enhanced through the use of various passive cooling techniques. The referred strategies are the most appropriate to the performance of the building in these climatic zones. Outside these areas, the use of air conditioning would be required.

It can be seen that, despite the variety in climatic profiles, the strategy with the greatest impact is natural ventilation (2 – highlighted in light blue), followed by night ventilation, (3 – dark blue), and, especially in the dry season, thermal inertia (4 – light pink), and evaporative cooling (5 – green).

The need for cooling is predominant, though in some cases there is a small period in which passive heating is also required (yellow zone). This occurs

FIGURE 16.91 Interior of a stack ventilation tower in a house in Sao Vicente.

FIGURE 16.92 The hot air must be flushed to the outside to prevent its accumulation at ceiling height.

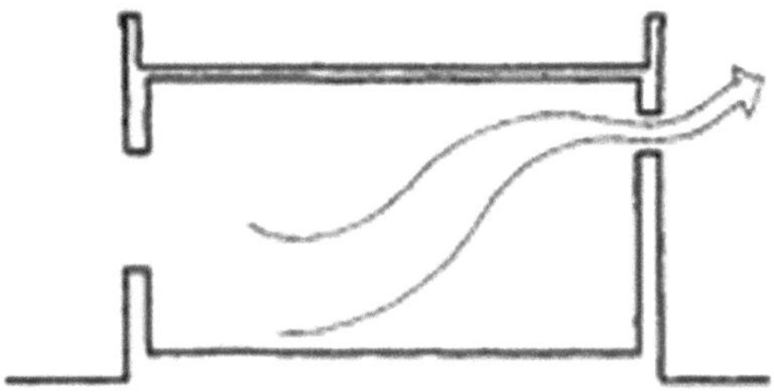

FIGURE 16.93 When the air intake openings are smaller than the air outlet, there is greater efficiency in the suction of cooler air, which flushes hot air out.

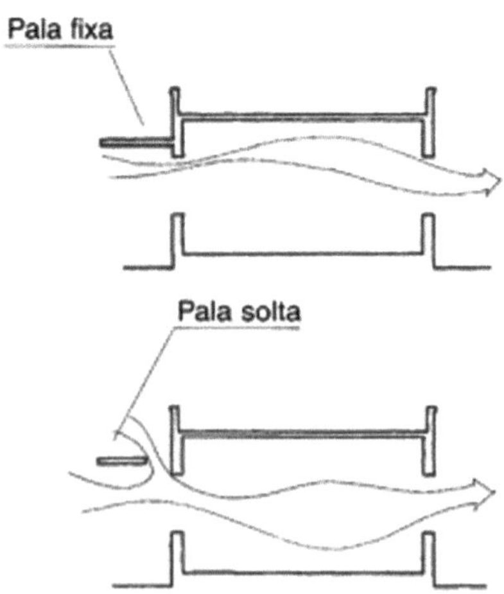

FIGURE 16.94 A blade away from the wall increases air intake.

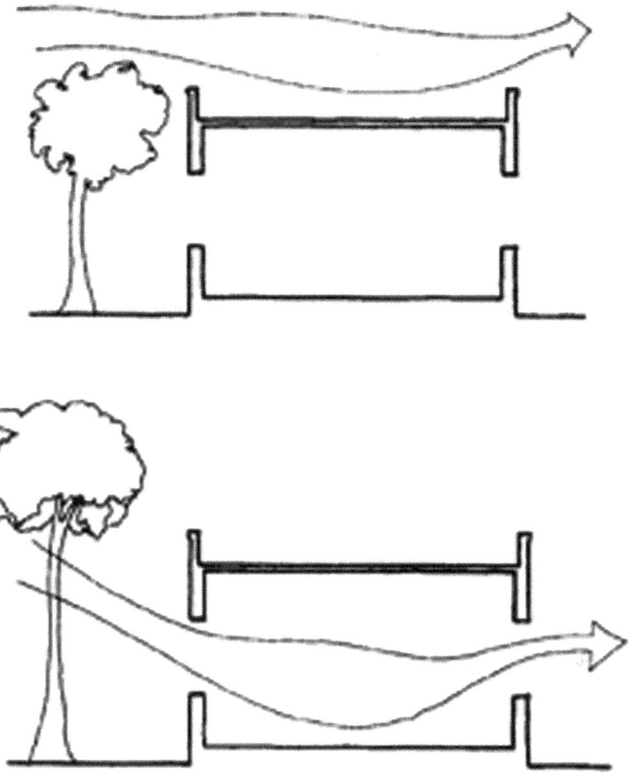

FIGURE 16.95 The breeze rises with low trees; with tall trees the breeze comes down and cools the room.

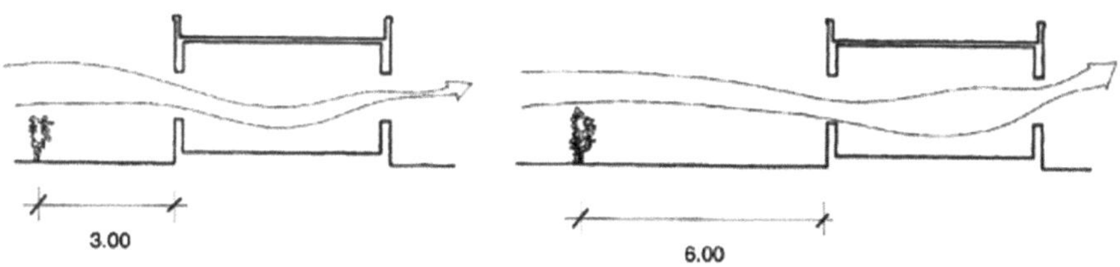

FIGURE 16.96 The greater the distance between the building and trees, the stronger will be the incoming breeze.

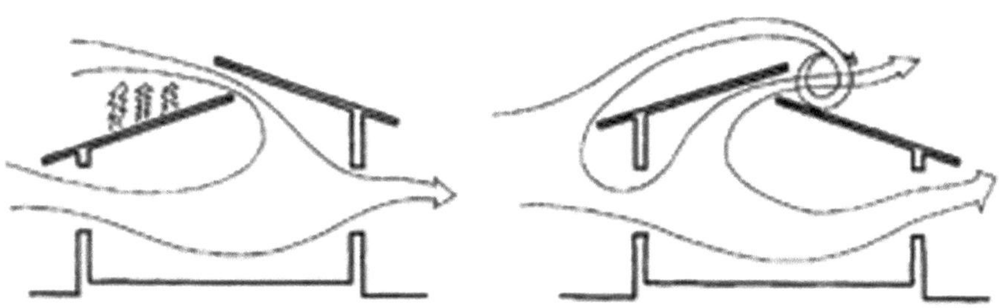

FIGURE 16.97 In the first scheme, the skylight is poorly located, because the warm air gets into the roof of the building. In the second scheme, there is a good position – the hot air of the compartments can be flushed through the skylight.

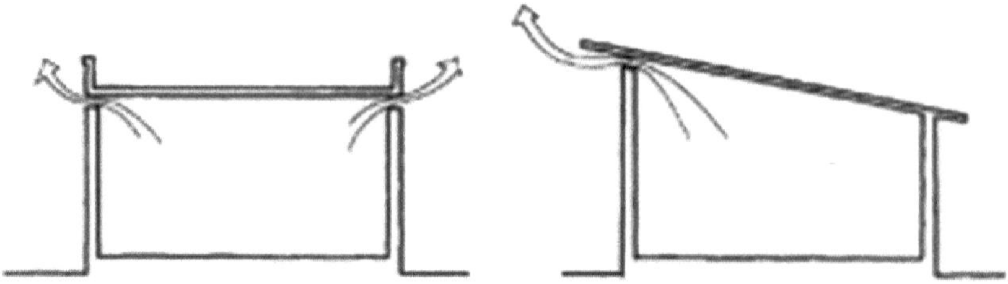

FIGURE 16.98 In the case of pitched roofs, the opening should be in the higher wall.

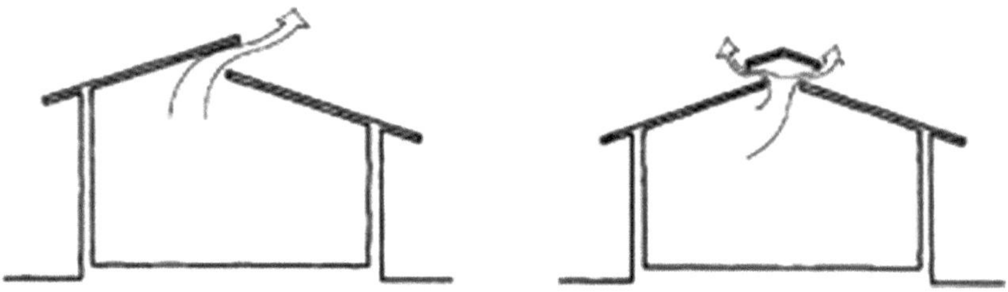

FIGURE 16.99 Two resources to force air movement through the opening in the ceilings.

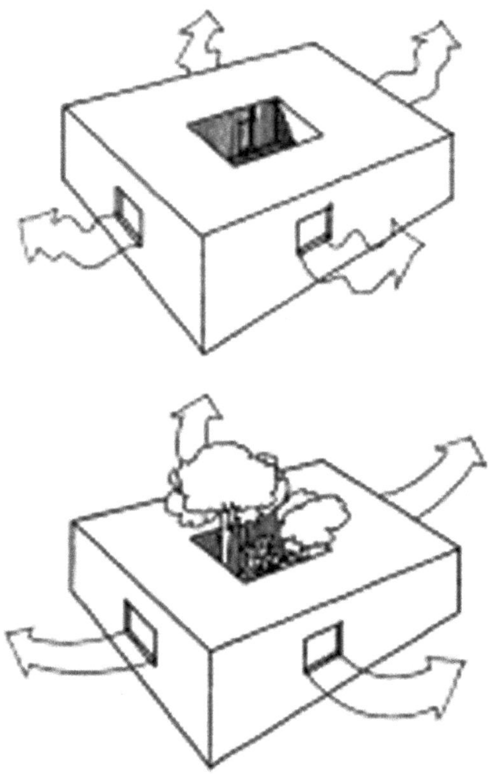

FIGURE 16.100 The inclusion of courtyards brings additional benefits to the climatization of the house. Fresh air from the courtyard circulates in the compartments. If the patio has plants, cooling will improve. In areas where there are few trees, a situation very common in Luanda and in the islands of Cape Verde, the house can be cooled using a courtyard to create shaded zones, where the air is cooler. The use of the courtyard facilitates more openings in the façade, for ventilation of indoor rooms.

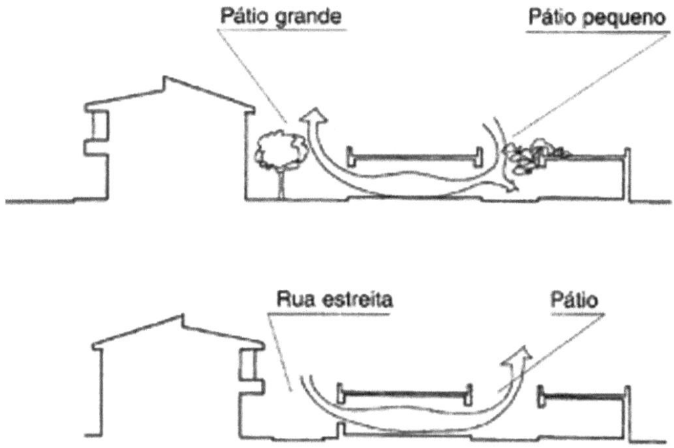

FIGURE 16.101 The movement of fresh air can also be produced through the addition of two courtyards, one smaller than the other. The air from the smaller patio, which is more shaded, is cooler than that of the larger patio. Thus, hot air rises, causing cool air to penetrate better in the compartments between the two courtyards.

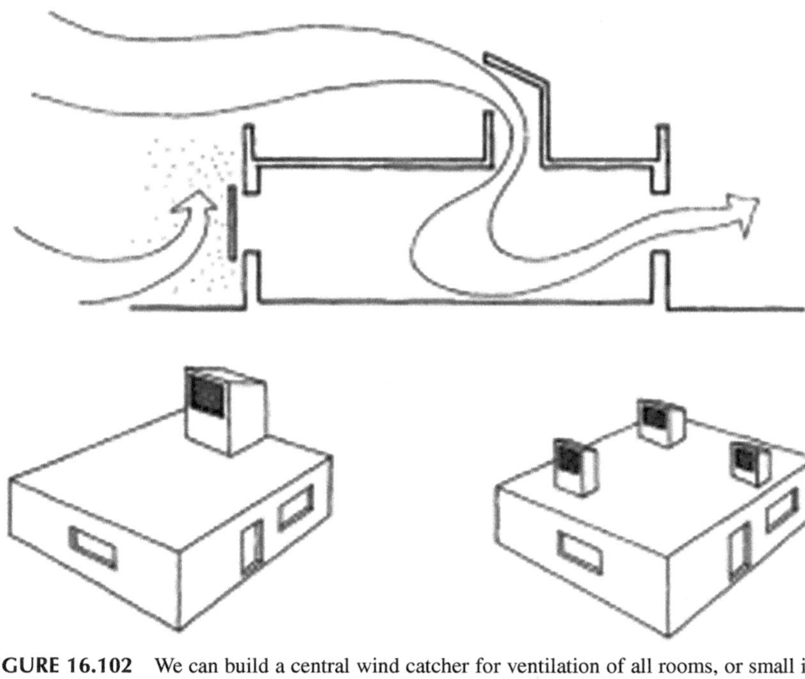

FIGURE 16.102 We can build a central wind catcher for ventilation of all rooms, or small individual catchers (wind towers). One way to get cool, clean air to the inside of a building is to use wind catchers that allow the recycling of the warm, stale air. The greater the height of the wind catcher, the cooler is the breeze. Wind catchers also prevent the entry of dust carried by the wind. In Angola and Cape Verde, the direction of the breeze is more or less constant, which makes this solution extremely effective.

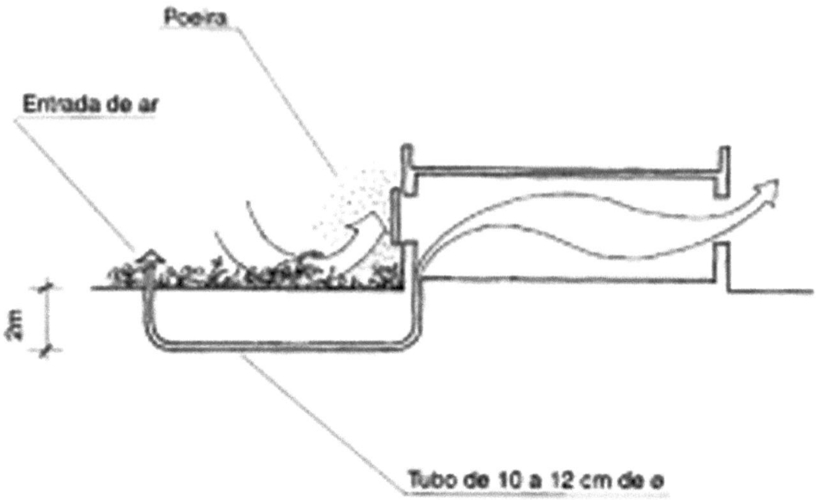

FIGURE 16.103 Scheme of a building ventilated through the underground. One can lower the temperature inside the house through an underground ventilation system. This technique consists of passing the air below ground through a tube, about two meters deep, to make the air cooler. It is important that the tube is that deep to get cool air. The tube is driven to the room that needs cooling. The capture is done in a cool area, shaded by trees or plants. The outlet opening of the tube, inside the room, must be protected with a mosquito net to prevent entry of insects, and shutters with movable blades to control the air intake.

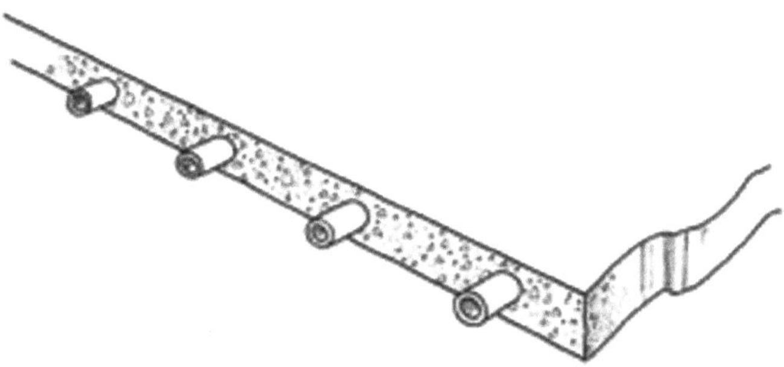

FIGURE 16.104 A detail of floor slab, ventilated with PVC pipes. The slabs may have channels for air circulation for cooling the house. These channels must have entry and exit to the outside so that air can circulate and be renewed. Openings must be protected against the entry of insects.

during the dry season, but heating can also be obtained passively (using solar energy), for example by proper orientation and window sizing. In very few cases, such as in Huambo and Ondjiva, in Angola, active heating (orange zone) may be required during a short period of the year – in these cases, solar thermal systems may prove useful.

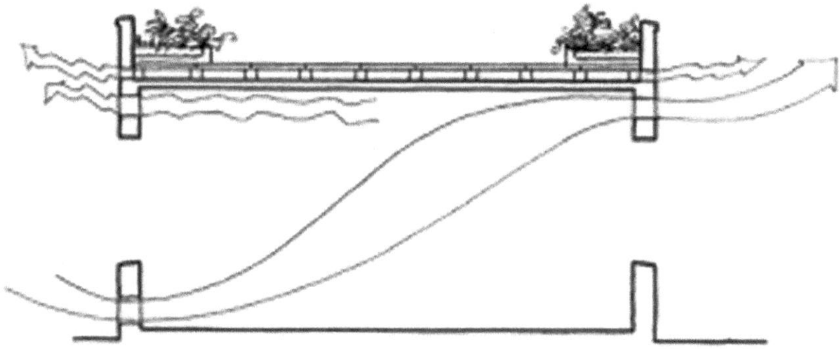

FIGURE 16.105 A scheme for a building with a ventilated roof. Most of the heat gain and loss occurs through the roof, which is more exposed to solar radiation. In hot and dry regions, the problem of rainwater infiltration is reduced, so roofs are generally flat or with a low slope. The most usual roof is a reinforced, massive, concrete slab, a poor solution in terms of energy efficiency. The massive concrete slabs absorb the sun's heat and are costly.

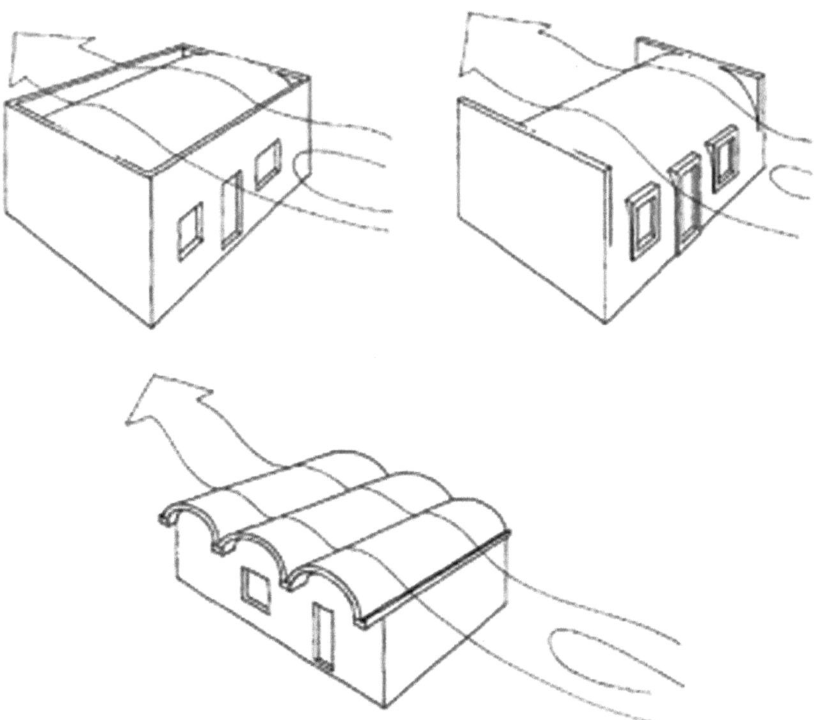

FIGURE 16.106 Recessed vault roof, half-round vault roof and pre-fabricated roof vault.

FIGURE 16.107 Tourist information building in Santo Antao, built using hollow concrete slabs, and incorporating openings on the roof for ventilation.

TABLE 16.9 Fan-Assisted Ventilation

	Description	Performance
FANS	The use of fans can improve the performance of natural ventilation techniques. Box, oscillating or ceiling fans increase internal air velocities and convection exchange, improving comfort by increasing the convective processes. They can be used particularly when opening windows may cause heat penetration, excessive air speeds, or security or noise problems. Fan-assisted mechanical ventilation systems involving ducts and special ventilation paths can also be used to improve air circulation through the building, being especially useful in the refurbishment of deep plan buildings, and in ventilation systems for cooling the building mass, such as night cooling.	The use of box, oscillating or ceiling fans alone can allow an increase of the indoor comfort temperature of 3–5°C, at 1m/s, say from 24°C to 28°C, much reducing cooling requirements. Ceiling fans can have a payback period of around just 3 years. The turbulent and variable air motion quality produced by fans also produces more comfortable effects than uniform air motion. A ceiling or desk fan is not annoying or draughty at 1m/s, and does not move loose papers. Fan-assisted mechanical ventilation systems involving ducts and special ventilation paths outside the occupied zone are used not for convective cooling of the body, but for the cooling of the building mass and fresh air provision. These systems can be much cheaper and less energy consuming than air conditioning.

FIGURE 16.108 New construction near Luanda, using massive materials – brick, stone and concrete, which provide thermal inertia to the building. This type of construction is adequate in hot climates with significant amplitudes between daytime and night time temperatures.

FIGURE 16.109 New construction (above) and traditional, popular construction (below) in Cape Verde, using massive materials.

FIGURE 16.110 Hostel designed by Alvaro Siza in Cidade Velha. New construction using materials with high thermal inertia (stone and concrete).

FIGURE 16.111 Examples of use of massive building materials in Guiné-Bissau.

FIGURE 16.112 Guiné-Bissau: the use of lightweight buildings may offer advantages in zones with low thermal amplitudes (below 8°C between night and day during most of the year), where thermal inertia has little effect. Building features like elevated structures, ventilated roofs and wind-catching verandas can be used to improve ventilation.

FIGURE 16.113 In Mozambique, the use of massive building materials, like earth, stone, bricks or concrete, can be observed throughout the country. Above: adobe and stone vernacular houses in Ilha de Moçambique. Below: brick and concrete construction in Maputo.

TABLE 16.10 Various Techniques that can be Used in Office Buildings to Optimize the Use of Thermal Mass

Thermal Mass	Description	Performance
Night ventilation	Massive elements, like walls, structure, slabs. Night ventilation of the thermal mass provides an efficient means of cooling the building. At night, when outside air temperature is considerably lower than inside, night ventilation is used to dissipate and exhaust the heat accumulated during the day in the building's mass into the low temperature atmospheric heat sink, preventing overheating the following day. The outside air is introduced to the building, either through special channels that contact with the building structure (allowing higher air speeds for convection), or through the windows. In night ventilation using windows, these can have openings on top for this type of ventilation. For reasons of safety and privacy, ground floor windows remain closed, and safety window screens are used.	Night cooling systems may be one of the most efficient passive cooling techniques. This system requires amplitudes of 8 to 10°C between day and night, ventilation rates of 10–25 ach/h, and the structure to be massive enough to store the cooling effect until the next day. It is most suitable when daytime temperatures are higher than 30°C and lower than 36°C. The storage mass can be cooled up to 2 or 3°C above the outdoor minimum. It is recommended for climatic regions such as in Cape Verde, Angola, and some zones in Mozambique (e.g. Tete, Lumbo, Lichinga, Quelimane and Maputo), and Guinea-Bissau (inland areas). The walls and the structure must be sufficiently exposed to the air stream, avoiding the use of false ceilings, and any other elements that could prevent this contact. This system does not usually require complex and costly actions – it may be enough to increase the thermal mass exposure, e.g. by removing false ceilings and opening existing windows; keeping in mind safety precautions, insect protection. To facilitate night ventilation, windows can be left opened overnight. This type of ventilation can be natural or assisted by fans
Chilled beams and chilled ceiling systems	Consist of an array of pipes embedded in the beams or ceiling slabs, through which cold water is circulated, lowering the temperature of the mass.	These systems have been successfully used in the refurbishment of office buildings, but still tend to be expensive.

FIGURE 16.114 Use of vegetation in the interior of a house in Mindelo: in addition to being pleasant, it slightly reduces the air temperature.

Also noteworthy is the fact that these passive strategies cover most of the climatic profile (dark blue area), of a significant number of locations, such as in Huige, Huambo and Onjiva in Angola, Sal in Cape Verde, or Tete, and Lichinga, in Mozambique, showing that, in theory, there is virtually no need for active air conditioning systems for cooling in these regions.

In other situations, such as in Beira and Quelimane in Mozambique, and particularly in Bissau or Bolama in Guiné-Bissau, the area located in the active zone (7 – where artificial cooling is needed), is significant: it corresponds to the hot and rainy season, with high temperature and humidity values that surpass the upper limits of comfort prescribed by Givoni. In these cases, one can use fans for example (a low-energy and economic system), or a mixed mode system, to increase upper comfort limit – or there is now an alternative technology to conventional air conditioning. This is the so-called solar HVAC: mechanical air conditioning systems in which the use of electricity from fossil fuels is replaced by solar energy, a renewable source, thereby reducing the negative impact on the environment, and also maintenance costs.

However, for tropical hot and humid zones like these in Guinea-Bissau and certain regions in Mozambique, existing comfort criteria may prove rather conservative and unrealistic, as they were based on empirical formulas and standards of comfort typical of cold or temperate climates. For example, field studies recently carried out in Guiné-Bissau show that with external temperatures of 39°C–43°C, and up to 75–80% RH%, people feel comfortable indoors at 31°C – hence the area where air conditioning would be needed would be reduced (in the presented graphs). This is an area where research is still necessary – to clarify the actual comfort requirements in hot and humid tropical regions, as in the presented cases – in order to avoid unnecessary energy consumption, with its serious economic and environmental consequences (see Figures 16.123 and 16.124).

FIGURE 16.115 Examples of use of vegetation in open spaces in Santiago and S. Vicente: as well as providing shade and contributing to the beauty of the place, the vegetation also contributes to a slight reduction of local temperature through the process of evapo-transpiration resulting from photosynthesis (evaporative cooling).

FIGURE 16.116 Examples of use of vegetation in open spaces in Luanda and Kuanza.

FIGURE 16.117 Guiné-Bissau: Examples of use of vegetation in open spaces, in Bafatá, Buba and Bissau.

FIGURE 16.118 Maputo, Mozambique – examples of the use of vegetation in open spaces. Trees generally have the effect of reducing the wind speed. However a row of trees with bare trunks for the lower 3 m of their height, may, if the foliage is dense above, deflect and enhance the breeze at ground level.

TABLE 16.11 Strategies to Reduce Internal Gains

	Description	Efficiency
ARTIFICIAL LIGHT	The use of artificial lighting is often excessive; either because the lighting levels are too high, the lighting systems are inefficient, or due to poor switching controls. Internal gains from artificial light can range from 6 to more than 20 W/m^2.	It is recommended to use task lighting, with low background illumination levels. High-efficacy light sources with low heat emissivity, like fluorescent lamps, should be used instead of incandescent tungsten or halogen lamps. Reduce the excessive indoor illumination levels: background lighting be reduced to 200 lux, and even to 150lux if task lights and VDU's are used. Extracting ventilation through the luminaries can also be used to reduce heat gains.
DAYLIGHT	The use of daylight can reduce substantially the cooling loads by delaying the use of artificial light. Daylight should be well distributed in the rooms, without glare and contrast.	It is estimated that for 1KWh saved for lighting in the cooling season, about 0.3KWh of electricity used by air conditioning is saved. Consider that the space area that can be effectively daylit is around 6m, corresponding to twice the floor-to-ceiling height. As a rule of thumb, high-level windows have a better performance than low-level windows and tall windows perform better than wide windows (as daylight goes deeper into the space). The use of clear (reflective) colors in wall painting and decoration, as well as light shelves also increases illumination levels. The use of skylights in the top floors can cause overheating during summer, as well as glare. Glare control is essential when using computers. Glare and contrast can be avoided by using splayed reveals, light shelves, prisms or reflectors, light ducts, or fiber optics.
COURTYARDS AND ATRIA	The introduction of courts and atria in buildings with a very deep plan could improve daylighting and ventilation, reducing energy consumption from artificial lighting and air conditioning (see also natural ventilation).	The introduction of glazed atria must be very carefully considered in warmer Mediterranean climates, as they often lead to problems of overheating. The daylit zone facing the atria to be considered is limited to the sky view zone (corresponding to about a 3 to 1 ratio between the height and width of the atrium). Courtyards are traditionally used in some countries in Africa.
OCCUPANTS AND OFFICE EQUIPMENT	Internal gains from occupants and office equipment like computers or photocopiers can produce annual heat gains in the range of 15 to 30W/m^2.	The reduction in internal gains can be achieved by locating the heat-generating equipment in special areas (e.g. computer room), with higher ventilation rates (or special climatization if required), serving as buffer spaces, and away from the occupants if possible. Internal gains from occupants can be reduced by avoiding high occupant density, through layout design.

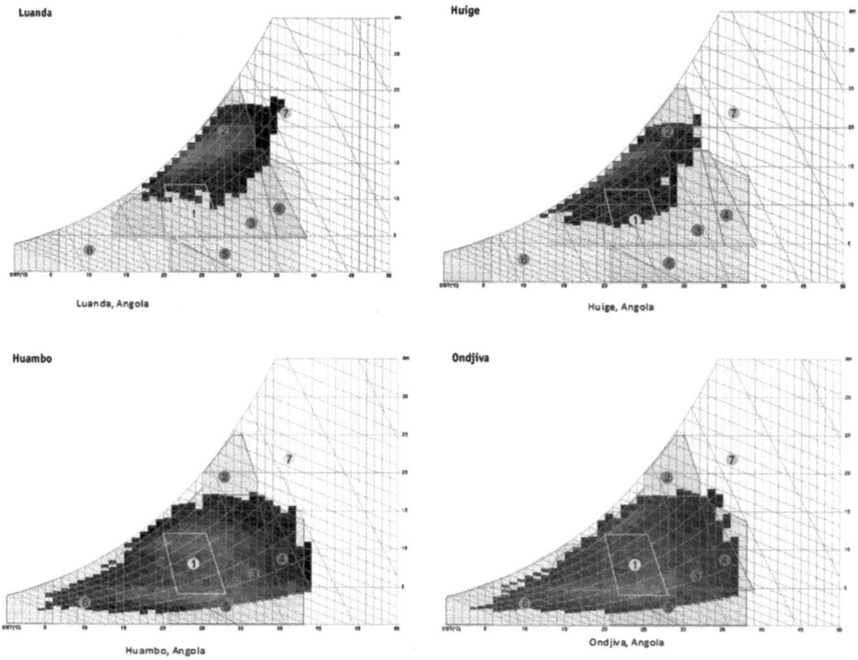

FIGURE 16.119 Psychometric diagram – for four cities in Angola, representative of the main climatic zones of the country. The dark blue area illustrates the climatic profile of the region. The diagram shows how the ASHRAE conventional summer comfort zone can be expanded through the use of various passive cooling techniques. The various areas shown on the graph were defined by Givoni (1969) and correspond to:

1. ASHRAE conventional summer comfort zone, used as standard for the use of air conditioning (yellow).
2. Zone of influence of daytime ventilation (light blue outline).
3. Zone of influence of night ventilation (blue outline).
4. Zone of influence of thermal inertia (pink contour). Includes zones 2 and 3.
5. Zone of influence of evaporative cooling (green outline). The evaporative cooling can also be used in zones 2, 3 and 4, for dry-bulb temperatures above 21°C.
6. Passive heating zone (dark yellow) and the zone of active heating (brown outline).
7. Zone where air conditioning is required (white background).

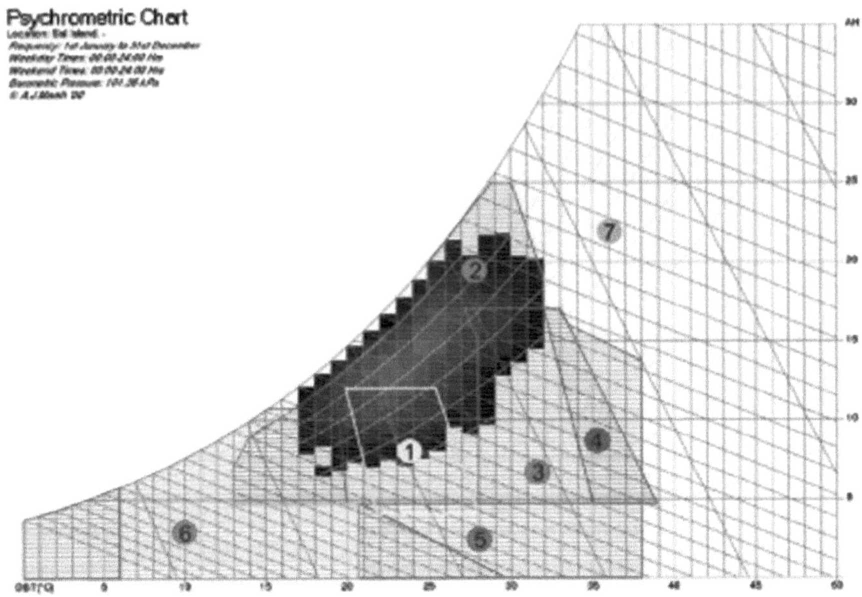

FIGURE 16.120 Psychometric diagram for the island of Sal, representative of the climate of the Cape Verde archipelago. The dark blue area illustrates the climatic profile of the region. The diagram shows how the ASHRAE conventional summer comfort zone can be expanded through the use of various passive cooling techniques. The various areas shown on the graph were defined by Givoni (1969) and correspond to:

1. ASHRAE conventional summer comfort zone, used as standard for the use of air conditioning (yellow).
2. Zone of influence of daytime ventilation (light blue outline).
3. Zone of influence of night ventilation (blue outline).
4. Zone of influence of thermal inertia (pink contour). Includes zones 2 and 3.
5. Zone of influence of evaporative cooling (green outline). The evaporative cooling can also be used in zones 2, 3 and 4, for dry-bulb temperatures above 21°C.
6. Passive heating zone (dark yellow) and the zone of active heating (brown outline).
7. Zone where air conditioning is required (white background).

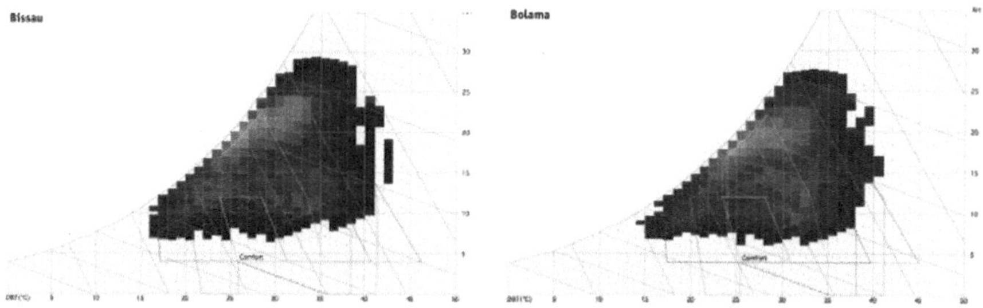

FIGURE 16.121 Psychometric diagram – for two cities in Guiné-Bissau, representative of the main climatic zones of the country. The dark blue area illustrates the climatic profile of the region. The diagram shows how the ASHRAE conventional summer comfort zone can be expanded through the use of various passive cooling techniques. The various areas shown on the graph were defined by Givoni (1969) and correspond to:

1. ASHRAE conventional summer comfort zone, used as standard for the use of air conditioning (yellow).
2. Zone of influence of daytime ventilation (light blue outline).
3. Zone of influence of night ventilation (blue outline).
4. Zone of influence of thermal inertia (pink contour). Includes zones 2 and 3.
5. Zone of influence of evaporative cooling (green outline). The evaporative cooling can also be used in zones 2, 3 and 4, for dry-bulb temperatures above 21 °C.
6. Passive heating zone (dark yellow) and the zone of active heating (brown outline).
7. Zone where air conditioning is required (white background).

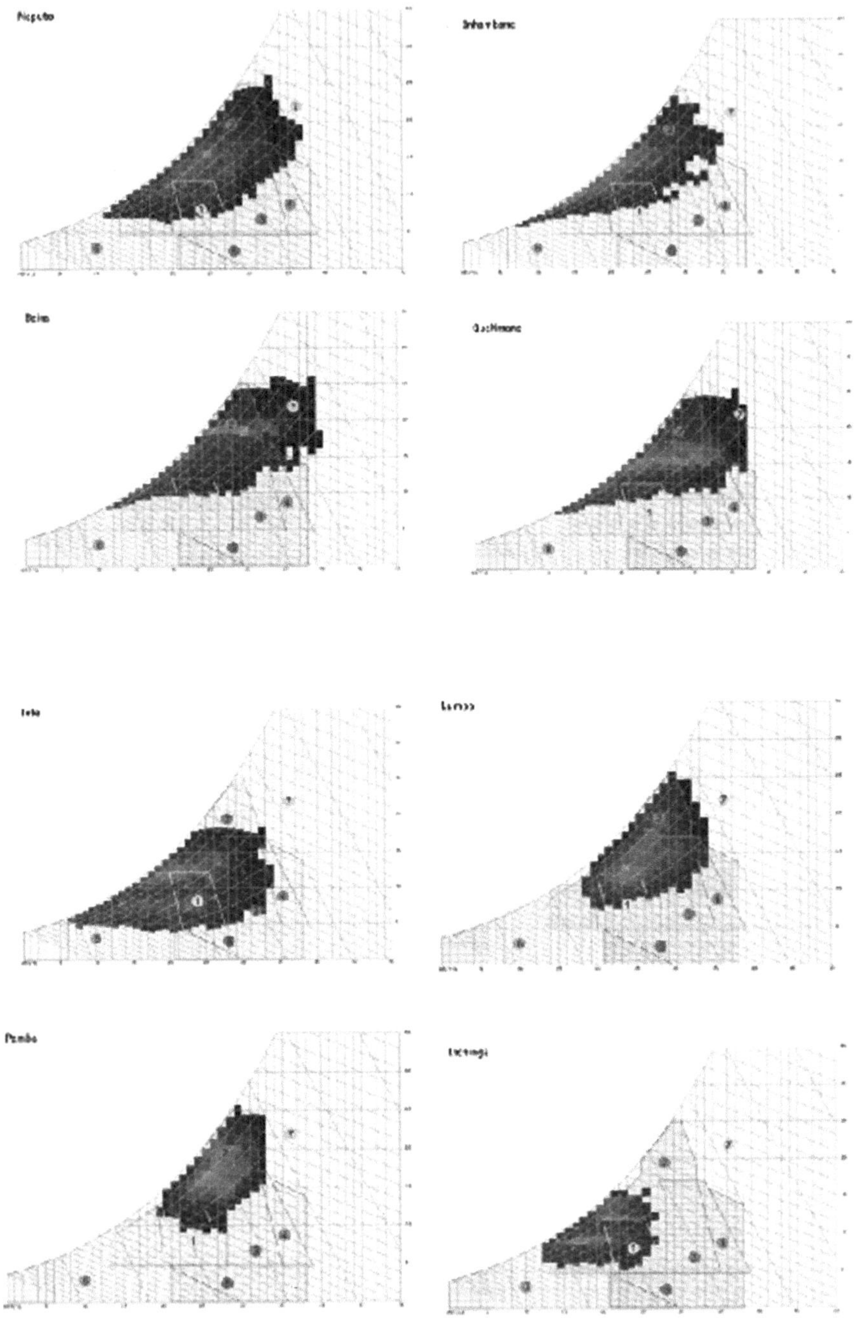

FIGURE 16.122 Psychometric diagram – for eight cities in Mozambique, representative of the main climatic zones of the country. The dark blue area illustrates the climatic profile of the region.

FIGURE 16.123 Luanda, Angola: the use of air conditioning can be avoided through a correct use of passive design, reducing damage to the environment and operation costs.

FIGURE 16.124 Use of air conditioning devices, city of Praia, Cape Verde.

Chapter 17

Mud to Skyscraper – Building Revolution in 50 Years in the Middle East

Ali Sayigh
World Renewable Energy Congress & Network, Brighton, UK

Chapter Outline

17.1 PORTABLE HOUSING: THE BEDOUIN TENT

In many parts of the world, in particular in the Gulf Cooperation Council, GCC, houses and dwellings used to be built from mud and clay baked in the sun, or adobe. Prior to the establishment of settled small towns and oases, the people of the region were nomads, moving through the desert depending on the seasons of the year, seeking pasture and water for their herds of camels and sheep following the minimum rainfall pattern. Their homes were tents constructed from wooden poles, and covered with camel or sheep knitted wool rugs. These were easy to erect, easy to dismantle and simple and light to carry. They relied mostly on the shading effect and prevailing north wind for cooling. In winter the fires were lit: to cook their food, and to provide coals to take into the tents to brew coffee and keep the tent warm. Examples of these tents are shown in Figure 17.1. Normally the family slept in one tent for security reasons. Saluki dogs were used for hunting deer and birds, with the help of falcons.

The Bedouin moved freely throughout the region, and until recent times were not hampered by the need for passports and international boundaries. Within a year they might travel from south of Arabia to east of Iraq or west of

Sustainability, Energy and Architecture. ISBN: 9780123972699

FIGURE 17.1 Portable housing; easy to assemble and pack, the Bedouin tent.

Syria, and back. Normally they pitched their tents one or two kilometers outside a town, in which they would briefly purchase stocks of rice, wheat, salt and the occasional fruit, or to sell their wool and animal skins.

17.2 MUD HOUSES AND COMFORT

As some of the Bedouins established settlements at oases and trading posts, they moved away from their tents to basic houses constructed from a combination of locally available wood and mud reinforced with straw (adobe). In Saudi Arabia and the Gulf region, the majority of buildings were made from adobe. These adobe houses were mostly single storey, with thick walls and flat roofs and a few openings for ventilation. All the rooms opened onto a courtyard containing a small well, and a palm tree at the center to create shade. Outside walls were decorated with traditional patterns. See Figure 17.2.

The inhabitants depended for comfort on utilizing shading, radiative cooling and to a lesser extent evaporative cooling. The flat roofs were used for sleeping on at night during the extreme heat of the summer months, where radiative cooling to the sky was achieved. In some parts of the Gulf, Iraq and Iran, extensive wind towers (*baudgeers*) were used. This concept was explained in another chapter of the book.

In southern parts of Iraq, the Marsh Arabs built their homes entirely from mud and reeds on platforms constructed on the shallow water of the marshes. Sadly the skills behind these buildings are being lost as the marshes are once again drying out. This is due to the decrease in water levels of the Tigris and Euphrates, resulting from extensive dam constructions in Turkey and Iran. These unique buildings are spacious, cool and attractive; see Figure 17.3.

The discovery of oil brought the opportunity to travel outside their traditional areas, and obtaining education abroad meant that the knowledge of and the desire for more sophisticated methods of construction prevailed. New construction methods gave rise to the need for cheap labor to be employed in building roads, the oil industry and other services. This in turn gave rise to a need for imported cement and steel, and the populations of these areas doubled or tripled in a few decades. Nowadays, for example, the UAE population is more than

FIGURE 17.2 Various mud houses in the GCC.

80% expatriates. These newcomers could not be housed in adobe houses. The construction industries flourished, but gave little consideration to comfort or environment. Traditional knowledge was discounted, thick walls and courtyards gave way to high rise condominiums with high levels of fenestration. Although the local people often continued to live in the decent, well-built houses, most of the expatriates lived in high rise buildings.

In the summer months of May, June, July, August and September, the temperature in the GCC at mid-day and in the shade reaches more than 50°C. These high rise buildings require cooling (air conditioning) even in the relatively mild winter months. These buildings are heat traps and if the supply of electricity fails, then the inhabitants go out in their air conditioned cars and travel around until the electricity is restored. Because of the boundless supply of electricity and wealth no consideration was given to the amount of energy consumed.

Architects, builders and property owners gave little consideration to the people who lived in these skyscrapers, and in the early years completely disregarded the cost of energy. Such development has not, in the long run, served the best interests of the region, considering that the average per capita consumption of electricity per annum is more than 18,000 kWh, compared an average of 6,000 kWh per capita in Europe.

One of the results of the rapid expansion of these towns into cities has been not only an increase in power consumption, but also that the demand for potable

FIGURE 17.3 The wind towers of Iran and the Marsh Arab of Southern Iraq.

and non-potable water has increased exponentially. One must not forget that these countries have few or no rivers or lakes and minimal rainfall. The cost per cubic meter of fresh water in the GCC is now more than US$3. All this development was achieved in less than 60 years (see Figure 17.4).

17.3 A NEW GENERATION OF BUILDINGS

Over the last 60 years, the young generation have become accustomed to a mostly western style of living, albeit within a very harsh environment and totally dependent on a limitless supply of cheap electricity. To maintain this

FIGURE 17.4 Skyscrapers in a country that used to be mud houses in 1970.

electricity, the authorities in the GCC have had to increase the supply by 20% each year. But this is not without its negative impacts: more and more motor vehicles; traffic jams; miles and miles of built roads and tens of thousands of concrete buildings. Because of the pressure from investors and developers, and the general lack of concern for the environment, architects have had to come up with increasingly extravagant buildings. This building bonanza has gone ahead in the absence of building rules and regulations. As a result, high rise buildings have been constructed with more than 70% of their façade made of glass and with only three meters between one building and the next. No wonder that we now face shortages in the supply of water and electricity.

17.4 WHAT IS THE SOLUTION?

Before it is too late, the authorities in this region must establish good architectural practice and properly enforced building codes. The teaching of architecture must change to encompass the need to reduce CO_2 emissions to counteract climate change and global warming. We knowing that 50% of the energy we consume is used in building and construction. For example, cement is one of the most energy intensive products. In 2011, China produced 1.5 bn tons of cement, India produced 0.5 bn tons, and the rest of the world produced 1.5 bn tons. Figure 17.5 shows a badly designed building with an almost 100% glass façade,

FIGURE 17.5 A very badly designed building with a 100% glass façade, and the BRE Building which is well ventilated, with a PV façade.

and another building of good design with a PV façade which needs minimum energy to run, the BRE (Building Research Establishment, UK).

17.5 ENERGY AND BUILDINGS

In the construction industries, the issue of energy was never a constraint. However this has changed with the knowledge of the harm caused by the excess emission of CO_2 into the atmosphere, climate change and global warming, and now it is a crucial issue in building design. Nowadays, the way one produces the building materials; the iron and steel, bricks and the glass, cement and wood, is vital, since the methods of production add to the total energy cost. In order to produce sustainable buildings it is necessary to consider wall thickness and insulation, building orientation, shading, ventilation and to reduce the reliance on air conditioning and other heavy electricity consumption devices. Had attention been paid to vernacular architecture and the use of indegenous materials,

FIGURE 17.6 Good practice in high rise buildings with proper shading, and a recent hotel in Dubai with wind towers.

some of these issues would never have arisen. An architect first must consider the climate in which he/she is designing. It is no longer acceptable to design a building suitable for one climate and adapt it for another climate by the use of electricity. For example, thick adobe walls kept houses cool in the summer and warm in winter, courtyards and verendas gave shading and wind towers were excellent ventilators. All of these contributed to indoor comfort. If this consideration alone had been addressed properly, then most of the GCC buildings would now be condemned as being too expensive to run and maintain. No doubt there are exceptions to the rule, with some buildings employing proper shading, using double skins, well ventilated with thick thermal mass, but they are few (see Figure 17.6).

17.6 FINAL REMARKS

As stated earlier, 50% or more of our total energy consumption globally is used in building construction. Now more than at any time previously architects, builders and building material producers must use approprite materials in their design. Adapt some of the techniques of vernacular architecture. The use of shading devices is essential in hot climates. This was well understood by our ancestors, whether shading in bazaar or in the building fabrics. This can lead to (10–15)° C temperature reductions. For example, Figure 17.7 shows three bazaars: Fes-Morocco, Al-Hameadia – Souk, Damascus and Hong Kong market – China.

Another important element to consider is the thermal mass of the building. Thick walls and structures are used to stabilize indoor temperatures, with very little variation from winter to summer. Figure 17.8 shows an example in Aqaba – Jordan, the Rock Palace in Yemen and Al-qarouan Mosque, Fes, in Morocco.

Additional features which made the buildings appealing to the inhabitants and visitors, as well as making them icons, was the use of traditional arts and design, which in most Islamic countries means the use of calligraphy. These were features of building in Andolusia – Spain of the middle centuries, and Iran (see Figure 17.9).

As final remarks, one should use optimization and personal touch in one's design in all parts of the building: façade, shutters, windows, materials,

FIGURE 17.7 Public places and their shading devices, Fes, Damascus and Hong Kong.

FIGURE 17.8 Thermal mass is essential in stabilizing indoors temperature, Al-Qarouan mosque, Rock Palace and Petra.

FIGURE 17.9 A decorative façade is essential in most public places to give peace and charm to the building.

vegetation and insulation. In inner cities, we must address the problems of transport and create adequate recreation parks, gardens and spaces for healthy living. We must introduce and abide by proper town planning and building codes suited for the locality in mind. We must address the issues of electricity supply, water supply, sewage and solid waste treatment and transport facilities.

Index

Note: Page numbers followed by "f" or "t" indicate figure or table respectively.

E

M

FSC
www.fsc.org
MIX
Papier | Fördert
gute Waldnutzung
FSC® C083411

Kolibri 360 GmbH
Ferdinand-Jühlke-Straße 7
99095 Erfurt, Deutschland
produktsicherheit@kolibri360.de